Physics of Neural Networks

Series Editors:
E. Domany J.L. van Hemmen K. Schulten

Springer
New York
Berlin
Heidelberg
Barcelona
Budapest
Hong Kong
London
Milan
Paris
Santa Clara
Singapore
Tokyo

Physics of Neural Networks

Models of Neural Networks
E. Domany, J.L. van Hemmen, K. Schulten (Eds.)

Models of Neural Networks II: Temporal Aspects of Coding and Information Processing in Biological Systems
E. Domany, J.L. van Hemmen, K. Schulten (Eds.)

Models of Neural Networks III: Association, Generalization, and Representation
E. Domany, J.L. van Hemmen, K. Schulten (Eds.)

Neural Networks: An Introduction
B. Müller, J. Reinhart

E. Domany J.L. van Hemmen
K. Schulten (Eds.)

Models of Neural Networks III

Association, Generalization, and Representation

With 67 Figures

Springer

Series and Volume Editors:

Professor Dr. J. Leo van Hemmen
Institut für Theoretische Physik
Technische Universität München
D-85747 Garching bei München
Germany

Professor Eytan Domany
Department of Electronics
Weizmann Institute of Science
76100 Rehovot
Israel

Professor Klaus Schulten
Department of Physics
and Beckman Institute
University of Illinois
Urbana, IL 61801
USA

Library of Congress Cataloging-in-Publication Data
Models of neural networks III / E. Domany, J.L. van Hemmen, K. Schulten, editors.
p. cm. — (Physics of neural networks)
Includes bibliographical references and index.
ISBN 0-387-94368-4 (alk. paper)
1. Neural networks (Computer science) — Mathematical models.
I. Domany, E. (Eytan). II. Hemmen, J.L. van (Jan Leonard). III. Schulten, K. (Klaus) IV. Series.
QA76.87.M59 1995
006.3 — dc20 95-14288

Printed on acid-free paper.

Production coordinated by Publishing Network and managed by Natalie Johnson; manufacturing supervised by Jeffrey Taub.
Typeset by Bytheway Typesetting, Norwich, NY.
Printed and bound by Braun-Brumfield, Inc., Ann Arbor, MI.
Printed in the United States of America.

9 8 7 6 5 4 3 2 1

ISBN 0-387-94368-4 Springer-Verlag New York Berlin Heidelberg
ISBN 3-540-94368-4 Springer-Verlag Berlin Heidelberg New York

Preface

One of the most challenging and fascinating problems of the theory of neural nets is that of asymptotic behavior, of how a system behaves as time proceeds. This is of particular relevance to many practical applications. Here we focus on *association*, *generalization*, and *representation*. We turn to the last topic first.

The introductory chapter, "Global Analysis of Recurrent Neural Networks," by Andreas Herz presents an in-depth analysis of how to construct a Lyapunov function for various types of dynamics and neural coding. It includes a review of the recent work with John Hopfield on integrate-and-fire neurons with local interactions.

The chapter, "Receptive Fields and Maps in the Visual Cortex: Models of Ocular Dominance and Orientation Columns" by Ken Miller, explains how the primary visual cortex may asymptotically gain its specific structure through a self-organization process based on Hebbian learning. His argument since has been shown to be rather susceptible to generalization.

Association long has been a key issue in the theory of neural nets. Local learning rules are quite convenient from the point of view of computer science, but they have a serious drawback: They do not see global correlations. In order to produce an extensive storage capacity for zero threshold, the couplings on the average should vanish. Accordingly, there is a deep truth behind Willshaw's slogan: "What goes up must come down." Meanwhile we have a zoo of local learning rules. In their chapter, "Associative Data Storage and Retrieval in Neural Networks," Palm and Sommer transform this zoo into a well-organized structure taking advantage of just a simple signal-to-noise ratio analysis.

Hebb's epoch-making book *The Organization of Behavior* appeared in 1949. It proposed one of the most famous local learning rules, viz., the Hebbian one. It was preceded by the 1943 paper of McCulloch and Pitts, which is quite notorious because of its formal logic. In "Inferences Modeled with Neural Networks," Carmesin takes up this lead and integrates it with the Hebbian approach, viz., ideas on assemblies and coherence. In so doing he provides a natural transition from "association" to "generalization."

Generalization means that, on the basis of certain known data, one extrapolates the meaning of a new set. There has been quite a bit of progress in formally understanding the process of generalization, and Opper and Kinzel's chapter "Statistical Mechanics of Generalization" summarizes this progress. It starts from scratch, assuming only some basic knowledge of statistical mechanics.

Bayes stands for *conditional* probabilities. For example, what is the probability of having sunshine on the American East coast tomorrow given that today's weather has no clouds? The sentence starting with "given that . . ." is a condition and the question entails an extrapolation. Adding one further condition, viz., that it is during the summer, the chance in question is about one. MacKay presents a careful and detailed exposition of the beneficial influence of "Bayesian Methods for Backpropagation Networks."

The last two chapters return to representation. Optical character recognition is well known as a playground of neural network ideas. The chapter "Penacée: A Neural Net System for Recognizing On-Line Handwriting," by Guyon et al., aims at making the underlying concepts also widely known. To this end, the setup is explained with great care. Their real-world examples show that an intelligently built but yet relatively simple structure can give rise to excellent performance.

Robotics has been in the realm of neural networks for a long time; and that is understandable. After all, we perform grasping movements ourselves with great ease. That is to say, our motor cortex allows us to do so. Cortical ideas also have permeated robotics. In their chapter "Topology Representing Networks in Robotics," Sarkar and Schulten present a detailed algorithm for the visually guided control of grasping movements of a pneumatic robot as they are performed by a highly hysteretic five-joint pneumatic robot arm. In so doing, they unfold a modified version of the manifold-representing network algorithm, a Kohonen-type approach. Here, too, governing asymptotic behavior is the algorithm's goal.

All of the chapters have one element in common: answering the question of how one can understand an algorithm or procedure theoretically. And that is what each volume of *Models of Neural Networks* is after.

The Editors

Contents

Contributors

J. Bromley AT&T Bell Labs, Room 4G-338, Holmdel, NJ 07733, USA

H.-O. Carmesin Institut für Theoretische Physik, Universität Bremen, D-28334 Bremen, Germany

I. Guyon AT&T Bell Labs, 955 Craston Road, Berkeley, CA 94708, USA

Andreas V.M. Herz Department of Zoology, University of Oxford, Oxford, OX1 3PS, England

Wolfgang Kinzel Physikalisches Institut, Universität Würzburg, D-97074 Würzburg, Germany

David J.C. MacKay Cavendish Laboratory, University of Cambridge, Madingley Road, Cambridge, CB3 0HE, England

N. Matić AT&T Bell Labs; presently at Synaptics, 2698 Orchard Parkway, San Jose, CA 95134, USA

Kenneth D. Miller Departments of Physiology and Otolaryngology, W.M. Keck Center for Integrative Neuroscience, and Sloan Center for Theoretical Neurobiology, University of California, San Francisco, CA 94143-0444, USA

Manfred Opper Physikalisches Institut, Universität Würzburg, D-97074 Würzburg, Germany

Günther Palm Abteilung Neuroinformatik, Fakultät für Informatik, Universität Ulm, Oberer Eselsberg, D-89081 Ulm, Germany

Kakali Sarkar Department of Physics/Beckman Institute, University of Illinois, Urbana, IL 61801, USA

M. Schenkel AT&T Bell Labs and ETH-Zürich, CH-8092 Zürich, Switzerland

Klaus Schulten Department of Physics/Beckman Institute, University of Illinois, Urbana, IL 61801, USA

Friedrich T. Sommer Institut für Medizinische Psychologic und Verhaltensneurobiologic der Universität Tübingen, Gartenstr. 29, D-72074 Tübingen, Germany

H. Weissman AT&T Bell Labs; presently at 12 Mordehai-Hetez St., Petah-Tikua, Israel

1

Global Analysis of Recurrent Neural Networks

Andreas V.M. Herz[1]

with 6 figures

Synopsis. This chapter reviews recurrent neural networks whose retrieval dynamics have been analyzed on a global level using Lyapunov functions. Discrete-time and continuous-time descriptions are discussed. Special attention is given to distributed network dynamics, models with signal delays, and systems with integrate-and-fire neurons. The examples demonstrate that Lyapunov's approach provides powerful tools for studying the retrieval of fixed-point memories, the recall of temporal associations, and the synchronization of action potentials.

1.1 Global Analysis — Why?

Information processing may be defined as the systematic manipulation of external data through the internal dynamics of some biological system or artificial device. In general, such a manipulation requires a highly nontrivial mapping between input data and output states. Important parts of this task can be accomplished with recurrent neural networks characterized by massive nonlinear feedback: Triggered by an appropriate external stimulus, such systems relax toward attractors that encode some a priori knowledge or previously stored memories.

Within this approach to information processing, understanding associative computation is equivalent to knowing the complete attractor structure of a neural network, that is, knowing what kind of input drives the network to which of its possibly time-dependent attractors. Understanding the computational properties of a recurrent neural network thus requires at least three levels of analysis: (1) What can be said about the existence and stability of fixed-point solutions? (2) Are there static attractors only, or are there also periodic limit cycles and aperiodic attractors, as would be

[1]Department of Zoology, University of Oxford, Oxford, OX1 3PS, England.

expected for generic nonlinear systems? (3) What is the structure of the basins of attraction?

Questions about the precise time evolution between the initial network state and the final output define a fourth level of analysis. Though less important within the framework of attractor neural networks, these questions are highly relevant for systems that extract information "en route" without waiting for the arrival at some attractor [1]. At a fifth level of analysis, one might finally be interested in questions concerning the structural stability of a given network, that is, its robustness under small changes of the evolution equations.

With regard to the computational capabilities of a neural network, questions about the type of attractor and the structure of basins of attraction are of paramount importance. These questions deal with *global* properties of the network dynamics. Accordingly, they cannot be answered using local techniques only: A linear stability analysis of fixed-point solutions, the first level of analysis, may reveal helpful knowledge about the network behavior close to equilibria, but it never can be used to rule out the existence of additional time-dependent attractors that may dominate large parts of the network's state space. Due to computational constraints, numerical simulations can offer limited additional information only.

Highly simplified network models provide a partial solution in that they often permit the application of global mathematical tools. However, such formal networks are characterized by bold approximations of biological structures. In the manner of good caricatures, they may nevertheless capture features that are also essential for more detailed descriptions.

One of the global mathematical tools is Lyapunov's "direct" or "second method" [2]. In the present context, it may be described as follows. Let the vector $x = (x_1, \ldots, x_N)$ denote the state variables of a neural network. These variables change in time according to some evolution equation, for example, a set of coupled differential equations $(d/dt)x_i = f_i(x)$ if time is modeled as a continuous variable t. A solution will be denoted by $x(t)$. If there exists an auxiliary scalar state function $L(x)$ that is bounded below and nonincreasing along all trajectories, then the network has to approach a solution for which $L(t) \equiv L(x(t))$ does *not* vary in time.[2] The global dynamics can then be visualized as a downhill march on an "energy landscape" generated by L. In this picture, every solution approaches the bottom of the valley in which it was initialized.

[2] Special care has to be taken with respect to unbounded solutions and continuous families of solutions with equal L. Note at this point that, in the present chapter, formal rigor often will be sacrificed for transparency of presentation. A mathematically rigorous introduction to Lyapunov functions can be found in the monograph of Rouche, Habets, and Laloy [3]. It also contains — apart from a large number of interesting theorems and proofs — some fascinating examples that illuminate possible pitfalls due to imprecise definitions.

The asymptotic expression for $L(t)$ and the equation $(d/dt)L(t) = 0$ contain valuable information about the very nature of the attractors — the first and second levels of analysis. Notice in particular that a solution that corresponds to a local minimum of the Lyapunov function has to be asymptotically stable, that is, it attracts every solution sufficiently close to it.

As an example, consider a gradient system

$$\frac{dx_i}{dt} = -\frac{\partial L(x)}{\partial x_i}. \tag{1.1}$$

Using the chain rule, the time derivative of L is given by

$$\frac{d}{dt}L(t) = \sum_{i=1}^{N} \frac{\partial L}{\partial x_i}\frac{dx_i}{dt} = -\sum_{i=1}^{N}\left(\frac{dx_i}{dt}\right)^2. \tag{1.2}$$

The last expression is negative unless $x(t)$ is a fixed-point solution. It follows that, if $L(x)$ is bounded below, the system has to relax to an equilibrium.

The most important feature of Lyapunov's direct method cannot be overemphasized: The method does not require any knowledge about the precise time evolution of the network; the mere existence of a bounded function that is nonincreasing along every solution suffices to characterize the system's long-time behavior. As a consequence, one can analyze the long-time dynamics of a feedback network without actually solving its equations of motion. Furthermore, most Lyapunov functions studied in this chapter play a role similar to that of the Hamiltonian of a conservative system: For certain stochastic extensions of the deterministic time evolution, the network dynamics approach a Gibbsian equilibrium distribution generated by the Lyapunov function of the noiseless dynamics. This has allowed the application of powerful techniques from statistical mechanics and has led to quantitative results about the performance of recurrent neural networks far beyond the limits of a local stability analysis. The existence of a Lyapunov function is thus of great conceptual as well as technical importance.

Lyapunov's method suffers, however, from one serious flaw: No systematic technique is known to decide whether a dynamical system admits a Lyapunov function or not. Finding Lyapunov functions requires experience, intuition, and luck. Fortunately, a wealth of knowledge on both practical and theoretical issues has been accumulated over the years.

The present chapter is intended as an overview of neural network architectures and dynamics where Lyapunov's method has been successfully employed to study the global network behavior. A general framework for modeling the dynamics of biological neural networks is developed in Sec. 1.2. This framework allows for a classification of various dynamical schemes found in the literature and facilitates the formal analysis presented in later sections.

Recurrent networks that relax to fixed-point attractors only have been used as auto-associative memories for static patterns. Section 1.3 reviews convergence criteria for a number of prototypical networks: The Hopfield model [4], the Little model [5], systems with graded-response neurons [6, 7], iterated-map networks [8], and networks with distributed dynamics [9, 10]. A statistical mechanical analysis of networks with block-sequential dynamics and results about the convergence to fixed points in networks with signal delays conclude the section.

Neural networks with signal delays can be trained to learn pattern sequences. Such systems are analyzed in Sec. 1.4. It is shown that, with a discrete-time evolution, these networks can be mapped onto "equivalent networks" with block-sequential updating and no time delays. This connection allows for a quantitative analysis of the storage of temporal associations in time-delay networks. Next, the time evolution of a single neuron with delayed feedback and continuous-time dynamics is discussed. Two different Lyapunov functions are presented. The first shows that, under certain conditions, all solutions approach special periodic attractors; the second demonstrates that, under less restrictive conditions, the system relaxes to oscillating solutions that need not be periodic.

The pulselike nature of neural activity has frequently been modeled using (coupled) threshold elements that discharge rapidly when they reach a trigger threshold. With uniform positive couplings, some networks composed of such *integrate-and-fire neurons* approach globally synchronized solutions where all neurons fire in unison. With more general coupling schemes, the systems approach phase-locked solutions where neurons only exhibit locally synchronized pulse activity. Section 1.5 presents Lyapunov functions for such a class of integrate-and-fire models. An additional proof shows that the phase-locked solutions are reached in minimal time.

1.2 A Framework for Neural Dynamics

Starting with a brief description of the anatomy and physiology of single neurons, this section introduces a general framework for modeling neural dynamics.

1.2.1 Description of Single Neurons

Neurons consist of three distinct structures: dendrites, a cell body, and an axon. *Dendrites* are thin nerve fibers that form highly branched structures called *dendritic trees*. They extend from the central part of a neuron, called the *cell body* or *soma*, which contains the cell nucleus. The *axon*, a single long fiber, projects from the soma and eventually branches into *strands* and *substrands*. Located along the axon and at its endings are *synapses* that connect one (*presynaptic*) neuron to the dendrites and/or cell bodies of other (*postsynaptic*) neurons [11].

Neurons communicate via an exchange of electrochemical signals. At rest, a cell is held at a negative potential relative to the exterior through selective ion pumps in the cell membrane. If the potential at the soma exceeds a firing threshold due to incoming signals, a strong electrochemical pulse is generated. This excitation is called an *action potential* or *spike*. It is propagated along the axon by an active transport process that results in a solitonlike pulse of almost constant size and duration [12]. Following the generation of a spike, the membrane potential quickly drops to a subthreshold value. After the event, the neuron has to recover for a short time of a few milliseconds before it can become active again. This time interval is called the *refractory period*.

At synapses, action potentials trigger the release of *neurotransmitters*, which are chemical substances that diffuse to the postsynaptic cell where they bind to receptors. This process leads to changes of the local membrane properties of the postsynaptic neuron, causing either an increase or decrease of the local potential. In the first case, the synapse is called an *excitatory synapse*; in the second case, an *inhibitory synapse*. Through (diffusive) transport processes along the dendritic tree, an incoming signal finally arrives at the soma of the postsynaptic neuron where it makes a usually minute contribution to the membrane potential.

How can one construct a mathematical framework for neural dynamics that may be used to analyze large networks of interconnected neurons?

Let me begin with the description of neural output activity. A spike is an all-or-none event and thus may be modeled by a binary variable as was pointed out by McCulloch and Pitts [13]. It will be denoted by $S_i = \pm 1$, where i enumerates the neurons. This specific representation emphasizes the resemblance between McCulloch-Pitts neurons and Ising spins.[3] Following the conventional notation, $S_i = 1$ means that cell i is firing an action potential, and $S_i = -1$ means that the cell is quiescent.

In an alternative formulation, a quiescent cell is denoted by $S_i = 0$. Both representations are equivalent if the network parameters are transformed appropriately. In the integrate-and-fire models that are discussed in this chapter, the duration of action potentials is set to 0 for simplicity. To obtain a nonvanishing pulse integral, a spike is modeled by a Dirac δ-function, so that, formally speaking, one is dealing with a $0/\infty$ representation of action potentials.

---

[3]The Ising model [14] provides an extremely simple and elegant description of ferromagnets and has become one of the most thoroughly studied models in solid-state physics. The formal similarity between certain extensions of this model, namely, spin glasses, and neural networks such as the Hopfield model has stimulated the application of statistical mechanics to neural information processing (see also Sec. 1.3.6).

An action potential is generated if the membrane potential u_i exceeds a firing threshold u_{thresh}. Since the trigger process operates without significant time lags, spike generation (in the ± 1-representation) may be written

$$S_i(t) = \mathrm{sgn}[u_i(t) - u_{\mathrm{thresh}}], \tag{1.3}$$

where $\mathrm{sgn}(x)$ denotes the signum function.

In most of the models that will be analyzed in this chapter, the membrane potential u_i is not reset after the emission of an action potential. An important exception are networks with integrate-and-fire neurons whose precise reset mechanism is discussed in Sec. 1.2.3.

Some cortical areas exhibit pronounced coherent activity of many neurons on the time scale of interspike intervals, that is, 10 – 100 ms [15, 16, 17]. Modeling this phenomenon requires a description of output activity in terms of single spikes, for example, by using integrate-and-fire neurons.[4] In other cases, the exact timing of individual action potentials does not seem to carry any relevant information. One then may switch to a description in terms of a coarse-grained variable, the short-time-averaged firing rate V. Unlike the binary outputs of McCulloch-Pitts neurons, the firing rate is a continuous variable. The firing rate varies between 0 and a maximal rate V_{max}, which is determined by the refractory period. Within a firing-rate description, model neurons are called *analog neurons* or *graded-response neurons*.

In such a real-valued representation of output activity, the threshold operation (1.3) is replaced by an s-shaped ("sigmoid") transfer function to describe the graded response of the firing rate to changes of the membrane potential,

$$V_i(t) = g_i[u_i(t)] \tag{1.4}$$

with $g_i : \mathbb{R} \to [0, V_{\mathrm{max}}]$. The functions g_i can be obtained from neurophysiological measurements of the response characteristic of a cell under quasi-stationary conditions.

Once generated by a neuron, say neuron j, an action potential travels as a sharp pulse along the axon and arrives at a synapse with neuron i after some time lag τ_{ij}. The delay depends on the distance traveled by the signal and its propagation speed, and may be as long as 10 – 50 ms. It follows that the release of neurotransmitter at time t does not depend on the present presynaptic activity but that it should be modeled by some function whose argument is the earlier activity $S_j(t - \tau_{ij})$. Diffusion across the synaptic cleft adds a distributed delay that is usually modeled by an integral kernel with a single hump.

What remains in the modeling process is the formalization of the dendritic and somatic signal processing. The force driving the membrane poten-

---

[4]Alternative approaches are discussed in the contribution of Gerstner and van Hemmen in this volume [18].

tial u_i up or down will be called the *local field* and denoted by h_i. Formally, the local field can always be written as a power series of the synaptic input currents. The exact form of the coefficients depends on the microscopic cell properties.

Dendrites and cell bodies are complex extended objects with intricate internal dynamics. This implies that, within any accurate microscopic description, even the dendrites and soma of a *single* cell have to be represented by a large number of parameters and dynamical variables [19, 20].[5] However, such a detailed approach cannot be pursued to analyze the time evolution of large networks of highly interconnected neurons as they are found in the cerebral cortex, where a neuron may be connected with up to 10,000 other cells [21].

The theory of formal neural networks offers a radical solution to this fundamental problem. Following a long tradition in statistical physics, the theory is built on the premise that detailed properties of single cells are not essential for an understanding of the *collective* behavior of large systems of interacting neurons: "Beyond a certain level complex function must be a result of the interaction of large numbers of simple elements, each chosen from a small variety." [22]. This point of view invites a long and controversial debate about modeling the brain and, more general, modeling complex biological systems. Such a discussion is beyond the scope and intention of the present chapter. Instead, I will cautiously adopt this position as a powerful working hypothesis whose neurobiological foundations require further investigation.[6] The advantage is obvious: Under the assumption that the function of large neural networks does not depend on microscopic details of single cells, and knowing that, in general, *many* incoming signals are necessary to trigger an action potential, it is sufficient to consider just the first terms of the power series defining the local field h_i. For the rest of this chapter, I will use the simplest approach and take only linear terms into account. The local field then may be written as

$$h_i(t) = \sum_{j=1}^{N} \int_0^{\tau_{\max}} J_{ij}(\tau) V_j(t-\tau) d\tau + I_i^{\text{ext}}(t). \tag{1.5}$$

For two state neurons, the term $V_j(t-\tau)$ is replaced by $S_j(t-\tau)$. The weight $J_{ij}(\tau)$ describes the influence of the presynaptic activity of neuron j at time $t-\tau$ on the local field of neuron i at time t. Input currents due to external stimuli are denoted by $I_i^{\text{ext}}(t)$.

---

[5]The argument applies to axons as well, but due to the emergent simplicity of axonal signal transport — action potentials are characterized by a dynamically stabilized, fixed pulse shape — a macroscopic description in terms of all-or-none events is justified.

[6]Unexpected support for this viewpoint comes from elaborate computer simulations of the dynamics of single cerebellar Purkinje cells [23].

The temporal details of signal transmission are reflected in the functional dependence of $J_{ij}(\tau)$ on the delay time τ. Axonal signal propagation corresponds to a discrete time lag; diffusion processes across the synapses and along the dendrites result in delay distributions with single peaks. Distributed time lags with multiple peaks may be used to include pathways via interneurons that are not explicitly represented in the model. A synapse is excitatory if $J_{ij}(\tau) > 0$ and inhibitory if $J_{ij}(\tau) < 0$. Self-couplings $J_{ii}(\tau)$ that are strongly negative for small delays may be used to model refractoriness [24, 25].[7] In network models without synaptic and dendritic delays, the local field h_i is identical to the total synaptic input current to neuron i, which often is denoted by I_i in the neural network literature.

As shown in this section, there are three main variables to describe the activity of single neurons — the membrane potential u_i, the output activity V_i or S_i, and the local field h_i. These three variables correspond to the three main parts of a neuron — soma, axon, and dendritic tree. The strongly nonlinear dependence of V_i or S_i on u_i captures the "decision process" of a neuron — to fire or not to fire. This decision is based on some evaluation of the weighted average h_i of incoming signals. To close the last gap in the general framework, one has to specify the dynamical relation between the membrane potential u_i and the local field h_i.

If there are no transmission delays, Eqs. (1.3)–(1.5) contain only a single time argument and no time derivatives, that is, they do not describe any *dynamical* law. It follows that the relation between u_i and h_i has to be formulated as an evolution equation. If one opts for a description where time is treated as a discrete variable, the evolution equation will be a difference equation; otherwise, a differential equation. As a first approximation, both types of dynamical descriptions may be linear since the main source for nonlinear behavior, namely, spike generation, is already described by Eq. (1.3) or (1.4).

1.2.2 Discrete-Time Dynamics

Within a discrete-time approach, time advances in steps of fixed length, usually taken to be unity. To obtain a consistent description, all signal delays should be nonnegative integers. Accordingly, the temporal integral $\int_0^{\tau_{\max}} J_{ij}(\tau)S_j(t-\tau)d\tau$ in Eq. (1.5) is replaced by a sum $\sum_{\tau=0}^{\tau_{\max}} J_{ij}(\tau)S_j(t-\tau)$.

In a discrete-time model, the most straightforward dynamic relation between u_i and h_i is the shift operation

$$u_i(t+1) = h_i(t). \tag{1.6}$$

[7] In some sense, the same is achieved in integrate-and-fire models where the membrane potential is explicitly reset after spike generation.

At a first glance, this dynamical relation neglects any inertia of the membrane potential caused by a nonzero transmembrane capacitance. According to Eq. (1.6), the membrane potentials are just time-shifted copies of the local fields. Inertia could be included on the single-neuron level by an additive term $\alpha u_i(t)$ on the right-hand side of Eq. (1.6); however, a similar effect can be obtained through a proper choice of the update rule for the overall network, as will be discussed at the end of this section.

For two state neurons, Eqs. (1.3), (1.5), and (1.6) may be combined to yield the single-neuron dynamics

$$S_i(t+1) = \mathrm{sgn}[h_i(t)], \tag{1.7}$$

where

$$h_i(t) = \sum_{j=1}^{N} \sum_{\tau=0}^{\tau_{\max}} J_{ij}(\tau) S_j(t-\tau) + I_i^{\mathrm{ext}}(t). \tag{1.8}$$

The term u_{thresh} has been absorbed in I_i^{ext} without loss of generality. In passing, note that, in the exceptional case $h_i(t) = 0$, it is advisable to supplement Eq. (1.7) by the convention $S_i(t+1) = S_i(t)$ for (purely technical) reasons that will become apparent in Sec. 1.3.1.

For analog neurons, Eqs. (1.7) and (1.8) are replaced by

$$V_i(t+1) = g_i[h_i(t)] \tag{1.9}$$

and

$$h_i(t) = \sum_{j=1}^{N} \sum_{\tau=0}^{\tau_{\max}} J_{ij}(\tau) V_j(t-\tau) + I_i^{\mathrm{ext}}(t). \tag{1.10}$$

The membrane potential u_i no longer appears in Eqs. (1.7)–(1.10) as the single-neuron description has been reduced from three to two variables — output activity and local field. Either one might be used as a state variable.

Neurotransmitters are released in small packages by a stochastic mechanism that includes spontaneous release at times when no spikes arrive at a synapse [26, 27]. This phenomenon, known as *synaptic noise*, is the most important source of stochasticity in neural signal transmission.

If one takes synaptic noise into account, the local field becomes a fluctuating quantity $h_i + \nu_i$, where ν_i denotes the stochastic contributions. The probability of spike generation then is equal to the probability that the local field exceeds the firing threshold. For two state neurons, this probability may be written as

$$\mathrm{Prob}[S_i(t+1) = +1] = f[h_i(t)], \tag{1.11}$$

where Prob denotes probability and $f : \mathbb{R} \to [0,1]$ is a monotone increasing function.

A careful analysis of synaptic transmission reveals that, under the assumption of linear dendritic processing, the stochastic variable ν_i is distributed according to a Gaussian probability distribution [22, 28]. In that case, Eq. (1.11) can be approximated by

$$\text{Prob}[S_i(t+1) = \pm 1] = \tfrac{1}{2}\{1 \pm \tanh[\beta h_i(t)]\}, \tag{1.12}$$

where $T \equiv \beta^{-1}$ is a measure of the noise level. In the limit as $T \to 0$, one recovers the deterministic threshold dynamics (1.7). In the physics literature, the update rule (1.12) is known as *Glauber dynamics* [29]. It was invented as a heat-bath algorithm for the Ising model [14] and has become an important tool for analyzing the collective properties of many-particle systems.

Equations (1.7)–(1.10) describe the time evolution of *individual neurons.* This leaves a number of options for the updating process at the level of the *overall network* [10].

First, there is the question of how many neurons may change their state at a time. Theoretical investigations of recurrent networks with discrete-time dynamics have almost exclusively focused on two cases: parallel dynamics (PD) and sequential dynamics (SD). In the former case, all neurons are updated in perfect synchrony, which has led to the name *synchronous dynamics.* In the latter case, only one neuron is picked at each time to evaluate its new state — *one-at-a-time updating* — while the activities of all other neurons remain constant. Parallel updating and sequential updating are two extreme realizations of discrete-time dynamics. Intermediate schemes will be called *distributed dynamics* (DD) and include block-sequential iterations where the network is partitioned into fixed clusters of simultaneously updated neurons.

Next, there is the question of how groups (of one or more neurons) are selected at each time step. One may have a fixed partition of the network, or one may choose random samples at each time step. Alternatively, one may study selective mechanisms such as a *maximum-field* or *greedy dynamics* [30]. Here, the neuron with the largest local field opposite to its own activity is updated.[8]

Network dynamics are said to be *fair sampling* if, on an intermediate time scale, no neuron is skipped for the updating process on average. The terminology emphasizes the similarity with the idea of "fairness" used by the computer science community [31]. On a conceptual level, fair sampling assures that all neurons have a chance to explore the part of phase space accessible to them through their single-neuron dynamics. Most computa-

[8]The network dynamics of integrate-and-fire neurons also may be viewed as a selective update process: Only those neurons whose local fields are larger than the threshold are active for the duration of an action potential. After that time, both output S_i and membrane potential u_i are reset to their rest values.

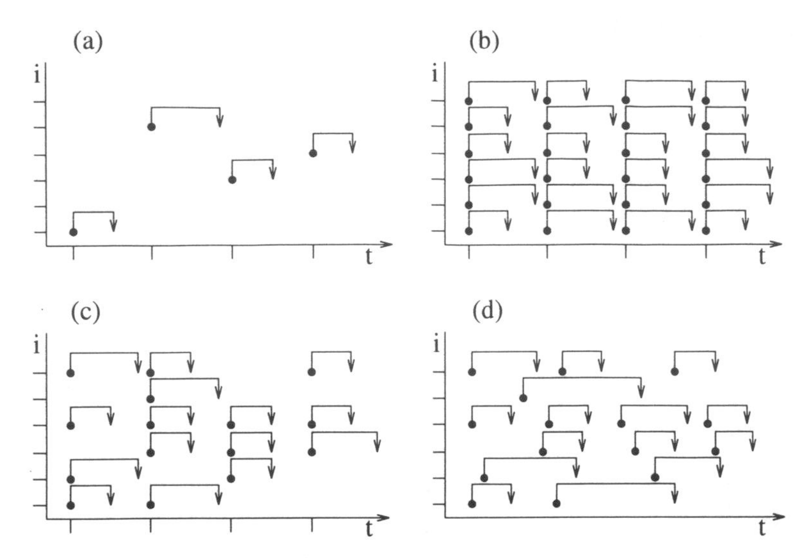

Fig. 1.1. Schematic representation of discrete-time updating schemes. Horizontal axes represent time, ticks on the vertical axes label the neurons. Delays due to transmission and computation times are indicated by the finite duration of the updating "event" for a given neuron. Clocked networks have ticks on the time axis. (a) One-at-a-time or sequential dynamics (SD); (b) synchronous or parallel dynamics (PD); (c) distributed dynamics (DD): still clocked, but with arbitrary update groups at each time step; (d) fully asynchronous dynamics including overlapping delays.

tionally useful iteration schemes are of this type. All updating schemes with a fixed partition or a random selection process are fair sampling. Exceptions may only occur in pathological situations within selective algorithms.

Finally, there is the question of whether signal delays may or may not overlap, as is illustrated in Fig. 1.1. The latter case is of utmost importance for the storage and retrieval of pattern sequences, as will be discussed in Sec. 1.4.

Summarizing the above discussion, updating rules for networks with distributed discrete-time dynamics may be categorized according to the following five criteria:

1. Description of output activity: (a) discrete; (b) continuous.

2. Single-neuron dynamics: (a) deterministic; (b) stochastic.

3. Size of group to be updated at each time step:

   (a) all neurons — parallel dynamics (PD);
   (b) some neurons — distributed dynamics (DD);

(c) one neuron — sequential dynamics (SD).

4. Selection of the update group at each time step: (a) fixed partition; (b) random sample; (c) selective choice.

5. Handling of delays: (a) overlapping not allowed; (b) overlapping allowed.

Most discrete-time descriptions appearing in the literature can be classified by these five criteria. For instance, Caianiello's model [32] uses McCulloch-Pitts neurons (rule 1a) and includes a broad distribution of transmission delays (rule 5b). All neurons are updated at the same time (rules 3a and 4a) according to a deterministic threshold operation (rule 2a). The Little model [5] differs from Caianiello's approach in that it describes single neurons as stochastic elements (rule 2b) with instantaneous interactions only (rule 5a). In the Hopfield model [4], neurons are updated one at a time (rule 3c), again without signal delays (rule 5a).

If neurons are picked in a random order, there is a nonzero chance that a neuron will be skipped during an elementary cycle of the network dynamics. On the level of macroscopic order parameters, this leads to an effective inertia comparable to that generated by an additive term $\alpha u_i(t)$ in Eq. (1.6).[9]

In closing this section, we introduce some helpful notation: Networks with deterministic parallel dynamics, continuous neurons, and no transmission delays (rules 1b, 2a, 3a, 4a, and 5a) will be called *iterated-map networks* (IM); those with (a broad distribution of) transmission delays and a deterministic parallel dynamics (rules 2a, 3a, 4a, and 5b) will be referred to as *time-delay networks* (TD).

1.2.3 Continuous-Time Dynamics

The step size in a discrete-time description is usually identified with the duration of an action potential. This implies on the one hand that such a description cannot accommodate the time resolution required to study the synchronization of action potentials.[10] On the other hand, the feedback delay implicitly built into any discrete-time description may lead to dynamical artefacts such as spurious oscillations. To avoid both problems, one may alternatively study networks with continuous-time dynamics.

---

[9] For a derivation of the evolution equations of macroscopic order parameters, see for example, reference [33].

[10] Decreasing the step size leads to a complication in the mathematical formulation because one is forced to introduce effective delayed interactions if one wants to assure that action potentials last for multiple elementary time steps.

Graded-Response Neurons

Membrane potentials of real neurons are subject to leakage currents due to the finite resistivity of biological membranes. Once charged by a short input current modeled by the local field $h_i(t)$, the membrane potential $u_i(t)$ of cell i relaxes to some rest value that is set to 0 for simplicity.

The physics of charging and leakage is best captured by the linear first-order differential equation

$$C\frac{d}{dt}u_i(t) = -R^{-1}u_i(t) + h_i(t). \tag{1.13}$$

Here, C denotes the input capacitance of a neuron and R is its trans-membrane resistance. Model neurons whose membrane potential changes according to the differential equation (1.13) will be called *graded-response neurons* (GR).

Inserting Equation (1.5) into (1.13), the time evolution of graded-response neurons may be written as

$$C\frac{d}{dt}u_i(t) = -R^{-1}u_i(t) + \sum_{j=1}^{N}\int_0^{\tau_{\max}} J_{ij}(\tau)V_j(t-\tau)d\tau + I_i^{\text{ext}}(t), \tag{1.14}$$

where, as in Sec. 1.2.1, the output activity V_j depends on the membrane potential u_j through the nonlinear response characteristic (1.4).

Similar to the discrete-time dynamics considered in Sec. 1.2.2, one of the original three variables to describe neural activity has become superfluous. In Sec. 1.2.2, the membrane potential $u_i(t)$ was expressed through the (time-shifted) local field $h_i(t-1)$; now, the local field $h_i(t)$ has been replaced by the membrane potential $u_i(t)$ and its time derivative $\dot{u}_i(t)$.

Integrate-and-Fire Neurons

Below the firing threshold, (leaky) integrate-and-fire neurons operate in the same way as graded-response neurons [Eq. (1.13)]. However, when the membrane potential of a cell reaches the threshold u_{thresh}, the cell produces an action potential and resets its potential to u_{reset}. For convenience, units can be chosen such that $u_{\text{thresh}} = 1$ and $u_{\text{reset}} = 0$.

Assuming vanishing signal delays and action potentials of negligible duration, the local field $h_i(t)$ of neuron i then is given by

$$h_i(t) = \sum_j J_{ij}f_j(t) + I_i^{\text{ext}}(t), \tag{1.15}$$

where the instantaneous firing rate $f_j(t)$ is a sum of Dirac δ-functions,

$$f_j(t) = \sum_n \delta(t - t_j^n), \tag{1.16}$$

and the t_j^n are the times at which neuron j generates an action potential. Throughout the remaining sections on integrate-and-fire neurons, the external input $I_i^{\text{ext}}(t)$ is assumed to be constant in time, $I_i^{\text{ext}}(t) = I_i^{\text{ext}}$.

The general behavior of the system is now as follows. While none of the neurons is producing an action potential, Eq. (1.13) can be integrated to yield

$$u_i(t) = [u_i(t_0^+) - RI_i^{\text{ext}}]e^{-(t-t_0)/RC} + RI_i^{\text{ext}} \qquad \text{for } t \geq t_0, \tag{1.17}$$

where t_0 denotes the last firing time. When the potential u_j of neuron j reaches 1 (the threshold), it drops instantaneously to 0. At the same time, the potential u_i of each neuron i to which j makes a synapse is increased by J_{ij}.

Because the durations of action potentials and synaptic currents have been set equal to 0, the description given so far contains an ambiguity. To which value should neuron i be reset if at time t an action potential is produced by cell j, if the synapse from j to i is excitatory, $J_{ij} > 0$, and if $u_i(t^-) > 1 - J_{ij}$? In this case, the action potential will raise u_i above 1, and cell i should generate its action potential during the flow of synaptic current produced by the synapse J_{ij}. When synaptic (and dendritic) time constants of the nerve cells to be modeled are longer than the duration of action potentials, what should actually happen in the model is that cell j should fire when its potential reaches $u_{\textbf{thresh}} = 1$, and the synaptic current from synapse J_{ij} that arrives after i fires should be integrated to yield a positive potential (relative to $u_{\textbf{reset}}$) afterward. Thus, if cell j fires first and at time t, and that event evokes a firing of neuron i, then, after both action potentials have been generated, the two membrane potentials should be

$$u_j(t^+) = J_{ji} \tag{1.18}$$

and

$$u_i(t^+) = u_i(t^-) + J_{ij} - 1. \tag{1.19}$$

The first equation represents the fact that j fired first when $u_j = 1$ was reset to 0, and when neuron i subsequently generated its action potential, this changed the potential of j to J_{ji}. The second equation represents the fact that i fired second, reduced its potential by 1 when it did so, but received the synaptic current J_{ij} when neuron j fired.

The updating rule can be generalized to a large network of neurons by the following algorithm. As the potentials all increase with time, a first neuron j reaches $u_j = 1$. Reset that potential to 0. Then change the potential of each neuron i by J_{ij}. If, following this procedure, some of the potentials become greater than 1, pick the neuron with the largest potential, say, neuron k, and decrease its potential by 1.[11] Then change the potential of each neuron

[11] If several neurons exhibit the same maximum potential, one may use some fixed, random, or selective update order to pick one of them.

l by J_{lk}. Continue the procedure until no membrane potential is greater than 1. Then "resume the flow of time," and again let each potential u_i increase according to Eq. (1.17).

This deterministic algorithm preserves the essence of the idea that firing an action potential carries a neuron from u_{thresh} to u_{reset}, and effectively apportions the synaptic current into a part that is necessary to reach threshold and a part that raises the potential again afterward. Because the firing of one neuron can set off the instantaneous firing of others, this model can generate events in which many neurons are active simultaneously.

When synaptic (and dendritic) time constants are shorter than the duration of an action potential, all contributions from the synaptic current that arrive during spike generation are lost, and Eq. (1.19) should be replaced by $u_i(t^+) = 0$. Generalizing from these two extreme cases, Eq. (1.19) becomes

$$u_i(t^+) = \gamma[u_i(t^-) + J_{ij} - 1] \tag{1.20}$$

with $0 \leq \gamma \leq 1$.

For models with $\gamma = 1$, the order in which the neurons are updated in an event in which several neurons fire at once does not matter as long as $J_{ij} \geq 0$. For these cases, any procedure for choosing the updating sequence of the neurons at or above threshold will yield the same result because the reset is by a fixed negative amount (here: -1) regardless of whether immediately prior to reset $u_i = 1$ or $u_i > 1$.

If, in addition to choosing $\gamma = 1$, the limit $R \to \infty$ is considered, one is dealing with perfectly integrating cells. For a network of such neurons, the cumulative effects of action potentials and slow membrane dynamics commute if $J_{ij} \geq 0$. This makes the model formally equivalent to a class of *Abelian avalanche* models [34, 35]. Closely related earthquake models and (discrete-time) "sandpile models" relax to a critical state with fluctuations on all length scales, a phenomenon known as *self-organized criticality* [36].

The similarity between the microscopic dynamics of such model systems and networks of integrate-and-fire neurons has led to speculations about a possible biological role of the stationary self-organized critical state [37, 38, 39]. However, whereas for earthquakes, avalanches, and sandpiles the main interest is in the properties of the stationary state, for neural computation it is the convergence process itself which does the computation and is thus of particular interest. Furthermore, computational decisions must be taken rapidly, and in any event the assumption of constant input from other cortical areas implicit in all models breaks down at longer times [40, 41].

1.2.4 Hebbian Learning

The previous sections focused on the dynamics of neural activity. Synaptic efficacies were treated as time-independent parameters. Real synapses, however, are often modifiable. As was postulated by Hebb [42], their strengths may change in response to correlated pre- and postsynaptic activity: "When an axon of cell A is near enough to excite cell B and *repeatedly* or *persistently* takes part in firing it, some growth process or metabolic change takes place in one or both cells such that A's efficiency, as one of the cells firing B, is increased."

Hebbian plasticity has long been recognized as a key element for associative learning [43].[12] How should it be implemented in a formal neural network that might include transmission delays?

Hebbian learning is local in both space and time: Changes in synaptic efficacies depend only on the activity of the presynaptic neuron and the evoked postsynaptic response. Within the present framework, presynaptic activity is described by the axonal output V_j or S_j. Which neural variable should be chosen to model the postsynaptic response?

Neurophysiological experiments demonstrate that postsynaptic spiking is *not* required to induce long-term potentiation (LTP) of synaptic efficacies — "a critical amount of postsynaptic depolarization is normally required to induce LTP in active synapses, but sodium spikes do not play an essential role in the LTP mechanism" [45]. This result implies that the postsynaptic response is best described by the local field h_i — it represents the dendritic potential and is not influenced by the detailed dynamics of the cell body (u_i) or the spike-generating mechanism (V_i or S_i).

Let us now study a discrete-time system where delays arise due to the finite propagation speed of axonal signals, and focus on a connection with delay τ between neurons j and i. Originally, Hebb's postulate was formulated for excitatory synapses only, but, for simplicity, it will be applied to all synapses of the model network.

A presynaptic action potential that arrives at the synapse time t was generated at time $t - \tau$. Following the above reasoning, $J_{ij}(\tau)$ therefore should be altered by an amount that depends on $V_j(t-\tau)$ and $h_i(t)$, most simply, their product

$$\Delta J_{ij}(\tau) \propto h_i(t) V_j(t-\tau) \Delta t. \tag{1.21}$$

The bilinear expression (1.21) does not cover saturation effects. They could be modeled by an additional decay term $- \alpha J_{ij}(\tau)\Delta t$ on the right-hand side of Eq. (1.21).

The combined equations (1.3)–(1.5) and (1.21) describe a "double dynamics," where both neurons and synapses change in time. In general, such a system of coupled nonlinear evolution equations cannot be analyzed using Lyapunov's direct method, although there are some interesting counterexamples [46]. To simplify the analysis, one usually splits the network operation into two phases — learning and retrieval. For the learning phase, one frequently considers a *clamped* scheme, where neurons evolve according

[12] Various hypotheses about the microscopic mechanisms of synaptic plasticity are the subject of an ongoing discussion [44].

to external inputs only, $h_i(t) = I_i^{\text{ext}}(t)$. Once the learning sessions are over, the $J_{ij}(\tau)$ are kept fixed.

In the following, we focus on deterministic discrete-time McCulloch-Pitts neurons in a clamped scheme with $I_i^{\text{ext}}(t) = \pm 1$. This simplification implies that $S_j(t+1) = I_j^{\text{ext}}(t)$. Starting with a *tabula rasa*, $J_{ij}(\tau) = 0$, one obtains after P learning sessions, labeled by μ and each of duration D_u,

$$J_{ij}(\tau) = \varepsilon(\tau) N^{-1} \sum_{\mu=1}^{P} \sum_{t_\mu=1}^{D_\mu} I_i^{\text{ext}}(t_\mu) I_j^{\text{ext}}(t_\mu - 1 - \tau) \equiv \varepsilon(\tau) \tilde{J}_{ij}(\tau). \quad (1.22)$$

The parameters $\varepsilon(\tau)$ model morphological characteristics of the axonal delay lines, and N^{-1} is a scaling factor useful for the theoretical analysis. Note that an input sequence should be offered $\tau_{\max}$ time steps before the learning session starts so that all variables in Eq. (1.22) are well defined. According to Eq. (1.22), synapses act as microscopic feature detectors during the learning sessions: They measure and store correlations of the taught sequences in both space (i, j) and time (τ). This leads to a resonance phenomenon where connections with delays that approximately match the time course of the external input receive maximum strength. Note that these connections are also the ones that would support a stable sequence of the same duration. Thus, due to a subtle interplay between external stimulus and internal architecture (distribution of τ's), the Hebb rule (1.22), which prima facie appears to be instructive in character, exhibits in fact pronounced selective characteristics [47].

An external stimulus encoded in a network with a *broad* distribution of transmission delays enjoys a rather multifaceted representation. Synaptic couplings with delays that are short compared to the typical time scale of single patterns within the taught sequence are almost symmetric in the sense that $J_{ij}(\tau) \approx J_{ij}(\tau)$. These synapses encode the individual patterns of the sequence as *unrelated static objects*. On the other hand, synapses with transmission delays of the order of the duration of single patterns of the sequence are able to detect the transitions between patterns. The corresponding synaptic efficacies are asymmetric and establish various temporal relations between the patterns, thereby representing the complete sequence as *one dynamic object*.

Note that the interplay between neural and synaptic dynamics, and in particular the role of transmission delays, has been a subject of intensive research [32, 42, 48, 49]. The full consequences for the learning and retrieval of temporal associations have, however, been explored only recently.

As a special case of Eq. (1.22), consider the Hebbian learning of static patterns, $I_i^{\text{ext}}(t_\mu) = \xi_i^\mu$, offered during learning sessions of equal duration $D_\mu = D$ to a network with a uniform delay distribution. For mathematical convenience, the distribution is taken to be $\varepsilon(\tau) = D^{-1}$. In this case, Eq.

(1.22) yields synaptic strengths that are independent of the delay τ,

$$J_{ij}(\tau) = J_{ij} = N^{-1} \sum_{\mu=1}^{P} \xi_i^\mu \xi_j^\mu, \tag{1.23}$$

and symmetric,

$$J_{ij} = J_{ji}. \tag{1.24}$$

The synaptic symmetry (1.24) plays a key role in the construction of Lyapunov functions, as will be shown in the following sections.

Another kind of symmetry arises if all input sequences $I_i^{\text{ext}}(t_\mu)$ are cyclic with equal periods $D_\mu = D$. If one defines patterns ξ_{ia}^μ by $\xi_{ia}^\mu = I_i^{\text{ext}}(t_\mu = a)$ for $0 \leq a < D$, one obtains from Eq. (1.22)

$$\tilde{J}_{ij}(\tau) = N^{-1} \sum_{\mu=1}^{P} \sum_{a=0}^{D-1} \xi_{ia}^\mu \xi_{i,a-1-\tau}^\mu. \tag{1.25}$$

Note that the synaptic strengths are now in general asymmetric. They do, however, obey the symmetry $\tilde{J}_{ij}(\tau) = \tilde{J}_{ij}(D - (2 + \tau))$. For all networks whose a priori weights $\varepsilon(\tau)$ satisfy $\varepsilon(\tau) = \varepsilon(D - (2 + \tau))$, this leads to an *extended synaptic symmetry* [50, 51],

$$J_{ij}(\tau) = J_{ij}(D - (2 + \tau)), \tag{1.26}$$

extending the previous symmetry (1.24) in a natural way to the temporal domain. This type of synaptic symmetry allows the construction of a Lyapunov function for time-delay networks, as will be explained in Sec. 1.4.1.

1.3 Fixed Points

This section focuses on the storage of static patterns in networks with instantaneous interactions. It will be shown that, under certain conditions for the model parameters, various network dynamics exhibit the same long-time behavior: They relax to fixed points only.

Feedback networks with fixed-point attractors can be made potentially useful devices for associative computation as soon as one knows how to embed desired activity patterns as attractors of the dynamics. In such circumstances, an initial state or "stimulus" lying in the basin of attraction of a stored "memory" will spontaneously evolve toward this attractor. Within a biological context, the arrival at the fixed point may be interpreted as a cognitive event, namely, the "recognition of the stimulus."

The hypothesis that the brain utilizes fixed-point attractors to perform associative information processing has led to quantitative predictions [52] that are in good agreement with neurophysiological measurements [53]. However, even if the hypothesis was refuted in its literal sense, it would nevertheless continue to provide an important conceptual tool to think about neural information processing.

1.3.1 SEQUENTIAL DYNAMICS: HOPFIELD MODEL

Hopfield's original approach [4] is based on McCulloch-Pitts neurons with discrete-time dynamics, instantaneous interactions, and constant external stimuli. Neurons are updated one at a time, either according to a deterministic threshold operation (1.7) or probabilistic Glauber dynamics (1.12). In the original model neurons are chosen in a random sequential manner, but in simulations the update order is often fixed in advance, corresponding to a quenched random selection. Within the classification scheme of Sec. 1.2.2, the Hopfield model is thus characterized by rules 1a, 3c, and 5a.

If the single-neuron dynamics are deterministic, the time evolution of the network is a special realization of Eqs. (1.7) and (1.8) and may be written as

$$S_k(t+1) = \mathrm{sgn}[h_k(t)], \tag{1.27}$$

where k is the index of the neuron updated at time t and

$$h_k(t) = \sum_j J_{kj} S_j(t) + I_k^{\mathrm{ext}}. \tag{1.28}$$

All other neurons remain unchanged, $S_j(t+1) = S_j(t)$ for $j \neq k$.

What can be said about the global dynamics generated by Eqs. (1.27) and (1.28)? Consider the quantity

$$L_{\mathrm{SD}} = -\frac{1}{2} \sum_{i,,j=1}^{N} J_{ij} S_i S_j - \sum_{i=1}^{N} I_i^{\mathrm{ext}} S_i. \tag{1.29}$$

The change of L_{SD} in a single time step, $\Delta L_{\mathrm{SD}}(t) \equiv L_{\mathrm{SD}}(t+1) - L_{\mathrm{SD}}(t)$, is

$$\begin{aligned} \Delta L_{\mathrm{SD}}(t) = &- \frac{1}{2} \sum_{i,,j=1}^{N} J_{ij} [S_i(t+1) S_j(t+1) - S_i(t) S_j(t)] \\ &- \sum_{i=1}^{N} I_i^{\mathrm{ext}} [S_i(t+1) - S_i(t)]. \end{aligned} \tag{1.30}$$

Assume again that neuron k is updated at time t. The difference $\Delta S_j(t) \equiv S_j(t+1) - S_j(t)$ equals 0 or ± 2 if $j = k$ and vanishes otherwise. For the special case where the synaptic efficacies satisfy the symmetry condition

(1.24), one obtains

$$\begin{aligned}\Delta L_{\mathrm{SD}}(t) &= \Delta S_k(t)J_{kk}S_k(t) - \Delta S_k(t)\left[\sum_{j=1}^{N} J_{kj}S_j(t) + I_k^{\mathrm{ext}}\right] \\ &= -\frac{1}{2}J_{kk}[\Delta S_k(t)]^2 - \Delta S_k(t)h_k(t). \end{aligned} \tag{1.31}$$

According to Eq. (1.27) and the remark following Eq. (1.8), neuron k does not change its state if $h_k(t)S_k(t) \geq 0$. If this condition is not fulfilled, the neuron flips and $\Delta S_k(t) = 2S_k(t+1)$. The change of L_{SD} then may be written as

$$\begin{aligned}\Delta L_{\mathrm{SD}}(t) &= -2[J_{kk} + S_k(t+1)h_k(t)] \\ &= -2[J_{kk} + |h_k(t)|]. \end{aligned} \tag{1.32}$$

The last line follows from the evolution equation (1.27) and the identity $|a| = a\ \mathrm{sgn}(a)$. Equation (1.32) proves that L_{SD} is nonincreasing along every solution if the self couplings J_{ii} are nonnegative.[13] As a finite sum of finite terms, L_{SD} is bounded. If $J_{ii} \geq 0$ for all neurons, $L_{\mathrm{SD}}(t)$ has to approach a limit as $t \to \infty$. Furthermore, $\Delta L_{\mathrm{SD}}(t)$ vanishes only if the neuron updated at time t does not change its state.[14] This proves that the Hopfield network relaxes to fixed-point solutions only. According to Eqs. (1.27) and (1.28), these equilibria satisfy

$$S_i = \mathrm{sgn}\left[\sum_j J_{ij}S_j + I_i^{\mathrm{ext}}\right] \qquad \text{for all } i. \tag{1.33}$$

The results obtained may be summarized as follows:

If the synaptic efficacies J_{ij} satisfy the symmetry condition (1.24), and if the self- interactions J_{ii} are nonnegative, then the dynamics of the Hopfield model [Eqs. (1.27) and (1.28)] admit the Lyapunov function (1.29) and converge to fixed points (1.33) only.

Let me clarify a potentially confusing point. For neural networks with McCulloch-Pitts neurons, the state space consists of the corners of an N-dimensional hypercube $\{-1,+1\}^N$, also known as *Hamming space.* In this *discrete* space, the smallest state change possible is a single-spin flip, $S_i \to -S_i$. As a consequence, the system may converge to fixed points that are *not* stable with respect to activity changes of single neurons, in the sense that

---

[13] This condition is satisfied in Hopfield's original model, where all self-couplings are set to 0.

[14] For zero self-coupling J_{kk}, and in the exceptional case $h_k(t) = 0$, $\Delta L_{\mathrm{SD}}(t)$ vanishes for any update rule, even if one chooses $S_k(t+1) = -S_k(t)$ if $h_k(t) = 0$. However, if one sets $S_k(t+1) = S_k(t)$ as mentioned in Sec. 1.2.1, $\Delta L_{\mathrm{SD}}(t) = 0$ implies $\Delta S_k(t) = 0$, as desired.

a single-spin flip made to a fixed-point solution could actually lower L. For instance, consider a network where, for some neuron i, the self-interaction J_{ii} dominates possible contributions from other neurons, $J_{ii} < \sum_{j \neq i} |J_{ij}|$. In such a case, the initial value of S_i will never be changed, independent of its sign. The earlier results about network convergence continue to hold; that is, the system evolves towards fixed-point solutions only, but those are not necessarily local minima of L in the discrete-space sense.

1.3.2 Parallel Dynamics: Little Model

The Little model [5] uses the most simple discrete-time dynamics conceivable: It is a network of McCulloch-Pitts neurons, updated in parallel using instantaneous interactions only (rules 1a, 3a, 4a, and 5a). Within a deterministic description of single neurons (rule 2a), the time evolution of the network is given by

$$S_i(t+1) = \text{sgn}[h_i(t)] \qquad \text{for all} \quad i, \tag{1.34}$$

where

$$h_i(t) = \sum_j J_{ij} S_j(t) + I_i^{\text{ext}}. \tag{1.35}$$

Except for the update order, Eqs. (1.34) and (1.35) are identical to Eqs. (1.27) and (1.28). Accordingly, the fixed-point solutions of the Little model are the same as those of the Hopfield model, given by Eq. (1.33). Are there additional time-dependent attractors?

For simplicity, only the case $I_i^{\text{ext}} = 0$ will be analyzed in this section. Nonzero inputs will be treated in Secs. 1.3.4 and 1.3.5. As in Sec. 1.3.1, we focus on networks with symmetric couplings and study the time evolution of a suitable auxiliary function:

$$L_{\text{PD}} = -\sum_{i=1}^N |h_i| = -\sum_{i=1}^N h_i \ \text{sgn}(h_i). \tag{1.36}$$

If one evaluates this expression along a solution generated by the network dynamics (1.34) and (1.35), one obtains

$$\begin{aligned} L_{\text{PD}}(t) &= -\sum_{i=1}^N h_i(t) S_i(t+1) \\ &= -\sum_{i,j=1}^N J_{ij} S_j(t) S_i(t+1). \end{aligned} \tag{1.37}$$

Using the synaptic symmetry in Eq. (1.24), the last line also may be written as

$$L_{\text{PD}}(t) = -\sum_{j=1}^N S_j(t) h_j(t+1). \tag{1.38}$$

The difference $\Delta L_{\mathrm{PD}}(t) \equiv L_{\mathrm{PD}}(t+1) - L_{\mathrm{PD}}(t)$ is then

$$\begin{aligned}\Delta L_{\mathrm{PD}}(t) &= -\sum_{i=1}^{N} |h_i(t+1)| + \sum_{i=1}^{N} S_i(t) h_i(t+1) \\ &= -\sum_{i=1}^{N} [S_i(t+2) - S_i(t)] h_i(t+1), \end{aligned} \tag{1.39}$$

where Eq. (1.34) has been used to obtain the last equation.

Like L_{SD}, the function L_{PD} is bounded. Evaluated along any solution of Eqs. (1.34) and (1.35), L_{PD} is nonincreasing because the right-hand side of Eq. (1.39) is nonpositive; the product $S_i(t)h_i(t+1)$ is $\pm h_i(t+1)$ and thus smaller or at most equal to $|h_i(t+1)|$. Consequently, $\Delta L_{\mathrm{PD}}(t)$ has to approach 0 as $t \to \infty$. $\Delta L_{\mathrm{PD}}(t)$ vanishes only if the system settles into a state with $S_i(t+2) = S_i(t)$ for all i, that is, a fixed-point solution [Eq. (1.33)] or a limit cycle of period two. In the latter case, some neurons switch between firing and quiescence at every time step while all other neurons remain in one activity state:

Assume that the synaptic couplings J_{ij} satisfy the symmetry condition (1.24). Then the dynamics of the Little model [Eqs. (1.34) and (1.35)] admit the Lyapunov function (1.36) and converge to fixed points (1.33) or period-two oscillations.

As will be shown in Sec. 1.3.5, the oscillating solutions can be excluded under additional assumptions for the synaptic couplings.

1.3.3 Continuous Time: Graded-Response Neurons

This section deals with the continuous-time dynamics of neural networks composed of analog neurons without signal delays. The network dynamics in Eq. (1.14) reduce to a set of coupled ordinary differential equations,

$$C\frac{d}{dt}u_i = -R^{-1}u_i + \sum_{j=1}^{N} J_{ij}V_j + I_i^{\mathrm{ext}}, \tag{1.40}$$

where

$$V_i = g_i(u_i). \tag{1.41}$$

Since the dynamical variables u_i and V_i in Eq. (1.40) are taken at equal times, all temporal arguments have been omitted.

The input–output relation g_i will be called *sigmoid* if it is increasing, differentiable, and grows in magnitude more slowly than linearly for large positive or negative arguments. The maximum slope of g_i will be referred to as the *gain* γ_i of neuron i. The nonlinearity is often modeled by a hyperbolic tangent, $g_i(u_i) = \frac{1}{2}[1 + \tanh(\gamma_i u_i)]$. In the high-gain limit $\gamma_i \to \infty$, one obtains a 0/1 representation of neural activity. It can be mapped onto Ising spins [14] through the identification $S_i = 2V_i - 1$.

Cohen and Grossberg [6] and Hopfield [7] studied the global behavior of networks with graded- response neurons, sigmoid response functions, and symmetric synapses. They used Lyapunov functions of the form

$$L_{\mathrm{GR}} = -\frac{1}{2}\sum_{i,j=1}^{N} J_{ij}V_iV_j - \sum_{i=1}^{N} I_i^{\mathrm{ext}}V_i + \sum_{i=1}^{N} R^{-1}G_i(V_i), \tag{1.42}$$

where the functions $G_i(V_i)$ are given by

$$G_i(V_i) = \int_0^{V_i} g_i^{-1}(x)dx. \tag{1.43}$$

The last expression is well defined because sigmoid nonlinearities are strictly monotone by definition. Since sigmoid functions grow less than linearly for large absolute arguments, the functions $G_i(V_i)$ increase faster than V_i^2 as $V_i \to \pm\infty$. The function L_{GR} is therefore bounded below.

Let us compute the time derivative of L_{GR} along a solution of the network dynamics. Using the synaptic symmetry in Eq. (1.24), one obtains

$$\begin{aligned} \frac{d}{dt}L_{\mathrm{GR}}(t) &= -\sum_{i=1}^{N}\left[\sum_{j=1}^{N} J_{ij}V_j + I_i^{\mathrm{ext}} - R^{-1}u_i\right]\frac{dV_i}{dt} \\ &= -\sum_{i=1}^{N} C^{-1}\frac{du_i}{dt}\frac{dV_i}{dt} \\ &= -\sum_{i=1}^{N} C^{-1}\left(\frac{du_i}{dt}\right)^2\frac{dg_i}{du_i} \le 0. \end{aligned} \tag{1.44}$$

The formula proves that the function L_{GR} is nonincreasing along every trajectory. The time derivative vanishes only at equilibria, which are given by

$$V_i = g_i\left[R\sum_j J_{ij}V_j + RI_i^{\mathrm{ext}}\right], \tag{1.45}$$

or at network states, where $dg_i/du_i = 0$ for all i. If, however, the latter states do not satisfy Eq. (1.45), the system will continue to evolve according to Eqs. (1.40) and (1.41). The final result may be stated as follows:

Suppose that the synaptic efficacies in a network of graded-response neurons [Eqs. (1.40) and (1.41)] respect the symmetry condition (1.24) and that the input–output relations are sigmoid. Then the network dynamics admit the Lyapunov function (1.42) and relax to fixed-point solutions (1.45) only.

A comparison of the Lyapunov function L_{GR} with the Lyapunov function L_{SD} provides some hints about how to construct Lyapunov functions for systems with sigmoid input–output characteristics: The additional term $\sum_i R^{-1}G_i(V_i)$ dominates the quadratic term $-\frac{1}{2}\sum_{i,j} J_{ij}V_iV_j$ for large V_i if the g_i are sigmoid. Consequently, the function L_{GR} is bounded below even if the V_i are not.[15] Furthermore, the term $\sum_i R^{-1}G_i(V_i)$ is constructed in such a way that its partial derivative with respect to V_i supplies the term $R^{-1}u_i$, which makes it possible to insert the evolution equation (1.40) into Eq. (1.44). Similar ideas will be applied in Secs. 1.3.4 and 1.3.5 to analyze discrete-time networks with sigmoid nonlinearities.

1.3.4 Iterated-Map Networks

Feedback networks with deterministic analog elements and synchronous discrete-time updating have been studied for a long time [32, 48, 49]. For vanishing signal delays and fixed inputs, the network dynamics, Eqs. (1.9) and (1.10), become

$$V_i(t+1) = g_i[h_i(t)] \qquad \text{for all } i, \tag{1.46}$$

where

$$h_i(t) = \sum_{j=1}^{N} J_{ij}V_j(t) + I_i^{\text{ext}}. \tag{1.47}$$

Systems described by Eqs. (1.46) and (1.47) have been called *iterated-map networks* [8]. Their fixed points coincide with those of graded-response networks [Eq. (1.45)] once one sets $R = 1$.

If the input–output functions g_i are threshold functions, $g_i(u_i) = \text{sgn}(u_i)$, one recovers the Little model, Eqs. (1.34) and (1.35). This connection indicates that one may find a Lyapunov function for iterated-map networks by combining appropriate parts of the Lyapunov function for the Little model with that for networks of graded-response neurons.

Let us follow the approach of Marcus and Westervelt [8] and study the time evolution of the function

$$\begin{aligned} L_{\text{IM}}(t) &= -\sum_{i,j=1}^{N} J_{ij}V_i(t)V_j(t-1) - \sum_{i=1}^{N} I_i^{\text{ext}}[V_i(t) + V_i(t-1)] \\ &\quad + \sum_{i=1}^{N}[G_i(V_i(t)) + G_i(V_i(t-1))], \end{aligned} \tag{1.48}$$

where $G_i(V_i)$ is defined as in Eq. (1.43).

Apart from a global time shift, the first term in Eq. (1.48) corresponds to L_{PD}, as can be seen from Eq. (1.37); the other terms should be com-

[15]It should be noted that, if a Lyapunov function is not globally bounded below, it still might be used for a local analysis.

pared with the second and third terms in Eq. (1.42). Notice that, unlike $L_{\rm PD}$ in Eq. (1.36), the function $L_{\rm IM}$ is written as an explicitly time-dependent function with temporal arguments t and $t-1$. In principle, one could use the evolution equations (1.46) and (1.47) and replace $V_i(t)$ by $g_i\left[\sum_{j=1}^{N} J_{ij}V_j(t-1) + I_i^{\rm ext}\right]$ to obtain a description that involves a single time argument only. However, since we mainly are interested in the evaluation of $L_{\rm IM}$ along trajectories, the shorter definition in Eq. (1.48) suffices.

Under the assumption of synaptic symmetry in Eq. (1.24), the temporal difference $\Delta L_{\rm IM}(t) \equiv L_{\rm IM}(t+1) - L_{\rm IM}(t)$ is

$$\Delta L_{\rm IM}(t) = -\sum_i h_i(t)\Delta_2 V_i(t) + \sum_i [G_i(V_i(t+1)) - G_i(V_i(t-1))], \quad (1.49)$$

where

$$\Delta_2 V_i(t) \equiv V_i(t+1) - V_i(t-1) \quad (1.50)$$

is the change of V_i over two time steps.

The right-hand side of Eq. (1.49) is 0 if $\Delta_2 V_i(t) = 0$ for all i. Let us analyze the case where $\Delta_2 V_i(t) \neq 0$ for at least some i. For sigmoid g_i, g_i^{-1} is single-valued and increasing. Consequently, G_i is strictly convex. Through a Taylor expansion of $G_i(V_i(t-1))$ around $V_i(t+1)$, one obtains

$$G_i(V_i(t+1)) - G_i(V_i(t-1)) < \Delta_2 V_i(t) G_i'(V_i(t+1)). \quad (1.51)$$

For an illustration of the inequality, see the left part of Fig. 1.2.

Inserting the identity

$$G_i'(V_i(t+1)) = g_i^{-1}(V_i(t+1)) = h_i(t) \quad (1.52)$$

and Eq. (1.51) into Eq. (1.49), one arrives at the expression

$$\Delta L_{\rm IM}(t) \leq 0, \quad (1.53)$$

where the strict inequality holds if $\Delta_2 V_i(t) \neq 0$ for at least one neuron.

As was demonstrated in the last section, the functions $G_i(V_i)$ increase faster than V_i^2 for large $|V_i|$. This result implies that $L_{\rm IM}$ is bounded below. As is shown by Eq. (1.53), the function $L_{\rm IM}$ strictly decreases along any solution of Eqs. (1.46) and (1.47) unless $\Delta_2 V_i(t) = 0$ for all neurons. The derivation may be summarized in the following way:

Assume that the synaptic efficacies in an iterated-map network [Eqs. (1.46) and (1.47)] are symmetric [Eq. (1.24)] and that the nonlinearities are sigmoid. Then the network dynamics admit the Lyapunov function (1.48) and relax to fixed-point solutions (1.45) or period-two oscillations.

In closing this section, let us briefly discuss antisymmetric synaptic couplings,

$$J_{ij} = -J_{ji}. \quad (1.54)$$

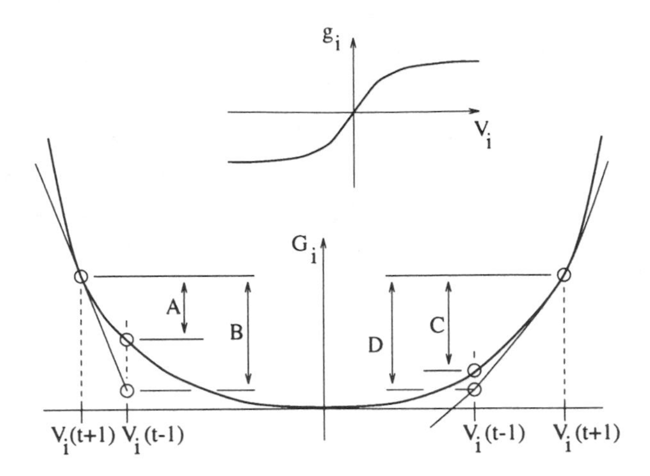

Fig. 1.2. Illustration of the inequalities (1.51) and (1.58) for a sigmoid input–output function $g_i(V_i)$. The convex function $G_i(V_i)$ is defined in Eq. (1.43). The straight line on the left-hand side and the parabola on the right-hand side are tangent to $G_i(V_i)$. The inequality (1.51) is the statement $A < B$, and the inequality (1.58) is the statement $C < D$.

The derivation of Sec. 1.3.2 for the Little model (with no external input) shows that, if Eq. (1.54) holds, one obtains

$$\Delta L_{\mathrm{PD}}(t) = -\sum_{i=1}^{N} [S_i(t+2) + S_i(t)] h_i(t+1). \tag{1.55}$$

In this case, the network approaches solutions that satisfy $S_i(t+2) = -S_i(t)$, that is, special limit cycles with period four [54].

It is left as an exercise to verify the same result for iterated maps without external input. Here, an additional condition is required, namely, that the input–output characteristics have to be odd functions, $g_i(V_i) = -g_i(-V_i)$. The interested reader may also try to construct Lyapunov functions for more general systems. In particular, he or she could look at two problems: (1) What kind of time-varying external stimuli can be incorporated into the Lyapunov function of the Little model if one focuses on antisymmetric couplings? (2) Are there Lyapunov functions for neural networks with McCulloch-Pitts neurons, antisymmetric couplings, and sequential dynamics with fixed update order?

1.3.5 Distributed Dynamics

In this section discrete-time updating schemes are considered that generalize beyond the Hopfield and Little models on both the single-neuron and network levels. Neurons are described by continuous variables with deterministic single-cell dynamics, that is, they fall into classes 1b and 2a in the scheme of Sec. 1.2.2. McCulloch-Pitts neurons with stochastic Glauber dynamics are discussed in Sec. 1.3.6. For the network dynamics, all choices of rules 3 and 4 are allowed that are fair sampling and do not lead to overlapping delays (rule 5a). The network dynamics are thus defined by a set of coupled nonlinear discrete-time equations:

$$V_i(t+1) = \begin{cases} g_i\left(\sum_{j=1}^{N} J_{ij}V_j(t) + I_i^{\text{ext}}\right) & \text{if } i \text{ is in } U(t), \\ V_i(t) & \text{otherwise.} \end{cases} \quad (1.56)$$

Here, $U(t)$ denotes the group of neurons updated at time t. The distributed dynamics, Eq. (1.56), reduce to block-sequential algorithms studied by Goles-Chacc et al. [9] if one considers McCulloch-Pitts neurons and fixed update groups $U_k, k = 0, 1, \ldots, K-1$ with $U(t) = U_{t(\text{modulo } K)}$.

There are a number of reasons to study partially parallel network dynamics such as Eq. (1.56). First, one may achieve a better understanding of the essential ingredients needed to construct feedback networks that possess fixed-point attractors only. Second, distributed dynamics map naturally on the architecture of parallel computers or computer networks. Third, the evolution equations (1.56) extend iterative methods that have been developed within the computer science community to solve nonlinear systems of equations [55, 56, 57, 58] to systems with *noncontracting* functions and multiple solutions.

What can be said about the long-time behavior of neural networks with distributed dynamics? As in Secs. 1.3.1 – 1.3.4, let us assume that the synaptic couplings are symmetric [Eq. (1.24)] and that the input–output characteristics are sigmoid. Consider again the Lyapunov function of networks with graded-response neurons in Eq. (1.42). The function now will be called L_{DD} to distinguish its discrete-time evolution from the continuous-time evolution of Sec. 1.3.3.

The only neurons that may change their state at time t belong to the update group $U(t)$. Accordingly, $\Delta V_i(t) \equiv V_i(t+1) - V_i(t)$ vanishes for all other neurons. Using the symmetry [see Eq. (1.24)] of the synaptic couplings, the change $\Delta L_{\text{DD}}(t) = L_{\text{DD}}(t+1) - L_{\text{DD}}(t)$ is given by

$$\begin{aligned} \Delta L_{\text{DD}}(t) &= -\frac{1}{2} \sum_{i \in U(t)} \sum_{j \in U(t)} J_{ij} \Delta V_i(t) \Delta V_j(t) - \sum_{j=1}^{N} \sum_{i \in U(t)} J_{ij} V_j(t) \Delta V_i(t) \\ &\quad - \sum_{i \in U(t)} I_i^{\text{ext}} \Delta V_i(t) + \sum_{i \in U(t)} [G_i(V_i(t+1)) - G_i(V_i(t)). \quad (1.57) \end{aligned}$$

Since the functions $g_i(V_i)$ are assumed to be sigmoid, the auxiliary functions $G_i(V_i)$ are again strictly convex. Expanding $G_i(V_i(t))$ to second order around $V_i(t+1)$ and replacing the coefficient of the quadratic term with the smallest possible value, that is, γ_i^{-1}, the following upper bound can be established (see also the right part of Fig. 1.2):

$$G_i(V_i(t+1)) - G_i(V_i(t)) \leq \Delta V_i(t) G_i'(V_i(t+1)) - \tfrac{1}{2}[\Delta V_i(t)]^2 \gamma_i^{-1}. \quad (1.58)$$

Equality holds if and only if $V_i(t+1) = V_i(t)$. Inserting Eqs. (1.52) and (1.58) into Eq. (1.57) gives

$$\Delta L_{\mathrm{DD}}(t) \leq -\frac{1}{2} \sum_{i \in U(t)} \sum_{j \in U(t)} (J_{ij} + \delta_{ij}\gamma_i^{-1}) \Delta V_i(t) \Delta V_j(t). \quad (1.59)$$

To facilitate further discussion, let us define $W(t)$ as the number of neurons in the group $U(t)$ and symmetric matrices $\mathbf{U}(t)$ of dimension $W(t) \times W(t)$ as submatrices of the connection matrix $\mathbf{J}$, which are given by the synaptic strengths of those neurons that are updated at time t. For the Hopfield model, Eqs. (1.27) and (1.28), where updating is one-at-a-time, $W(t) = 1$ for all t, and $\mathbf{U}(t)$ reduces to the self-interaction term J_{ii}, where i denotes the neuron being updated at time t. For the Little model, Eqs. (1.34) and (1.35), or iterated-map analog networks, Eqs. (1.46) and (1.47), the matrix is identical to $\mathbf{J}$ itself. As is obvious from these limiting cases, the structure of the set of matrices $\mathbf{U}(t)$ encodes the global dynamics.

The maximum neuron gain in the update group $U(t)$ will be denoted by $\gamma(t)$ and the minimum eigenvalue of the matrix $\mathbf{U}(t)$ by $\lambda_{\min}[\mathbf{U}(t)]$. Since, for arbitrary symmetric matrices $\mathbf{A}$ and $\mathbf{B}$, $\lambda_{\min}[\mathbf{A} + \mathbf{B}] \geq \lambda_{\min}[\mathbf{A}] + \lambda_{\min}[\mathbf{B}]$, a sufficient condition for $\Delta L(t) \leq 0$ is given by

$$\lambda_{\min}[\mathbf{U}(t)] \geq -\gamma(t)^{-1}. \quad (1.60)$$

If the above condition holds for all t, $L_{\mathrm{DD}}(t)$ is strictly decreasing as long as $V_i(t+1) \neq V_i(t)$ for at least some i in the update group $U(t)$. As before, the function L_{DD} is bounded below. The network therefore relaxes asymptotically to a state where L does not vary in time if all directions in the space spanned by the neural activities are explored, that is, if the updating scheme is fair sampling. Since equality in Eqs. (1.58) and (1.59) holds only if $V_i(t+1) = V_i(t)$, all solutions of Eq. (1.56) with time-independent L_{DD} are fixed-point solutions [10]. The result may be stated as follows:

Suppose the following three conditions hold: 1) the updating rule is fair sampling, 2) the neuron transfer functions are sigmoid, and 3) the symmetric connection matrix satisfies Eq. (1.60) for all times. Then the distributed dynamics (1.56) admit the Lyapunov function (1.42) and converge to fixed points only.

For iterated-map networks, $\mathbf{U}(t)$ is constant in time and equals the set of all neurons. The criterion $\lambda_{\min}[\mathbf{J}] \geq -\gamma(t)^{-1}$ provides a sufficient condition to exclude two cycles that exist in the general case as shown in Sec. 1.3.4: Lowering the neuron gain eliminates spurious oscillatory modes.

Neural networks with discrete elements correspond to the limit $\gamma_i \to \infty$, where Eq. (1.60) reduces to $\lambda_{\min}[\mathbf{U}(t)] \geq 0$. This implies in particular that there are no two cycles possible in the Little model if the whole connection matrix is nonnegative definite. The general remark from Sec. 1.3.1 about the convergence to solutions that are *not* minima of $L_{\rm DD}$ still holds in the discrete-neuron limit. This atypical behavior is, however, only possible because the g_i are piecewise constant functions in models with discrete neurons. For the generic case of continuous input–output characteristics, the network will always settle in a minimum as long as the initial conditions do not coincide with an unstable fixed-point of Eq. (1.56).

The convergence criterion in Eq. (1.60) is less restrictive for smaller update groups than for larger ones because

$$\lambda_{\min}[\mathbf{U}_1] \geq \lambda_{\min}[\mathbf{U}_2] \qquad \text{if } U_1 \subset U_2. \tag{1.61}$$

Note that Eq. (1.61) implies that the stability criterion for a fully parallel network, where $\lambda_{\min}[\mathbf{J}] \geq -\gamma^{-1}$, is a sufficient condition for Eq. (1.60) and thus is sufficient to assure that the system (1.56) will converge to a fixed point for *any* fair sampling updating scheme.

Formula (1.61) has direct consequences for possible applications. Consider a high-dimensional optimization task such as the traveling salesman problem. It may be mapped onto a neural network architecture which then defines a fixed connection matrix $\mathbf{J}$ [59]. The computational time needed to find a good solution can be reduced easily on a parallel computer by increasing the size of the update groups. However, the bounds given by Eq. (1.60) have to be met in order to assure convergence to fixed points, and will limit the maximal size of the update groups. The goal of large updating groups will be achieved in an optimal way if one can form update groups of weakly or noninteracting neurons. All submatrices $\mathbf{U}(t)$ will have small off-diagonal elements in that case, and their eigenvalues will be close or identical to the diagonal elements, that is, the bounds in Eq. (1.60) are largely independent of the size of the update groups. In principle, the search for optimal partitions of the above kind is itself a difficult optimization problem, but many applications exhibit an intrinsic structure (for example, predominantly short-range interactions) that naturally leads to good choices for the updating groups.

1.3.6 Network Performance

The results obtained thus far demonstrate that the long-time behavior of neural networks with symmetric synaptic couplings is surprisingly robust with respect to alterations of model details at both the level of single neurons and the level of the overall network dynamics. All systems studied relax to fixed-point solutions under appropriate additional conditions on the synaptic efficacies and the input–output characteristics.

Various prescriptions for the storage of static patterns as fixed-point attractors have been discussed in the literature [22, 60, 61]. In what follows, we concentrate on the Hebbian learning rule [Eq. (1.23)]. A statistical mechanical analysis of performance measures, such as storage capacity and retrieval quality, can be carried out most readily for networks with McCulloch-Pitts neurons and block-sequential dynamics. It also will be assumed that the network can be partitioned into n fixed update blocks of equal size W such that there are no interactions within a group [10]. As was emphasized before, such a situation can be arranged for many applications that map onto diluted or geometrically structured networks. In the limiting case $W = 1$, one recovers the Hopfield model.

To simplify the analysis, neurons are labeled by a double index S_{ia}. The first index, $1 \leq i \leq W$, refers to the position within an update group, while the second $1 \leq a \leq n$ labels the update group. The same notation applies to stored patterns ξ_{ia}^{μ}, where the additional index $\mu, 1 \leq \mu \leq p$, labels the patterns. With these conventions, the Hebb rule in Eq. (1.23) becomes

$$J_{ij}^{ab} = \begin{cases} \frac{1}{W(n-1)} \sum_{\mu=1}^{p} \xi_{ia}^{\mu} \xi_{jb}^{\mu} & \text{if } a \neq b, \\ 0 & \text{if } a = b. \end{cases} \tag{1.62}$$

The normalization factor N^{-1} in Eq. (1.23) has been changed to $[W(n-1)]^{-1}$ to guarantee the correct scaling behavior of L_{DD} in the thermodynamic limit $N \to \infty$.

Statistical mechanics may be used to analyze the emergent properties of feedback neural networks once it has been shown that, under a stochastic update rule, the network relaxes to a Gibbsian equilibrium distribution generated by the Lyapunov function of the deterministic dynamics [22, 60, 62]. For Glauber dynamics [Eq. (1.12)] and a one-at-a-time or a parallel updating scheme, such a relation exists as can be shown using the principle of detailed balance [28].

Although L_{DD} is identical to L_{SD} for two-state neurons, a block-sequential realization of Glauber dynamics need not approach a Gibbsian equilibrium distribution. However, in the special case of vanishing connection strength within all update groups [Eq. (1.62)], neurons "do not know" about the state of other neurons in the same group. Thus there is no formal difference between the block-sequential rule considered here and serial updating, where neurons change their state in consecutive order: Every set of W successive updates of the latter dynamics is identical to one time step in the former case.

In what follows, we focus on the retrieval of unbiased random patterns where $\xi_{ia}^{\mu} = \pm 1$ with equal probability and study networks at a finite storage level $\alpha \equiv p/N$. The case of large cluster size, $W \to \infty$, with the number n of update groups kept finite will be analyzed; n has to be at least equal to 2 because, according to Eq. (1.62), all neurons would be disconnected otherwise. Following the replica-symmetric theory of Amit, Gutfreund, and Sompolinsky [63], a fixed number s of patterns is singled out, and it is assumed that the network is in a state highly correlated with these "condensed" memories. The remaining patterns are described collectively by a noise term. Notice that, for coupling matrices of the form in Eq. (1.62), both the overlaps m and spin-glass parameters q have to be defined as order parameters *on the level of the update groups.* For retrieval solutions, this requirements leads to the Ansatz

$$m^{\mu}_{\sigma a} \equiv W^{-1} \sum_{i=1}^{W} \xi^{\mu}_{ia} S^{\sigma}_{ia} = m \delta_{\mu,1} \tag{1.63}$$

and

$$q^{\rho\sigma}_{ab} \equiv W^{-1} \sum_{i=1}^{W} S^{\rho}_{ia} S^{\sigma}_{ib} = \delta_{ab}[\delta_{\rho\sigma}(1-q)+q] \tag{1.64}$$

for a k-fold replicated network, $1 \leq \rho,\ \sigma \leq k$. The resulting fixed-point equations are

$$m = \langle\langle \tanh[T^{-1}\{m + \sqrt{\alpha r} z\}] \rangle\rangle \tag{1.65}$$

and

$$q = \langle\langle \tanh^2[T^{-1}\{m + \sqrt{\alpha r} z\}] \rangle\rangle, \tag{1.66}$$

where

$$r \equiv \frac{q}{[1 - T^{-1}(1-q)]^2} - \frac{q(n-1)}{[n - 1 + T^{-1}(1-q)]^2}. \tag{1.67}$$

Double angular brackets represent an average with respect to both the condensed patterns and the normalized Gaussian random variable z [10].

Equations (1.65)–(1.67) closely resemble their counterparts for the Hopfield model [63] and become identical to them in the limit of large n. On a formal level, the same holds for $n = 1$, but, as was explained before, this case does not correspond to a physical situation. For a general number of update groups there exists a first-order phase transition at $T = 0$ between the retrieval state and a spin-glass phase as α is varied. The critical storage level is denoted by α_c and the corresponding overlap by m_c.

The relative information content I_R, measured *per synapse* and *relative* to that of the Hopfield model,

$$I_R(n) \equiv \frac{I_n(\text{block-sequential})}{I(\text{random-sequential})} = \frac{n \cdot \alpha_c(n)}{(n-1) \cdot \alpha_c(\text{Hopfield})}, \tag{1.68}$$

is a third performance measure. A comparison between various network architectures in terms of all three measures is given in Table 1.1.

The performance of block-sequential updating schemes is quantitatively similar to that of the Hopfield model where $\alpha_c = 0.138$ and $m_c = 0.97$ [63]: The capability to retrieve stored random patterns is slightly lower when measured in terms of patterns per neuron, as is indicated in the second column of Table 1.1, and slightly higher when measured in terms of patterns per synapse, as is shown in the last column. Notice, in particular, that the information content increases with decreasing network connectivity, that is, for small n.

Table 1.1. Numerical solution of the saddle-point equations at $T = 0$, Displayed are the storage capacity α_c, the retrieval overlap m_c, and the relative information content I_R as functions of the number n of update groups.

n	α_c	m_c	I_R
2	0.100	0.93	1.45
3	0.110	0.95	1.20
4	0.116	0.96	1.12
5	0.120	0.96	1.09

(1.69)

The results demonstrate that feedback networks can be used to store large amounts of information: The number of patterns (each of size N) that can be memorized grows linearly with N, so that the information stored per synapse remains at a constant value of roughly 0.1 bits per synapse.[16] Stored patterns can be retrieved from noisy or incomplete data as long as the storage level remains below the critical level α_c. Compared to sequential or fully synchronous update schemes, partially parallel schemes offer a potentially large advantage in terms of computational costs when implemented on a parallel computer allowing for a speedup that may be as large as the number of processors without sacrificing network stability.

1.3.7 Intermezzo: Delayed Graded-Response Neurons

The dynamical description of Sec. 1.3.3 neglects any time lags due to finite propagation velocities of neural signals. As a first step toward the general formulation (1.14), one may study models where the communication time between neurons is modeled by one fixed delay τ,

$$C\frac{d}{dt}u_i(t) = -R^{-1}u_i(t) + \sum_{j=1}^{N} J_{ij}V_j(t-\tau) + I_i^{\text{ext}}(t) \tag{1.70}$$

with

$$V_i = g_i(u_i). \tag{1.71}$$

A mathematical analysis of this model is quite complicated. Because of the discrete delay, the initial condition for each neuron has to be specified as a function over a time interval of length τ. Consequently, Eqs. (1.70) and (1.71) describe an infinite-dimensional dynamical system even in the scalar case ($N = 1$), which will be discussed in detail in Sec. 1.4.2.

Obviously, fixed-point solutions of Eqs. (1.70) and (1.71) do not depend on the time lag and are thus identical with those of the original model

[16] This number is increased significantly by more elaborate learning rules [64].

without delays, described by Eqs. (1.40) and (1.41). However, equilibria that are stable without delays may become unstable for large enough time lag, as can be verified through a local stability analysis [65].

Global results about Eqs. (1.70) and (1.71) have been obtained under conditions that exclude nontrivial fixed-point solutions. A proof based on a Lyapunov functional shows that in this case there are no limit cycles either [66].

The lack of stronger global analytical results illustrates the limits of Lyapunov's direct method. It is often very hard or impossible to find a Lyapunov function for a *given* dynamical system under conditions that admit interesting applications — multiple fixed points in the present example. On the other hand, there are many cases where one can find Lyapunov functions as soon as one enlarges the class of systems studied. In the present case, one could replace the single discrete lag in Eq. (1.70) by a distributed delay such as the one used in Eq. (1.14). At a first glance, this seems to complicate the analysis even further. However, there exist nontrivial delay distributions for which the dynamics generated by Eq. (1.14) admit global Lyapunov functionals [67].

The remark applies also to systems with synaptic couplings $J_{ij}(\tau)$ that are of the form $J_{ij}\varepsilon(\tau)$, where $\varepsilon(\tau)$ satisfies a linear ordinary differential equation in τ. For instance, if $\tau_{\max} = \infty$ and $\varepsilon(\tau) = \exp(-\tau)$, one may rewrite the dynamical equations as a set of $2N$ ordinary differential equations. The example demonstrates that, unlike networks with discrete time lags, networks with distributed delays need not represent infinite-dimensional dynamical systems. Models with delay distributions that are "reducible" in this sense have been studied extensively in the applied mathematics literature [68]. For a neurobiologically motivated system of two limit-cycle oscillators with reducible signal delay, a Lyapunov function is given in reference [69].

1.4 Periodic Limit Cycles and Beyond

Natural stimuli provide information in both space and time. Recurrent neural networks with delayed feedback can be programmed to recognize and generate such pattern sequences or "temporal associations" [70, 71, 72, 73, 74, 75, 76].[17] Recurrent networks with a broad distribution of signal delays and a Hebbian learning rule such as Eq. (1.22) are well suited to *learn* pattern sequences as well [47, 77, 78, 79, 80, 81]. These systems are characterized by a high degree of compatibility between the network architecture, the task of learning spatio-temporal associations, and the learning algorithm. As in networks with fixed-point attractors, an initial state or

[17]A detailed discussion can be found in reference [33].

"stimulus" lying in the basin of attraction of a stored "memory" will spontaneously evolve toward this attractor. In the present context, however, memories are spatio-temporal patterns of neural activity.

This section demonstrates that one can understand the computation of certain networks with signal delays as a downhill march on an abstract spatio-temporal energy landscape. The result allows the application of techniques developed in the last sections.

1.4.1 Discrete-Time Dynamics

Let us focus on a synchronous discrete-time dynamics with deterministic McCulloch-Pitts neurons. For vanishing external inputs, the network dynamics in Eqs. (1.7) and (1.8) become

$$S_i(t+1) = \mathrm{sgn}[h_i(t)] \qquad \text{for all } i \tag{1.72}$$

with

$$h_i(t) = \sum_{j=1}^{N} \sum_{\tau=0}^{\tau_{\max}} J_{ij}(\tau) S_j(t-\tau). \tag{1.73}$$

In the following, it is assumed that the synaptic couplings $J_{ij}(\tau)$ satisfy the extended symmetry $J_{ij}(\tau) = J_{ij}(D-(2+\tau))$. As was shown in Sec. 1.2.4, this symmetry arises if the network is taught cyclic pattern sequences of equal duration D.

The construction of a Lyapunov function for the retrieval dynamics in Eqs. (1.72) and (1.73) is facilitated by the following consideration: If the network has learned cyclic associations with common length D, every correct retrieval solution corresponds to a D-periodic limit cycle. D-periodic oscillatory solutions of a discrete-time network, however, can always be interpreted as static states in a *fictitious* system of size $D \times N$ [50, 51].

Let us consider such a "D-plicated" network with D columns and N rows. The neural activities are denoted by S_{ia}, where $1 \leq i \leq N$ and $0 \leq a \leq D$. To reproduce the synchronous dynamics of the original system, neurons S_{ia} with $a = t$ (modulo D) are updated at time t.

The time evolution of the new network is block-sequential: synchronous within single columns and sequential with respect to these columns. In terms of the original variables S_i, the new activities S_{ia} are therefore given by $S_{ia}(t) \equiv S_i(a + n_t)$ for $a \leq t$ (modulo D) and $S_{ia}(t) \equiv S_i(a + n_t - D)$ for $a > t$ (modulo D), where n_t is defined through $t \equiv n_t + t$ (modulo D). The update rule reads

$$S_{ia}(t+1) = \begin{cases} \mathrm{sgn}\left[\sum_{j=1}^{N} \sum_{b=0}^{D-1} J_{ij}^{ab} S_{jb}(t)\right] & \text{if } a = t(\text{modulo } D), \\ S_{ia}(t) & \text{otherwise.} \end{cases} \tag{1.74}$$

The synaptic couplings J_{ij}^{ab} are defined as

$$J_{ij}^{ab} = J_{ij}\left((b-a-1)(\text{modulo } D)\right). \tag{1.75}$$

Notice that the time evolution [Eq. (1.74)] of the equivalent fictitious system is the same as a block-sequential updating of a network with $D \times N$ McCulloch-Pitts neurons and block size N, as is illustrated in Fig. 1.3. Section 1.3.5 shows how to guarantee that such a system relaxes to fixed points only: through synaptic symmetry together with the condition $\lambda_{\min}[\mathbf{U}(t)] \geq 0$.

Synaptic symmetry in the fictitious system, $J_{ij}^{ab} = J_{ji}^{ba}$, is equivalent to the extended symmetry of Eq. (1.26) for the original couplings $J_{ij}(\tau)$. The second condition, $\lambda_{\min}[\mathbf{U}(t)] \geq 0$, is equivalent to $\lambda_{\min}[\mathbf{J}(D-1)] \geq 0$. This condition can be satisfied by setting $\tau_{\max} = D - 2$.

It is left as an exercise for the interested reader to show that the Lyapunov function L_{DD}, formulated for the equivalent fictitious system, may be rewritten in terms of the original time-delay network as

$$L_{\mathrm{TD}}(t) = -\frac{1}{2} \sum_{i,j=1}^{N} \sum_{a,\tau=0}^{D-1} J_{ij}(\tau) S_i(t-a) S_j\left(t - (a+\tau+1)(\text{modulo } D)\right). \tag{1.76}$$

One may once again calculate the difference $\Delta L_{\mathrm{TD}}(t) \equiv L_{\mathrm{TD}}(t+1) - L_{\mathrm{TD}}(t)$ and arrive, as expected, at

$$\Delta L_{\mathrm{TD}}(t) = -\sum_{i=1}^{N} \left[S_i(t+1) - S_i(t+1-D)\right] h_i(t) \leq 0. \tag{1.77}$$

The derivation may be summarized as follows:

Suppose that the synaptic efficacies of the time-delay network [Eqs. (1.72) and (1.73)] satisfy the extended symmetry condition (1.26). Then the retrieval dynamics are governed by the Lyapunov function (1.76). The network relaxes to a fixed-point solution or a limit cycle with $S_i(t) = S_i(t-D)$, that is, an oscillatory solution with the same period as that of the taught cycles or a period that is equal to an integer fraction of D.

Due to the equivalence of Eqs. (1.72) and (1.73) with a block-sequential update rule for the fictitious system, one may apply the quantitative analysis of Sec. 1.3.6 to time-delay networks that store temporal associations. There is, however, a slight technical difficulty that has to be handled properly. Storing one D-periodic pattern sequence in the original model corresponds to memorizing D static patterns of size $D \times N$ in the equivalent system, each shifted by one column (modulo D) with respect to the next pattern. This complication arises because every sequence may be occurring with its first pattern recalled at some time t, or at time $t+1$, or at time $t+2$, and so on. In the equivalent D-plicated system, each of these time-shifted cyclic temporal associations corresponds to a new pattern.

For generic temporal associations, the analysis becomes rather complicated due to nontrivial correlations between shifted copies of the same pattern. If, however, each pattern of a sequence lasts for one time step only, all relevant correlations are the same as if one had stored D unrelated

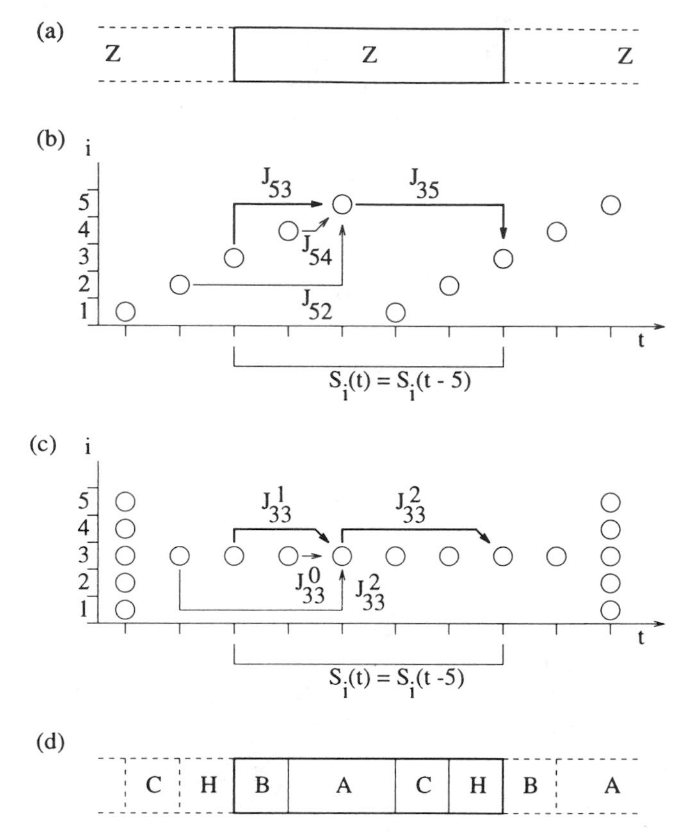

Fig. 1.3. Schematic drawing of the dynamics of a time-delay network (c and d) and its equivalent fictitious system with block-sequential time evolution (a and b). Horizontal axes represent time, vertical axes in (b) and (c) denote the index of neurons. (a) The pattern "Z" is retrieved in the fictitious network with five update groups that are represented in (b) by five neurons. (c) Time evolution of one neuron in a network with signal delays and discrete-time dynamics. The system recalls the cyclic pattern sequence "BAACH" as shown in (d).

Table 1.2. Influence of the weight distribution on the collective network properties. The storage capacity α_c, the critical overlap m_c, and the relative information content I_R are displayed for some choices of $\varepsilon(\tau)$ for $D = 4$.

τ	=	0	1	2	3	α_c	m_c	I_R
$\varepsilon(\tau)$	=	1/3	1/3	1/3	0	0.116	0.96	1.12
$\varepsilon(\tau)$	=	1/2	0	1/2	0	0.100	0.93	1.45
$\varepsilon(\tau)$	=	0	1	0	0	0.050	0.93	1.45

(1.78)

patterns. This implies that the results of Sec. 1.3.6 also cover the storage of pattern sequences where each pattern lasts for one unit of time.

As an example, take $D = 2$. With the maximal delay $\tau_{\max}$ set to $D - 2$, $\tau_{\max}$ is 0, and one has recovered the Little model. According to Table 1.1, $0.100N$ two-cycles of the form $\psi^\mu_{i1} \rightleftharpoons \psi^\mu_{i2}$ may be recalled as compared to $0.138N$ static patterns [82]: a 1.45-fold increase of the information content per synapse. At the same time, the retrieval overlap drops slightly from 0.97 to 0.93.

The performance of networks with distributed delays and $D = 4$ is displayed in Table 1.2.

As is shown in Table 1.2, the uniform distribution leads to the largest α_c but the smallest I_R. The other two networks have the same value of I_R as the (unique) $D = 2$ system due to the particular structure of their eigenvalue spectrum. Furthermore, one obtains $I_R = 1.45$ independently of D for all networks with a minimal connectivity where only one synapse links two neurons.[18] Simulation data show slightly higher values of α_c, possibly indicating effects of replica symmetry breaking as in the Hopfield model [63].

In passing, note that each cycle consists of D patterns so that the storage capacity for *single* patterns is $\overline{\alpha}_c = D\alpha_c$. During the recognition process, however, each pattern will trigger the cycle it belongs to and cannot be retrieved as a static memory.

If static patterns instead of temporal associations are learned, the synaptic strengths do not depend on the delay; see also Eq. (1.23). The synaptic couplings still satisfy the extended symmetry, and, with $\tau_{\max} = D - 2$, one recovers the Lyapunov function for networks with McCulloch-Pitts neurons and "multiple-time-step parallel dynamics" [83],

$$L_{\mathrm{MTS}}(t) = -\frac{1}{2} \sum_{i,j=1}^{N} J_{ij} \sum_{a=0}^{D-2} S_i(t-a) \sum_{b=0}^{D-2} S_j(t-b). \tag{1.79}$$

The evolution equations (1.72) and (1.73) may be generalized to analog

[18]This case is possible if D is an even number.

systems with periodic external inputs. Using the "cooking recipes" of Secs. 1.3.1–1.3.4, it is possible to construct a Lyapunov function for that case as well [84].

The learning rule in Eq. (1.26) also may be utilized to store cycles of correlated real-valued pattern sequences. Numerical studies have been performed for low-dimensional trajectories (small N) with high numbers of data points (large D). For many examples, good retrieval could be obtained without any need for highly time-consuming supervised learning schemes. However, algorithms of the latter kind facilitate the learning of more sophisticated real-world tasks. Here, Lyapunov functions are of great help since they permit the application of mean-field techniques [85] to a wide class of supervised learning strategies such as spatio-temporal extensions of the "Boltzmann Machine" concept [86] and contrastive-learning schemes [87].

In closing this section, let me mention that an analysis of the storage capacity along Gardner's approach [88] has been given in reference [89]. Analytical results on highly diluted systems with time lags have also been obtained [90].

1.4.2 Continuous-Time Dynamics

The global dynamics of certain networks with graded-response neurons and delayed interactions may be studied in a manner similar to that of Sec. 1.4.1 [67]. In the following, we focus on the simplest case, a single neuron (or a homogeneous assembly of neurons) coupled to itself through one inhibitory feedback loop with delay τ. Equation (1.14) reduces to

$$C\frac{d}{dt}u(t) = -R^{-1}u(t) - g[u(t-\tau)], \tag{1.80}$$

where g satisfies the condition

$$ug(u) > 0 \qquad \text{for } u \neq 0 \qquad \text{and} \qquad g(0) = 0. \tag{1.81}$$

Solutions of this seemingly simple scalar equation include a fixed point $u(t) = 0$ and, depending on the graph of g, periodic limit cycles and chaotic trajectories [91]. Such a diversity of temporal phenomena is possible since, due to the discrete delay, Eq. (1.80) describes an infinite-dimensional dynamical system as was already mentioned in Sec. 1.3.7.

Various aspects of the scalar delay differential equation (1.80) have been discussed in the mathematics literature. Most articles have concentrated on periodic solutions, in particular on those that are "slowly oscillating," that is, periodic solutions with zeros spaced at distances larger than the time lag τ. Results about their existence, uniqueness, and local stability have been obtained by Kaplan and Yorke [92], Nussbaum [93], and Chow and Walther [94], respectively.

The global analysis of Eq. (1.80) is simplified significantly if one neglects the transmembrane current $R^{-1}u(t)$ and if g is an odd sigmoid function. Without loss of generality, one may set $C = \tau = 1$ and study the evolution equation

$$\frac{d}{dt}u(t) = -g[u(t-1)]. \tag{1.82}$$

Consider the auxiliary function $L_{\mathrm{DDE}}(t)$,

$$\begin{aligned} L_{\mathrm{DDE}}(t) &= -\frac{1}{2}\int_0^1 \int_{t+\tau-1}^{t+1} \dot{u}(s)\dot{u}(s-\tau)ds\, d\tau \\ &+\frac{1}{2}\int_1^2 \int_{t+\tau-1}^{t+1} \dot{u}(s)\dot{u}(s-\tau)ds\, d\tau \\ &+\int_{t-1}^{t+1} G(\dot{u}(s))ds + \frac{1}{4}[u(t+1)+u(t-1)]^2, \end{aligned} \tag{1.83}$$

where $G(x)$ is defined as in Eq. (1.43).[19] For bounded nonlinearities g, all solutions of Eq. (1.82) are bounded. They are differentiable for $t > 1$. Consequently, $L_{\mathrm{DDE}}(t)$ is bounded below for $t > 2$. It follows that, for $t > 1$, the time derivative of $L_{\mathrm{DDE}}(t)$ along a solution of Eq. (1.82) is well defined and given by

$$\begin{aligned} \frac{d}{dt}L_{\mathrm{DDE}}(t) &= [\dot{u}(t+1)+\dot{u}(t-1)][u(t) - \tfrac{1}{2}u(t+1) - \tfrac{1}{2}u(t-1)] \\ &\quad +G(\dot{u}(t+1)) - G(\dot{u}(t-1)) \\ &\quad +\tfrac{1}{2}[u(t+1)+u(t-1)][\dot{u}(t+1)+\dot{u}(t-1)] \\ &= u(t)[\dot{u}(t+1)+\dot{u}(t-1)] + G(\dot{u}(t+1)) - G(\dot{u}(t-1)). \end{aligned} \tag{1.84}$$

Because the input–output characteristic is assumed to be an odd sigmoid function, g^{-1} is odd, single-valued, and monotone increasing. Consequently, the function G is even and strictly convex. In particular, the equality $G(\dot{u}(t-1)) = G(-\dot{u}(t-1))$ holds. Performing a Taylor expansion as in Eq. (1.51), one therefore obtains

$$\begin{aligned} G(\dot{u}(t+1)) - G(\dot{u}(t-1)) &\leq [\dot{u}(t+1)+\dot{u}(t-1)]g^{-1}(\dot{u}(t+1)) \\ &\leq -[\dot{u}(t+1)+\dot{u}(t-1)]u(t). \end{aligned} \tag{1.85}$$

Equality in Eq. (1.85) holds if and only if $\dot{u}(t+1) = -\dot{u}(t-1)$. Taking the evolution equation (1.82) and the strict monotonicity of g into account, the last equation also may be written $u(t) = -u(t-2)$.

---

[19] L_{DDE} has been introduced as an explicitly time-dependent function for simplicity and has been written in terms of both u and $\dot{u}$ for the same reason. However, the initial function may not be differentiable. This (purely technical) difficulty can be avoided if $\dot{u}(s)$ is replaced by $-g(u(s-1))$. L_{DDE} then may be properly defined as a functional in the space of continuous functions from the interval $[-2,0]$ to the real numbers [95].

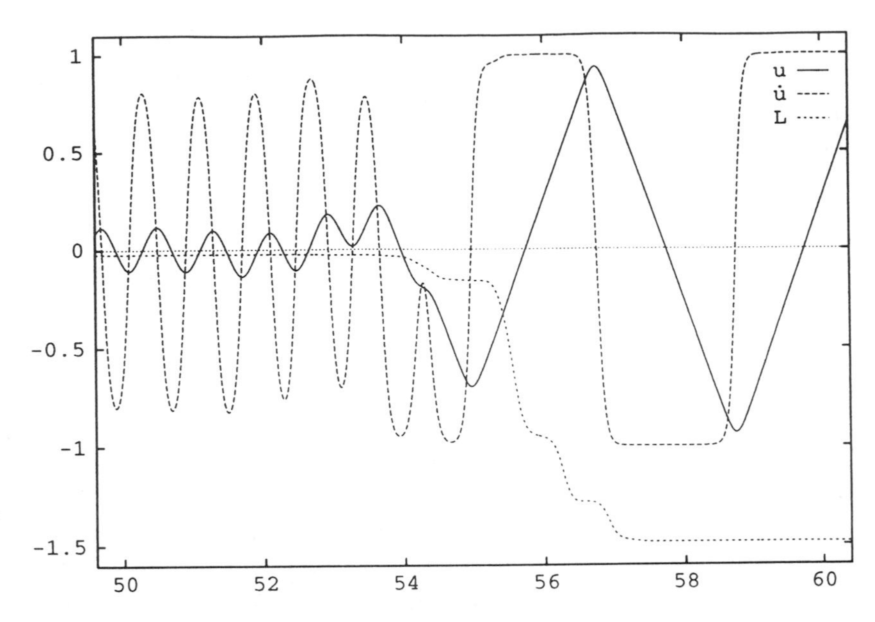

Fig. 1.4. Time evolution of a single neuron with delayed feedback according to the evolution equation (1.82). The input–output characteristic is $g(u) = \tanh(5u)$. The state variable u is plotted as a solid line, its derivative $\dot{u}$ as a dashed line, and the Lyapunov function L_{DDE} as a dotted line. Notice that $L_{\mathrm{DDE}}(t)$ approaches a constant value as required for a Lyapunov function, whereas u relaxes toward a periodic oscillatory solution with period four.

Inserting Eq. (1.85) into Eq. (1.84), one finally arrives at

$$\frac{d}{dt} L_{\mathrm{DDE}}(t) \le 0 \qquad \text{for } \; t \ge 2, \tag{1.86}$$

where equality holds if and only if $u(t) = -u(t-2)$.[20] An illustration is given in Fig. 1.4. According to Eq. (1.86), $L_{\mathrm{DDE}}(t)$ is nonincreasing along every solution for $t > 2$. The overall result may be summarized in the following way:

Suppose that the function g is odd, bounded, and sigmoid. Then the evolution equation (1.82) admits the Lyapunov function (1.83). Solutions of Eq. (1.82) converge either to the trivial fixed point $u = 0$ or to a periodic limit cycle that satisfies

$$u(t) = -u(t-2). \tag{1.87}$$

Notice that the period P of the limit cycles does *not* depend on the graph

[20]The curious reader is invited to compare this result and its derivation with that for the Little model with antisymmetric couplings in Eq. (1.54).

of g; according to Eq. (1.87), it is always given by $P = 4/(4k+1)$, where k is a nonnegative integer.[21] On the other hand, it is well known that, for the general equation (1.80), the period of a periodic solution is influenced by the ratio of RC to τ and the shape of g [96]. This fact implies that the above methods probably cannot be extended to study delay differential equations of the type in Eq. (1.80). There is, however, another way to analyze this equation [97]. To facilitate the discussion, let $t_i, i \in \mathbb{N}$ with $t_i < t_{i+1}$ denote the times of consecutive zero crossings $u(t) = 0$ of a solution of Eq. (1.80). One may then prove the following proposition:

Assume that the function g is bounded and satisfies the condition in Eq. (1.81). For every solution $u(t)$ of Eq. (1.80), the number $n(i)$ of zero crossings in the interval $[t_i - \tau, t_i)$ is a nonincreasing function of i.

This result means that a solution of Eq. (1.80) oscillates more and more slowly around 0 as time proceeds. For long times it approaches a solution with constant $n = n(i)$; possibly $n = 0$. In particular, if the system is initialized with a solution that has n zero crossings in the interval $[-\tau, 0)$, it can never reach an oscillation with more than n zero crossings in any one of the intervals $[t_i - \tau, t_i)$.

Let me briefly sketch the proof. The reader is also referred to Fig. 1.5. If g is bounded and satisfies the condition (1.81), solutions of Eq. (1.80) exist for all positive t and are continuous [98]. Assume without loss of generality that, at time $t_j, (d/dt)\,u(t_j) > 0$. According to Eqs. (1.80) and (1.81), this means that $u(t_j - \tau) < 0$ because $u(t_j) = 0$ by definition. The same argument may be used at time t_{j+1}, where it implies that $u(t_{j+1} - \tau) > 0$ because $u(t_{j+1}) = 0$ and $(d/dt)\,u(t_{j+1}) < 0$. Together with the continuity of $u(t)$, this implies that there is an odd number $k(j) \geq 1$ of zero crossings in the interval $[t_j - \tau, t_{j+1} - \tau)$.

Denote the number of zero crossings in the interval $[t_{j+1} - \tau, t_j)$ by $l(j)$.[22] It follows that $n(j) = l + k(j)$ and $n(j+1) = l(j) + 1$. Since $k(j) \geq 1$, both relations may be combined to the statement $n(j) \geq n(j+1)$, which proves the proposition.

The number of zero crossings in any interval is nonnegative — the function $n(i)$ is bounded below. Since it is nonincreasing along every solution of Eq. (1.80), it is an integer-valued Lyapunov function. Accordingly, solutions of Eq. (1.80) relax to solutions with constant $n(i)$. Notice that those solutions may be periodic but could — at least in principle — also be ape-

[21]Further results derived with the help of L_{DDE} can be found in reference [95]. One proof is well suited to highlight the potential of Lyapunov functions — once they are found: It can be shown that, for large enough $g'(0)$, the global minimum of L_{DDE} is always achieved on a slowly oscillating solution [otherwise on the trivial fixed point $u(t) = 0$]. This immediately implies that those solutions have to be asymptotically stable (except for global phase shifts), a conclusion that previously required elaborate analytical techniques.

[22]It is understood that $l(j) = 0$ if $t_{j+1} - \tau \geq t_j$.

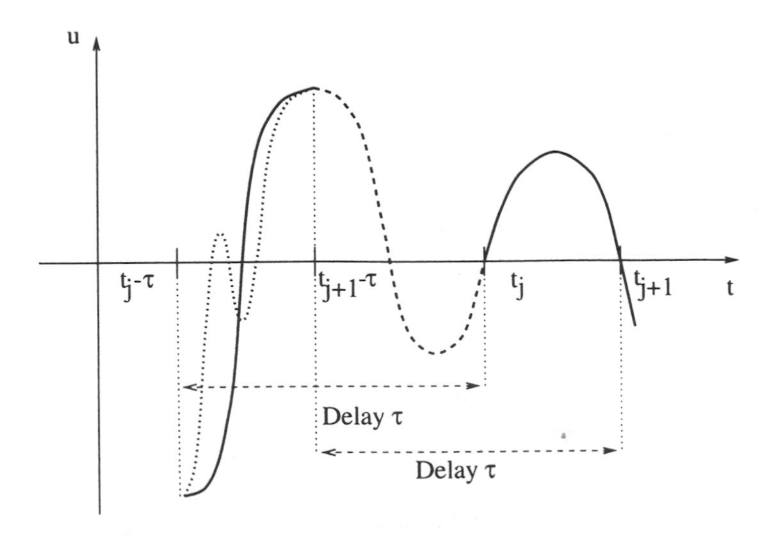

Fig. 1.5. Time evolution of a single graded-response neuron with delayed self-inhibition modeled by the delay differential equation (1.80). There are zero crossings of the solution $u(t)$ at time t_j, t_{j+1}, and at various earlier (and later) times. In the interval $[t_j - \tau, t_{j+1} - \tau)$, two possible solutions are drawn. They have one and three zero crossings, respectively.

riodic. This is a surprising result; it highlights the generality of Lyapunov's second method in a rather illuminating way.

1.5 Synchronization of Action Potentials

While it frequently may be the case that mean-firing rates are an adequate description of neural information, there are many instances where the detailed timing and organization of action potentials matter. An important example is given by the stimulus-dependent synchronization of action potentials [15, 16, 17].

Due to the inherent limitations of descriptions based on discrete-time dynamics or mean-firing rates, realistic synchronization processes are not captured by the networks discussed in Secs. 1.3 and 1.4. Synchronization processes may, however, be studied using networks with integrate-and-fire neurons, whose time evolution was introduced in Sec. 1.2.3.

Networks of that type often show globally synchronized neurons when all-to-all couplings are used.[23] Note that, throughout this section, terms

[23] Doubts about the structural stability of simple integrate-and-fire models have been raised because some model variants do not exhibit systemwide synchronization with all-to-all couplings [99, 100, 101].

such as "synchronized neurons" always refer to the time of spike generation. According to this definition, a periodic network state (also called a *phase-locked solution*) may or may not be "globally synchronized." A global analysis for networks described by Eqs. (1.13), (1.20), and the "absorbtion rule" $u_j(t^+) = 0$ [instead of Eq. (1.18)] has been given in reference [102]. With excitatory all-to-all couplings of equal strength, nonzero leakage currents, uniform external inputs, and a reset to 0 after spike generation ($\gamma = 0$), the size of the largest synchronized cluster is a nondecreasing function of time — a (discrete-valued) Lyapunov function! The proof then shows that such systems approach a globally synchronized solution where all neurons fire in unison.

Networks with more general nonuniform interaction admit richer dynamical behavior [25, 103, 104]. Equipped with excitatory finite-range couplings, one class of networks relaxes to phase-locked clusters of (locally) synchronized neurons [105, 106]. The shapes and relative phases of the clusters encode information about the initial stimulus. this result is in accordance with the hypothesis that synchronized cortical neurons are used to bind stimulus features together [107].

1.5.1 Phase Locking

Global results for locally coupled networks with integrate-and-fire neurons have been obtained in the limiting case $R \to \infty$ of perfectly integrating cells and uniform positive input currents $I_i^{\text{ext}} = I > 0$. In this situation, external information is encoded in the initial conditions $u_i(t = 0)$, not in the input currents. This choice is reminiscent of the experimental paradigm of *stimulus-induced oscillations* [15]. Due to the constant positive input current I, each model cell fires regularly if there is no further synaptic input from other cells. Thus, I^{-1} represents the spontaneous firing rate of an isolated neuron. By rescaling time, the capacitance C and input I in Eq. (1.13) can be taken as unity. The overall dynamics then may be summarized by the following update rules:

(i) Initialize the $u_i(t = 0)$ in [0, 1] according to the external stimulus.

(ii) If $u_i \geq 1$, and if neuron i is next in the update scheme, then

$$u_i \to u_i' = \gamma(u_i - 1) \tag{1.88}$$

and

$$u_j \to u_j' = u_j + J_{ji}. \tag{1.89}$$

(iii) Repeat step ii until $u_i < 1$ for all i.

(iv) If the condition of step ii does not apply, then

$$\frac{d}{dt}u_i = 1 \qquad \text{for all } i. \tag{1.90}$$

Under the condition that all neurons have the same total incoming synaptic strength,

$$\sum_j J_{ij} = A, \tag{1.91}$$

and the same total outgoing synaptic synaptic strength,

$$\sum_i J_{ij} = A, \tag{1.92}$$

none may prove that the simple function L_{IAF},

$$L_{\text{IAF}} = -\sum_i u_i, \tag{1.93}$$

that is, the total (negative) membrane potential, plays the role of a Lyapunov function for the system defined by steps i–iv, as is shown in reference [106]:

Assume that $\gamma = 1$ and that the synapses satisfy $J_{ij} \geq 0$ and the constraints in Eqs. (1.91) and (1.92) with $A < 1$. Then the dynamics generated by Eqs. (1.88)–(1.90) admit the Lyapunov function (1.93) and converge to cyclic solutions with period $P_{\text{IAF}} = 1 - A$. On periodic solutions, each neuron fires exactly once in a period.

Notice that synaptic symmetry has *not* been required! This distinguishes the present model from the networks discussed in the previous sections.

Depending on the initial conditions, the periodic solutions can contain events in which one neuron fires alone, and others in which many neurons fire in synchrony. In networks with excitatory short-range connections only, regions with small variability of the initial conditions are smoothed out and represented by locally synchronized clusters of neurons whose firing times encode the stimulus quality. Regions with high variability, on the other hand, give rise to spatially uncorrelated firing patterns. Through an appropriate choice of coupling strengths, more complex computations can be performed as demonstrated by numerical simulations [106].

In order to prove the proposition, let us first show that no neuron fires more than once in any interval of length P_{IAF}.

Lemma: Let $n_i(t,t')$ denote the number of times neuron i fires in $[t,t')$. If the conditions of the proposition hold, then $n_i(t, t + P_{\text{IAF}}) \leq 1$.

Starting at time t, if some neuron fires twice before $t + P_{\text{IAF}}$, then some neuron k must first fire twice, and at time $t' < t + P_{\text{IAF}}$. For that to happen, the total change in u_k from t to t' due to the synaptic currents and the external input must be greater than 1. Thus one requires that, for neuron k,

$$(t' - t)\frac{(1 - A)}{P_{\text{IAF}}} + \sum_j J_{kj} n_j(t,t') > 1. \tag{1.94}$$

However, by hypothesis $(t' - t) < P_{\text{IAF}}$, and, since k is the first neuron to fire twice, the number $n_j(t,t')$ of firings of each of the other neurons up to

t' is less than or equal to 1. For J_{ij} nonnegative, the left-hand side of Eq. (1.94) is less than $(1-A)+A=1$. The contradiction shows that k cannot have fired twice.

Returning to the proof of the proposition, let us consider the change of L_{IAF} in a time interval of length P_{IAF}, $\Delta L_{\text{IAF}}(t) \equiv L_{\text{IAF}}(t+P_{\text{IAF}})-L_{\text{IAF}}(t)$. It is

$$\Delta L_{\text{IAF}}(t) = -(1-A)N - \sum_{i,j} J_{ij} n_j(t, t+P_{\text{IAF}}) + \sum_i n_i(t, t+P_{\text{IAF}}). \quad (1.95)$$

The first term comes from the constant input current, the second term from the effect of the firing of other neurons, and the third term comes from i itself firing. Using the condition (1.91), one finds

$$\Delta L_{\text{IAF}}(t) = -(1-A)\left[N - \sum_i n_i(t, t+P_{\text{IAF}})\right]. \quad (1.96)$$

Due to the lemma, $n_i(t, t+P_{\text{IAF}}) \leq 1$ for all t. The change of L_{IAF} in each time interval P_{IAF} is thus nonpositive. Since L_{IAF} is bounded, the system performs a downhill march on the energy landscape generated by the Lyapunov function L_{IAF} — if the function is measured after time steps of length P_{IAF}. The difference $\Delta L_{\text{IAF}}(t)$ vanishes if and only if $n_i(t, t+P_{\text{IAF}}) = 1$ for all i, that is, on periodic solutions where every neuron fires exactly once in a time interval of length P_{IAF} [106].[24]

To avoid the unfamiliar evaluation of the Lyapunov function L_{IAF} at the discrete times $t+kP_{\text{IAF}}$, $k \in \mathbb{N}$, one may alternatively use the functional

$$\tilde{L}_{\text{IAF}} = \int_{-P_{\text{IAF}}}^{0} L_{\text{IAF}}(s)ds. \quad (1.97)$$

Along solutions, $\tilde{L}_{\text{IAF}}$ is differentiable with $(d/dt)\tilde{L}_{\text{IAF}}(t) = \Delta L_{\text{IAF}}(t - P_{\text{IAF}})$ for all $t \geq P_{\text{IAF}}$, so that the previous conclusions are reached again. For an illustration, see Fig. 1.6.

1.5.2 Rapid Convergence

The results of the previous section prove that specific networks of integrate-and-fire neurons approach phase-locked solutions. Numerical simulations of these and more general networks [102, 106, 108, 109, 110, 111] indicate that the convergence process takes place in a very short time — see also Fig. 1.6.[25] This observation can be substantiated under certain conditions [105, 106]:

---

[24] A related proof has been given in reference [35]. Notice also that a continuous set of stable (but not asymptotically stable) periodic solutions is reached.

[25] In general, clusters of locally synchronized neurons will slowly reorganize after the initial rapid convergence. The models analyzed in this chapter are an exception in that they do not show such slow relaxation phenomena.

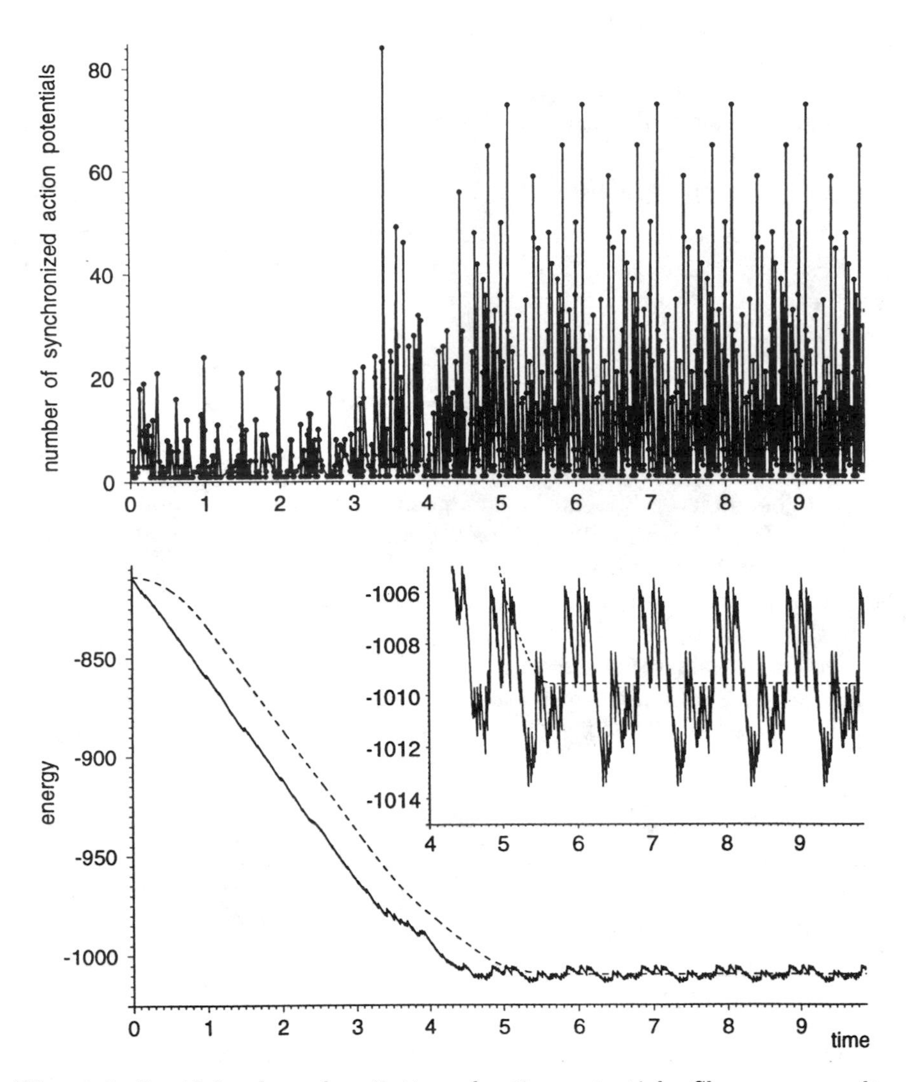

Fig. 1.6. Rapid local synchronization of action potentials. Shown are results from numerical simulations of a planar network with 40×40 integrate-and-fire neurons ($R^{-1} = 0$, $\gamma = 1$), periodic boundary conditions, and nearest-neighbor interactions of strength $J_{nn} = 0.24$. Each dot in the upper trace represents the number of simultaneous action potentials as a function of time. The lower trace depicts the time evolution of the Lyapunov function L_{IAF} (solid line) and the Lyapunov functional $\tilde{L}_{\text{IAF}}$ (dashed line). The inset verifies that, as predicted, the latter approaches a constant value.

Assume that the synapses satisfy $J_{ij} \geq 0$ *and the condition (1.91) with* $A < 1$. *Then all solutions of Eqs. (1.88)–(1.90) converge to cyclic solutions with period* $P_{\mathrm{IAF}} = 1 - A$. *The attractor is reached as soon as every neuron has fired once. On the periodic solution, each neuron fires exactly once in a period.*

Notice that, although the conditions on γ and on the sum of outgoing synaptic strengths have been dropped, the conclusions are now stronger than in the previous proposition. However, the proof given is *not* based on a Lyapunov function, so the concept of a downhill march on an energy landscape generated by the Lyapunov function no longer is available. The lack of a Lyapunov function might also be a drawback when stochastic extensions are considered in the future.[26]

Let $t_{\max}$ denote the first time every neuron has fired at least once. Some cells may have fired repeatedly before $t_{\max}$, depending on the parameter values and initial conditions. Let t_i denote the last time neuron i fires before $t_{\max}$, $t_{\min}$ the minimum of all these times t_i, and k a cell that fires at $t_{\min}$ for the last time without being triggered by other cells.

By definition, every cell discharges at least once in the interval $[t_{\min}, t_{\max}]$. This implies in particular that every neuron j from which cell k receives synaptic input emits one or more action potentials in that interval. Each spike adds J_{kj} to u_k. The total change of u_k in $[t_{\min}, t_{\max}]$ is thus equal to or greater than $A + t_{\max} - t_{\min}$. This number has to be smaller than 1 because, otherwise, neuron k would fire a second time in the interval $[t_{\min}, t_{\max}]$ in contradiction to the assumption. It follows that $t_{\max} - t_{\min} < P_{\mathrm{IAF}}$.

Going back to Sec. 1.5.1, one notices that the condition on the sum of *outgoing* synaptic strengths [Eq. (1.92)], although essential for the proof of the main proposition, is not required for the proof of the lemma: The lemma is also valid under the weaker conditions of the present section. Evaluated at time $t = t_{\max} - P_{\mathrm{IAF}}$ and combined with the previous results, the lemma implies that every cell fires exactly once in $[t_{\min}, t_{\max}]$ and no cell fires in $(t_{\max} - P_{\mathrm{IAF}}, t_{\min})$. Since $t_{\max} \leq 1$, the last result proves that, in finite time $t_{\max} - P_{\mathrm{IAF}}$, a limit cycle is approached in the sense that $u_i(t) = u_i(t + P_{\mathrm{IAF}})$ for $t \geq t_{\max} - P_{\mathrm{IAF}}$. The argument also shows that the attractor is reached as soon as every neuron has fired once.

The proof does not depend on the details of the reset mechanism. This means that it covers not only the present model with arbitrary $0 \leq \gamma \leq 1$,

---

[26]The sentence reflects the author's hope that it might be possible to construct simple stochastic dynamics of integrate-and-fire neurons such that the Lyapunov function of the noiseless dynamics determines a Gibbs distribution for the stochastic extension. Equilibrium statistical mechanics then could be applied to analyze the collective phenomena in networks of integrate-and-fire neurons in the same spirit as has been done for the neural network models discussed in Secs. 1.3 and 1.4. Regrettably, such evolution equations have not been found yet.

but also all schemes where a neuron i firing at time t is relaxed to some value between 0 and $u_i(t^-) - 1$. Perhaps surprisingly, this allows stochastic updatings during the transient phase.

In all model variants except from the limiting case $\gamma = 1$, cyclic solutions with period P_{IAF} and one spike per cycle cannot occur if a neuron is driven above threshold. In events with multiple neurons firing "at the same time," the potentials have to be fine-tuned such that, if neuron i is triggered by neuron $j, u_i(t^-) = 1 - J_{ij}$. This implies that, although *every firing sequence* of the model with $\gamma = 1$ can be realized in these models, the volume of all attractors is greatly reduced when measured in the space of the dynamical variables u_i.

1.6 Conclusions

The examples presented in this chapter demonstrate that Lyapunov's direct method has widespread applications within the theory of recurrent neural networks. With respect to the list of levels of analysis sketched in the Introduction, it has been shown that Lyapunov's method is most helpful on the second level, which deals with questions about the type of attractors possible in a neural network.

Combined with powerful techniques from statistical mechanics, Lyapunov's approach allows not only for a qualitative understanding of the global dynamics, but also for quantitative results about the collective network behavior. As was shown in Secs. 1.3, 1.4, and 1.5, Lyapunov's method applies to the retrieval of static patterns in networks with instantaneous interactions, to the recall of spatio-temporal associations in networks with signal delays, and to synchronization processes in networks of integrate-and-fire neurons.

There remain numerous interesting questions about the global dynamics of feedback neural networks. These include questions concerning the convergence of network models with discrete-time dynamics, symmetric couplings, and *overlapping* delays [see Fig. 1.1(d)]. Numerical simulations suggest that such systems relax to fixed-point solutions [112], but the analytic results from the computer science literature [55, 56, 57, 58] only cover the case where a single pattern is stored in the network.

With regard to networks with transmission delays, it would be interesting to know more about the global dynamics generated by Eqs. (1.70) and (1.71) under conditions that admit multiple fixed-point attractors. With a similar interest in mind, one could try to perform a statistical mechanical analysis of the system (1.72), (1.73) with delay-independent symmetric couplings [Eq. (1.23)] to study the influence of signal delays on the collective properties of networks that store static patterns.

In the proofs concerning integrate-and-fire neurons, synaptic strengths were assumed to be excitatory. There is, however, strong numerical evidence that inhibition does not change the overall results [106]. If the synaptic couplings continue to satisfy the condition (1.91) with $A < 1$, and if the network parameters are chosen such that there are no runaway solutions and no solutions with neurons that are permanently below threshold, then all simulations of the dynamics generated by Eqs. (1.88)–(1.90) approach periodic solutions of period $P_{\text{IAF}} = 1 - A$. For leaky integrate-and-fire models (finite R), the same is true, but the period is given by the period P_{LIAF} of the globally synchronized solution in such a system:

$$P_{\text{LIAF}} = RC[\ln(RI - A) - \ln(RI - 1)]. \tag{1.98}$$

This observation gives hope that further understanding of integrate-and-fire models is possible, although the mathematical situation is more complicated than in the cases discussed in Sec. 1.5. A convergence proof based on Lyapunov functions such as Eq. (1.93) is possible because *every* periodic solution of the model has the *same* period. This is not the case for models for finite R, as is shown by the following counterexample. Consider a spatio-temporal "checkerboard" pattern, where the "black" sites fire at even multiples of $\Delta/2$ and the "white" sites at odd multiples of $\Delta/2$. A self-consistent calculation of the firing pattern leads to an implicit equation for Δ:

$$Ae^{-\Delta/2RC} + RI\left[1 - e^{-\Delta/RC}\right] = 1. \tag{1.99}$$

Excepting from the limiting case $R \to \infty$, Δ differs from the period of the globally synchronized solution. A stability analysis verifies that the checkerboard pattern is unstable, but its mere existence indicates that it will be difficult to find Lyapunov functions for leaky integrate-and-fire models.

More generally, one may ask which conditions in the proofs of Secs. 1.3, 1.4, and 1.5 can be violated without changing the desired emergent network behavior. These questions deal with the structural stability of neural networks, the fifth level of analysis, and have to be answered if one wants to evaluate the biological relevance of specific networks. In order to keep the chapter within reasonable bounds, this topic has not been discussed here. A particularly important issue, the convergence of "conventional" recurrent neural networks (of the type studied in Sec. 1.3) without synaptic symmetry, has been studied extensively in the literature [113, 114]. In passing, let me note that one may always generate specific asymmetric networks through appropriate transformations of both the coupling matrix and dynamical variables of systems with symmetric interactions.

There are a number of other topics related to the main theme of this chapter that could not be included. Let me briefly mention two of these issues.

First, one may design dynamical systems such that they perform a downhill march on an energy landscape that encodes some optimization task [59]. Various biologically motivated examples can be found in the computer vision literature [115, 116].

Second, one may construct feedback networks that possess desired attractors but no spurious stable states [117, 118]. The construction of such artificial associative memories is greatly facilitated if one deliberately lifts modeling restrictions that otherwise would be naturally imposed by biological constraints.

Let me close with a general comment: "Associative computation" means that many different inputs are mapped onto few output states. The time evolution of a dynamical system that performs such a computation is characterized by a contraction in its state space, that is, it is dissipative.[27] This observation suggests that many dynamical systems that have been used as models for associative computation may admit Lyapunov functions. As was emphasized in Sec. 1.3.7, minor modifications of the models may be needed to satisfy technical requirements.

In view of the many Lyapunov functions already found, I would like to conclude with a remark from the monograph of Rouche, Habets, and Laloy [3]: "Lyapunov's second method has the undeserved reputation of being mainly of theoretical interest, because auxiliary functions are so difficult to construct. We feel this is the opinion of those people who have not really tried ..."

Acknowledgments. Most of the author's own results presented in this chapter were obtained in collaborations with Bernhard Sulzer, Charlie Marcus, John Hopfield, Leo van Hemmen, Reimer Kühn, and Zhaoping Li. Discussions with David MacKay and Hans-Otto Walther were of great help. The work on integrate-and-fire neurons was stimulated by a series of helpful conversations with John Rundle and benefitted from valuable comments by Tom Heskes and powerful computing resources provided by Klaus Schulten. Atlee Jackson, Ken Wallace, Reimer Kühn, Ron Benson, and Wulfram Gerstner contributed with critical remarks on earlier drafts of the manuscript. Burkhard Rost, John Hopfield, Li-Waj Tang, and Tanja Diehl were sources of inspiration when it came to understanding complex biological systems of various kinds.

It is a great pleasure to acknowledge the continuous support from the above people. I would also like to thank these colleagues and friends for sharing happy and exciting times. This work has been made possible through grants from the Deutsche Forschungsgemeinschaft, the Beckman Institute, and the Commission of the European Communities under the Human Capital and Mobility Programme.

---

[27]The threshold operation of a two-state neuron might be interpreted as a special realization of this contraction process.

REFERENCES

[1] M. Abeles (1991) *Corticonics: Neural Circuits of the Cerebral Cortex* (Cambridge University Press, Cambridge)

[2] J. LaSalle, S. Lefschetz (1961) *Stability by Ljapunov's Direct Method* (Academic Press, New York)

[3] N. Rouche, P. Habets, M. Laloy (1977) *Stability Theory by Liapunov's Direct Method* (Springer-Verlag, New York)

[4] J.J. Hopfield (1982) *Proc. Natl. Acad. Sci. USA* **79**:2554–2558

[5] W.A. Little (1974) *Math. Biosci.* **19**:101–120

[6] M.A. Cohen, S. Grossberg (1983) *IEEE Trans. SMC* **13**:815–826

[7] J.J. Hopfield (1984) *Proc. Natl. Acad. Sci. USA* **81**:3088–3092

[8] C.M. Marcus, R.M. Westervelt (1989) *Phys. Rev. A* **40**:501–504

[9] E. Goles-Chacc, F. Fogelman-Soulie, D. Pellegrin (1985) *Disc. Appl. Math.* **12**:261–277

[10] A.V.M. Herz, C.M. Marcus (1993) *Phys. Rev. E* **47**:2155–2161

[11] E.R. Kandel, J.H. Schwartz (1985) *Principles of Neural Science* (Elsevier, New York)

[12] A.L. Hodgkin, A.F. Huxley (1952) *J. Physiol.* (London), **117**:500–544

[13] W.C. McCulloch, W. Pitts (1943) *Bull. Math. Biophys.* **5**:115–133

[14] E. Ising (1925) *Z. Phys.* **31**:253

[15] R. Eckhorn (1988) *Biol. Cybern.* **60**:121–130

[16] C.M. Gray, W. Singer (1989) *Proc. Natl. Acad. Sci. USA* **86**:1698–1702

[17] G. Laurent, H. Davidowitz (1994) *Science* **265**:1872–1875

[18] W. Gerstner, J.L. van Hemmen (1994) In: *Models of Neural Networks II*, E. Domany, J.L. van Hemmen, K. Schulten (Eds.) (Springer-Verlag, New York), pp. 1–93.

[19] M.A. Wilson, J.M. Bower (1989) In: *Methods in Neuronal Modeling: From Synapses to Networks*, C. Koch, I. Segev (Eds.) (MIT Press, Cambridge), pp. 291–334

[20] R.D. Traub, R. Miles (1991) *Neural Networks of the Hippocampus* (Cambridge University Press, Cambridge)

[21] V. Braitenberg (1986) In: *Brain Theory*, G. Palm, A. Aertsen (Eds.) (Springer-Verlag, Berlin) pp. 81–96

[22] D.J. Amit (1989) *Modeling Brain Function: The World of Attractor Neural Networks* (Cambridge University Press, Cambridge)

[23] E. de Schutter, J.M. Bower (1994) *Proc. Natl. Acad. Sci. USA* **91**:4736–4740

[24] P.C. Bressloff (1991) *Phys. Rev. A* **44**:4005–4016

[25] W. Gerstner, J.L. van Hemmen (1992) *Biol. Cybern.* **67**:195–205; Network **3**:139–164

[26] B. Katz (1966) *Nerve, Muscle, and Synapse* (McGraw-Hill, New York)

[27] Y. Burnod, H. Korn (1989) *Proc. Natl. Acad. Sci. USA* **86**:352–256

[28] P. Peretto (1984) *Biol. Cybern.* **50**:51–62

[29] R.J. Glauber (1963) *J. Math. Phys.* **4**:294–307

[30] H. Horner (1988) In: *Computational Systems — Natural and Artificial*, H. Haken (Ed.) (Springer-Verlag, Berlin) pp. 118–132

[31] K.M. Chandy (1990) *Sci. Comp. Pr.* **14**:117–132

[32] E. Caianiello (1961) *J. Theor. Biol.* **1**:204–235

[33] R. Kühn, J.L. van Hemmen (1991) In: *Physics of Neural Networks*, E. Domany, J.L. van Hemmen, K. Schulten (Eds.) (Springer-Verlag, Berlin) pp. 213–280

[34] A. Gabrielov (1993) *Physica A* **195**:253–274

[35] A. Gabrielov, W.I. Newman, L. Knopoff (1994) *Phys. Rev. E* **50**:188–196

[36] P. Bak, C. Tang, K. Wiesenfeld (1987) *Phys. Rev. Lett.* **59**:381–384

[37] S. Dunkelmann, G. Radons (1994) In: *Proceedings of the International Conference on Artificial Neural Networks*, M. Marimnaro, P.G. Morasso (Eds.) (Springer-Verlag, London), pp. 867–871

[38] R.W. Kentridge (1994) In: *Computation and Neural Systems*, F.H. Eeckman, J.M. Bower (Eds.) (Kluwer, Netherlands), pp. 531–535

[39] M. Usher, M. Stemmler, C. Koch, Z. Olami (1994) *Neural Comput.* **6**:795–836

[40] J.J. Hopfield (1994) *Physics Today* **46**:40–46

[41] J.B. Rundle, A.V.M. Herz, J.J. Hopfield (1994) preprint

[42] D.O. Hebb (1949) *The Organization of Behavior* (Wiley, New York)

[43] J.H. Byrne, W.O. Berry (Eds.) (1989) *Neural Models of Plasticity* (Academic Press, San Diego, CA)

[44] F. Edwards (1991) *Nature* **350**:271–272

[45] T.H. Brown, A.H. Ganong, E.W. Kairiss, C.L. Keenan, S.R. Kelso (1989) In: *Neural Models of Plasticity*, J.H. Byrne, W.O. Berry (Eds.) (Academic Press, San Diego, CA), pp. 266–306

[46] D.W. Dong, J.J. Hopfield (1992) *Network* **3**:267–283

[47] A.V.M. Herz, B. Sulzer, R. Kühn, J.L. van Hemmen (1988) *Europhys. Lett.* **7**:663–669; (1989) *Biol. Cybern.* **60**:457–467

[48] S. Grossberg (1968) *J. Math. Anal. Appl.* **21**:643–694

[49] S.I. Amari (1972) *IEEE Trans. Comp. C* **21**:1197–1206

[50] Z. Li, A.V.M. Herz (1990) In: *Proceedings of the XI. Sitges Conference, "Neural Networks,"* L. Garrido (Ed.) (Springer-Verlag, Berlin), pp. 287–302

[51] A.V.M. Herz, Z. Li, J.L. van Hemmen, (1991) *Phys. Rev. Lett.* **66**:1370–1373

[52] D.J. Amit, N. Brunel, M.V. Tsodyks (1994) *J. Neurophysiol.* **14**:6445

[53] K. Sakai, Y. Miyashita (1991) *Nature* **354**:152–155

[54] H. Gutfreund, J.D. Reger, A.P. Young (1988) *J. Phys. A* **21**:2775–2797

[55] D. Chazan, W. Miranker (1969) *Lin. Alg. Appl.* **2**:199

[56] G.M. Baudet (1978) *J. Assoc. Comp. Mach.* **25**:226

[57] D. Mitra (1987) *SIAM J. Sci. Stat. Comput.* **8**:43–58

[58] P. Tseng, D.P. Bertsekas, J.N. Tsitsiklis (1989) *Siam J. Control* **28**:678–710

[59] J.J. Hopfield, D.W. Tank (1985) *Biol. Cybern.* **52**:141–152

[60] J. Hertz, A. Krogh, R.G. Palmer (1991) *Introduction to the Theory of Neural Computation* (Addison-Wesley, Redwood City, CA)

[61] E. Domany, J.L. van Hemmen, K. Schulten (Eds.) *Physics of Neural Networks* (Springer-Verlag, Berlin)

[62] J.L. van Hemmen, R. Kühn (1991) In: *Physics of Neural Networks* E. Domany, J.L. van Hemmen, K. Schulten (Eds.) (Springer-Verlag, Berlin), pp. 1–106

[63] D.J. Amit, H. Gutfreund, H. Sompolinsky (1985) *Phys. Rev. Lett.* **55**:1530–1533

[64] B.M. Forrest, D.J. Wallace (1991) In: *Physics of Neural Networks*, E. Domany, J.L. van Hemmen, K. Schulten (Eds.) (Springer-Verlag, Berlin), pp. 121–148

[65] C.M. Marcus, R.M. Westervelt (1989) *Phys. Rev.* **A**:347–359

[66] T.A. Burton (1993) *Neural Networks* **6**:677–680

[67] A.V.M. Herz (1992) In: *Proceedings, SFB Workshop, Riezlern 1991*, U. Krüger (Ed.) (SFB 185, Frankfurt), pp. 151–164

[68] N. MacDonald (1989) *Biological Delay Systems: Linear Stability Theory* (Cambridge University Press, Cambridge)

[69] E. Niebur, H.G. Schuster, D.M. Kammen (1991) *Phys. Rev. Lett.* **67**:2753–2756

[70] D. Kleinfeld (1986) *Proc. Natl. Acad. Sci.* USA **83**:9469–9473

[71] H. Sompolinsky, I. Kanter (1986) *Phys. Rev. Lett.* **57**:2861–2864

[72] H. Gutfreund, M. Mezard (1988) *Phys. Rev. Lett.* **61**:235–238

[73] D.J. Amit (1988) *Proc. Natl. Acad. Sci. USA* **85**:2141–2145

[74] U. Riedel, R. Kühn, J.L. van Hemmen (1988) *Phys. Rev. A* **38**:1105–1108

[75] D. Kleinfeld, H. Sompolinsky (1988) *Biophys. J.* **54**:1039–1051

[76] A.C.C. Coolen, C.C.A.M. Gielen (1988) *Europhys. Lett.* **7**:281–285

[77] K. Bauer, U. Krey (1990) *Z. Phys. B* **79**:461–475

[78] M. Kerszberg, A. Zippelius (1990) *Phys. Scr. T* **33**:54–64

[79] M. Bartholomeus, A.C.C. Coolen (1992) *Biol. Cybern.* **67**:285–290

[80] T.M. Heskes, S. Gielen (1992) *Neural Networks* **5**:145–152

[81] B. de Vries, J.C. Principe (1992) *Neural Networks* **5**:565–576

[82] J.F. Fontanari, R. Köberle (1987) *Phys. Rev. A* **36**:2475

[83] C.M. Marcus, R.M. Westervelt (1990) *Phys. Rev. A* **42**:2410–2417

[84] A.V.M. Herz (1991) *Phys. Rev. A* **44**:1415–1418

[85] C. Peterson, J.R. Anderson (1987) *Complex Systems* **1**:995

[86] D.H. Ackley, G.E. Hinton, T.J. Sejnowski (1985) *Cognitive Sci.* **9**:147–169

[87] P. Baldi, F. Pineda (1991) *Neural Comput.* **3**:526–545

[88] E. Gardner (1987) *Europhys. Lett.* **4**:481–485

[89] K. Bauer, U. Krey (1991) *Z. Phys. B* **84**:131–141

[90] G. Mato, N. Parga (1991) *Z. Phys. B* **84**:483–486

[91] M.C. Mackey, L. Glass (1977) *Science* **197**:287–289

[92] J.L. Kaplan, J.A. Yorke (1974) *J. Math. Anal. Appl.* **48**:317–324

[93] R.D. Nussbaum (1979) *J. Differential Equations* **34**:25–54

[94] S.-N. Chow, H.-O. Walther (1988) *Trans. Amer. Math. Soc.* **307**:127–142

[95] A.V.M. Herz (1995) *J. Differential Equations* **118**:36–53

[96] G.S. Jones (1962) *J. Math. Anal. Appl.* **4**:440–469

[97] J. Mallet-Paret (1988) *J. Differential Equations* **72**:270–315

[98] J.K. Hale (1977) *Theory of Functional Differential Equations* (Springer-Verlag, New York)

[99] L.F. Abbott, C. van Vreeswijk (1993) *Phys. Rev. E* **48**:1483–1490

[100] M. Tsodyks, I. Mitkov, H. Sompolinsky (1993) *Phys. Rev. Lett.* **71**:1280–1283

[101] C. van Vreeswijk, L.F. Abbott (1993) *SIAM J. Appl. Math.* **53**:253–254

[102] R.E. Mirollo, S.H. Strgatz (1990) *SIAM J. Appl. Math.* **50**:1645–1662

[103] L.F. Abbott (1990) *J. Phys. A* **23**:3835–3859

[104] A. Treves (1993) *Network* **4**:259–284

[105] A.V.M. Herz, J.J. Hopfield (1995) *Phys. Rev. Lett.* **75**:1222–1225

[106] J.J. Hopfield, A.V.M. Herz (1995) *Proc. Natl. Acad. Sci.* USA **92**:6655–6662

[107] C. von der Malsburg (1981) Internal Report 81-2, MPI for Biophysical Chemistry, Göttingen

[108] M. Usher, H. Schuster, E. Niebur (1993) *Phys. Rev. Lett.* **71**:1280–1283

[109] P. Bush, T. Sejnowski (1994) preprint

[110] D. Hansel, G. Mato, C. Meunier (1995) *Neural Computation* **7**:25–26

[111] M. Tsodyks, T. Sejnowski (1994) *Network* **6**:111–124

[112] G. Sawitzki (1989) *The NetWork Project*, StatLab, Universität Heidelberg. Republished on Apple Developer CD Series disk *IV* (1990); G. Sawitzki, R. Kühn, J.L. van Hemmen (private communication)

[113] M.W. Hirsch (1989) *Neural Networks* **2**:331–349

[114] E.K. Blum, X. Wang (1992) *Neural Networks* **5**:577–587

[115] T. Poggio, V. Torre, C. Koch (1985) *Nature* **317**:314–319

[116] A.L. Yuille (1989) *Biol. Cybern.* **61**:115–123

[117] C.M. Bachmann, L.N. Cooper, A. Dembo, O. Zeitouni (1987) *Proc. Natl. Acad. Sci. USA* **84**:7529–7531

[118] M.A. Cohen (1992) *Neural Networks* **5**:83–103

2

Receptive Fields and Maps in the Visual Cortex: Models of Ocular Dominance and Orientation Columns*

Kenneth D. Miller[1]

with 4 figures

Synopsis. The formation of ocular dominance and orientation columns in the mammalian visual cortex is briefly reviewed. Correlation-based models for their development are then discussed, beginning with the models of Von der Malsburg. For the case of semilinear models, model behavior is well understood: correlations determine receptive field structure, intracortical interactions determine projective field structure, and the "knitting together" of the two determines the cortical map. This provides a basis for simple but powerful models of ocular dominance and orientation column formation: ocular dominance columns form through a correlation-based competition between left-eye and right-eye inputs, while orientation columns can form through a competition between ON-center and OFF-center inputs. These models account well for receptive field structure but are not completely adequate to account for the details of cortical map structure. Alternative approaches to map structure, including the self-organizing feature map of Kohonen, are discussed. Finally, theories of the computational function of correlation-based and self-organizing rules are discussed.

2.1 Introduction

The brain is a learning machine. An animal's experience shapes the neural activity of its brain; this activity in turn modifies the brain, so that

---

*An earlier and briefer version of this chapter appeared in *The Handbook of Neural Networks* (M.A. Arbib, Ed.), The MIT Press, 1995, under the title "Models of Ocular Dominance and Orientation Columns." Reused by permission.

[1]Departments of Physiology and Otolaryngology, W.M. Keck Center for Integrative Neuroscience, and Sloan Center for Theoretical Neurobiology, University of California, San Francisco, CA 94143-0444, USA.

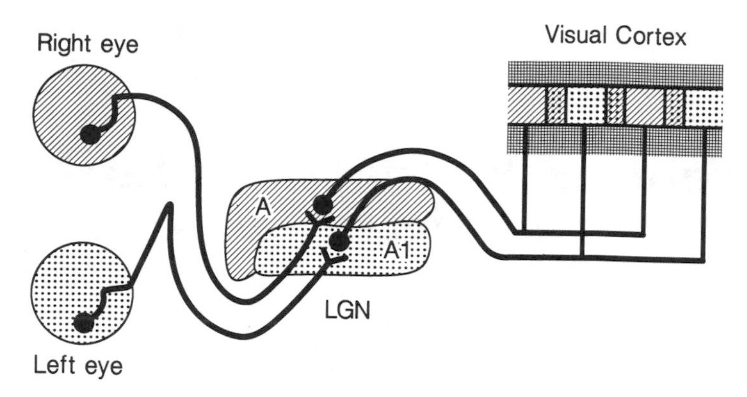

Fig. 2.1. Schematic of the mature visual system. Retinal ganglion cells from the two eyes project to separate layers of the lateral geniculate nucleus (LGN). Neurons from these two layers project to separate patches or stripes within layer 4 of the visual cortex (V1). Binocular regions (receiving input from both eyes) are depicted at the borders between the eye-specific patches. The cortex is depicted in cross-section, so that layers 1–3 are above and layers 5–6 below the LGN-recipient layer 4. Reprinted by permission from [42]. © 1989 by the AAAS.

the animal learns from its experience. This self-organization, the brain's reshaping of itself through its own activity (reviewed in [7, 14, 39, 51]), has long fascinated neuroscientists and modelers.

The classic example of activity-dependent neural development is the formation of ocular dominance columns in the cat or monkey primary visual cortex (reviewed in [44]). The cerebral cortex is the uniquely mammalian part of the brain. It is thought to form the complex, associative representations that characterize mammalian and human intelligence. The primary visual cortex (V1) is the first cortical area to receive visual information. It receives signals from the lateral geniculate nucleus of the thalamus (LGN), which in turn receives input from the retinas of the two eyes (Fig. 2.1).

To describe ocular dominance columns, several terms must be defined. First, the *receptive field* of a cortical cell refers to the area on the retinas in which appropriate light stimulation evokes a response in the cell, and also to the pattern of light stimulation that evokes such a response. Second, a *column* is defined as follows. V1 extends many millimeters in each of two, "horizontal" dimensions. Receptive field positions vary continuously along these dimensions, forming a *retinotopic* map, a continuous map of the visual world. In the third, "vertical" dimension, the cortex is about 2 mm in depth and consists of six layers. Receptive field positions do not significantly vary through this depth. Such organization, in which cortical properties are

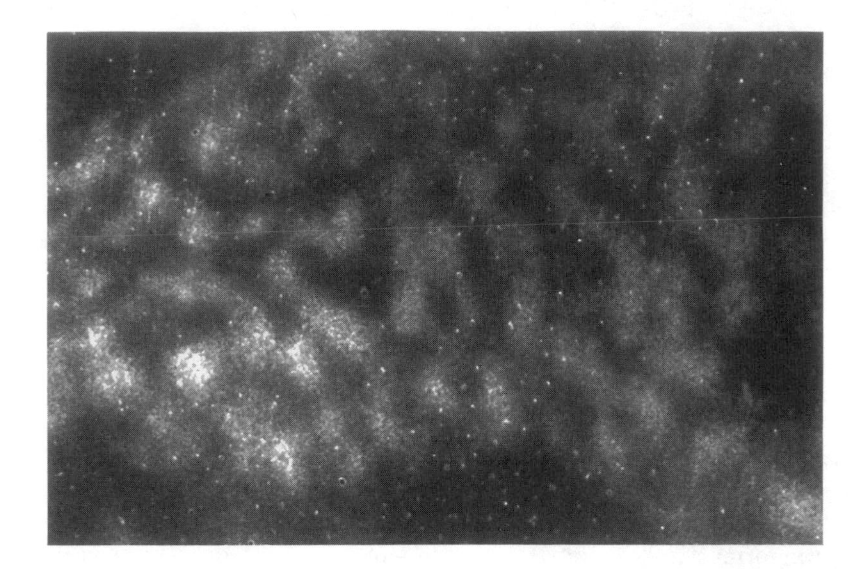

Fig. 2.2. Ocular dominance columns from cat V1. A horizontal cut through the layer 4 of V1 is shown. Terminals serving a single eye are labeled white. Dark regions at the edges are out of the plane containing LGN terminals. Region shown is 5.3 × 7.9 mm. Photograph generously supplied by Dr. Y. Hata.

invariant through the vertical depth of cortex but vary horizontally, is called *columnar* organization and is a basic feature of the cerebral cortex.

Third, *ocular dominance* must be defined. Cells in the LGN are *monocular*, responding exclusively to stimulation of a single eye (Fig. 2.1). LGN cells project to layer 4 of V1, where they terminate in alternating stripes or patches of terminals representing a single eye (Figs. 2.1 and 2.2). Most or, in some species, all layer-4 V1 cells are monocular. Cells in other layers of V1 respond best to the eye that dominates layer-4 responses at that horizontal location. Thus, V1 cells can be characterized by their *ocular dominance*, or eye preference. The stripes or patches of cortex that are dominated throughout the cortical depth by a single eye are known as *ocular dominance columns.*

The segregated pattern of termination of the LGN inputs to V1 arises early in development. Initially, LGN inputs project to layer 4 of V1 in an overlapping manner, without apparent distinction by eye represented. The terminal arbors of individual LGN inputs extend horizontally in layer 4 for distances as large as 2 mm (for comparison, a typical spacing between cortical cells is perhaps 20 μm). Subsequently, beginning either prenatally or shortly after birth, depending on the species, the inputs representing each eye become horizontally confined to the alternating, approximately 1/2-mm wide ocular dominance patches.

This segregation results from an activity-dependent competition between the geniculate terminals serving the two eyes (see discussion in [44]). The signal indicating that different terminals represent the same eye appears to be the correlations in their neural activities [54]. These correlations exist due both to spontaneous activity, which is locally correlated within each retina [36, 37, 38, 64], and to visually-induced activity, which correlates the activities of retinotopically nearby neurons within each eye and, to a lesser extent, between the eyes [26]. The segregation process is competitive. If one eye is caused to have less activity than the other during a critical period in which the columns are forming, the more active eye takes over most of the cortical territory [25, 52, 60]; but the eye with reduced activity suffers no loss of projection strength in retinotopic regions in which it lacks competition from the other eye [15, 16]. In summary, ocular dominance column formation is a simple system in which correlated patterns of neural activity sculpt the patterns of neural connectivity.

Orientation columns are another striking feature of visual cortical organization. Most V1 cells are orientation-selective, responding selectively to light/dark edges over a narrow range of orientations. The preferred orientation of cortical cells varies regularly and periodically across the horizontal dimension of the cortex and is invariant in the vertical dimension. The maturation of orientation selectivity is activity-dependent (e.g., [6, 11]). However, it has not yet been possible to test whether the initial development of orientation selectivity is activity-dependent. This is because some orientation selectivity already exists at the earliest developmental times at which visual cortical responses can be recorded [1, 4, 6, 20, 61], and it has not been possible to block visual system activity immediately before this time. Nonetheless, it has long been a popular notion that the initial development of orientation selectivity, like that of ocular dominance, may occur through a process of activity-dependent synaptic competition.

The inputs from LGN to V1 serving each eye are of two types: ON-center and OFF-center. Both kinds of cells have circularly symmetric, orientation-insensitive receptive fields and respond to contrast rather than uniform luminance. ON-center cells respond to light against a dark background, or to light onset; OFF-center cells respond to dark against a light background, or to light offset. In the cat, the orientation-selective V1 cells in layer 4 are *simple cells*: cells with receptive fields consisting of alternating oriented subregions that receive exclusively ON-center or exclusively OFF-center input (Fig. 2.3). As shall be discussed, one theory for the development of orientation selectivity is that, like ocular dominance, it develops through a competition between two input populations: in this case, a competition between the ON-center and the OFF-center inputs [41].

Fig. 2.3. Two examples of simple cell receptive fields (RFs). Regions of the visual field from which a simple cell receives ON-center (white) or OFF-center (dark) input are shown. Note: Ocular dominance columns (Fig. 2.2) represent an alternation, across the cortex, in the type of input (left- or right-eye) received by different cortical cells; while a simple-cell RF (this figure) represents an alternation across visual space in the type of input (ON- or OFF-center) received by a *single* cortical cell.

2.2 Correlation-Based Models

To understand ocular dominance and orientation column formation, two processes must be understood: (1) the development of *receptive field structure*: under what conditions do receptive fields become monocular (drivable only by a single eye) or orientation-selective? (2) the development of *periodic cortical maps* of receptive field properties: what leads ocular dominance or preferred orientation to vary periodically across the horizontal dimensions of the cortex, and what determines the periodic length scales of these maps? Typically, the problem is simplified by consideration of a two-dimensional model cortex, ignoring the third dimension in which properties such as ocular dominance and orientation are invariant.

One approach to addressing these problems is to begin with a hypothesized mechanism of synaptic plasticity, and to study the outcome of cortical development under such a mechanism. Most commonly, theorists have considered a *Hebbian synapse*: a synapse whose strength is increased when pre- and postsynaptic firings are correlated, and possibly decreased when they are anticorrelated. Other mechanisms, such as activity-dependent release and uptake of a diffusible modification factor, can lead to similar dynamics [42], in which synaptic plasticity depends on the correlations among the activities of the competing inputs. Models based on such mechanisms are referred to as *correlation-based models* [39].

2.2.1 The Von der Malsburg Model of V1 Development

Von der Malsburg [57, 59] first formulated a correlation-based model for the development of visual cortical receptive fields and maps. His model had two basic elements. First, synapses of LGN inputs onto cortical neurons were modified by a Hebbian rule that is *competitive*, so that some synapses were strengthened only at the expense of others. He enforced the competition by holding constant the total strength of the synapses converging on each cortical cell (conservation rule). Second, the cortical cells tended to be activated in *clusters*, due to intrinsic cortical connectivity, e.g., short-range horizontal excitatory connections and longer range horizontal inhibitory connections.

The conservation rule leads to competition among the inputs to a single target cell. Inputs that tend to be coactivated — that is, that have correlated activities — are mutually reinforcing, working together to activate the postsynaptic cells and thus to strengthen their own synapses. Different patterns that are mutually un- or anticorrelated compete, since the strengthening of some synapses means the weakening of others. Cortical cells eventually develop receptive fields that are responsive to a correlated pattern of inputs.

The clustered cortical activity patterns lead to competition between the different groups of cortical cells. Each input pattern comes to be associated with a cortical cluster of activity. Overlapping cortical clusters contain many coactivated cortical cells, and thus become responsive to overlapping, correlated input patterns. Adjacent, nonoverlapping clusters contain many anticorrelated cortical cells, and thus become responsive to un- or anticorrelated input patterns. Thus, over distances on the scale of an activity cluster, cortical cells will have similar response properties; while, on the scale of the distance between nonoverlapping clusters, cortical cells will prefer un- or anticorrelated input patterns. This combination of local continuity and larger scale heterogeneity leads to continuous, periodic cortical maps of receptive field properties.

In computer simulations, this model was applied to the development of orientation columns [57] and ocular dominance columns [59]. For orientation columns, inputs were activated in oriented patterns of all possible orientations. Individual cortical cells then developed selective responses, preferring one such oriented pattern, with nearby cortical cells preferring nearby orientations. For ocular dominance columns, inputs were activated in monocular patterns consisting of a localized set of inputs from a single eye. Individual cortical cells came to be driven exclusively by a single eye, and clusters of cortical cells came to be driven by the same eye. The final cortical pattern consisted of alternating stripes of cortical cells preferring a single eye, with the width of a stripe approximately set by the diameter of an intrinsic cluster of cortical activity.

In summary, a competitive Hebbian rule leads individual receptive fields to become selective for a correlated pattern of inputs. Combined with the idea that the cortex is activated in intrinsic clusters, this suggests an origin for cortical maps: coactivated cells in a cortical cluster tend to become selective for similar, coactivated patterns of inputs. These basic ideas are used in most subsequent models.

2.2.2 Mathematical Formulation

A typical correlation-based model is mathematically formulated as follows [57, 27, 40, 42]. Let $x, y, \ldots$ represent retinotopic positions in V1, and let $\alpha, \beta, \ldots$ represent retinotopic positions in the LGN. Let $S^{\mu}(x, \alpha)$ be the synaptic strength of the connection from α to x of the LGN projection of type μ, where μ may signify left-eye, right-eye, ON-center, OFF-center, etc. Let $B(x, y)$ represent the synaptic strength and sign of connection from the cortical cell at y to that at x. For simplicity, $B(x, y)$ is assumed to take different signs for a fixed y as x varies, but, alternatively, separate excitatory-projecting and inhibitory-projecting cortical neurons may be used. Let $a(x)$ and $a^{\mu}(\alpha)$ represent the activity of a cortical or LGN cell, respectively.

The activity $a(x)$ of a cortical neuron is assumed to depend on a linear combination of its inputs:

$$a(x) = f_1\left(\sum_{\mu,\alpha} S^{\mu}(x, \alpha) a^{\mu}(\alpha) + \sum_{y} B(x, y) a(y)\right). \tag{2.1}$$

Here, f_1 is some monotonic function such as a sigmoid or linear threshold.

A Hebbian rule for the change in feedforward synapses can be expressed as

$$\Delta S^{\mu}(x, \alpha) = A^{\mu}(x, \alpha) f_2\left[a(x)\right] f_3\left[a^{\mu}(\alpha)\right]. \tag{2.2}$$

Here, $A(x, \alpha)$ is an *arbor function* that expresses the number of synapses of each type from α to x; a minimal form is $A(x, \alpha) = 1$ if there is a connection from α to x, and $A(x, \alpha) = 0$ otherwise. A typical form for the functions f_2 and f_3 is $f(a) = (a - \langle a \rangle)$, where $\langle a \rangle$ indicates an average of a over input patterns. This yields a *covariance rule*: synaptic change depends on the covariance of postsynaptic and presynaptic activity.

Next, the Hebbian rule must be made *competitive*. This can be accomplished by conserving the total synaptic strength over the postsynaptic cell [57], which in turn may be done either subtractively or multiplicatively [43]. The corresponding equations are

$$\tfrac{d}{dt} S^{\mu}(x, \alpha) = \Delta S^{\mu}(x, \alpha) - \epsilon(x) A(x, \alpha) \qquad \text{(Subtractive)} \tag{2.3}$$

$$\tfrac{d}{dt} S^{\mu}(x, \alpha) = \Delta S^{\mu}(x, \alpha) - \gamma(x) S^{\mu}(x, \alpha) \qquad \text{(Multiplicative)}, \tag{2.4}$$

where

$$\epsilon(x) = \frac{\sum_{\kappa,\alpha} \Delta S^{\kappa}(x,\alpha)}{\sum_{\kappa,\alpha} A(x,\alpha)} \qquad \text{and} \quad \gamma(x) = \frac{\sum_{\kappa,\alpha} \Delta S^{\kappa}(x,\alpha)}{\sum_{\kappa,\alpha} S^{\kappa}(x,\alpha)}.$$

Either form of constraint ensures that $\sum_{\mu,\alpha}(d/dt)S^{\mu}(x,\alpha)=0$. Alternative methods have been developed to force Hebbian rules to be competitive [43].

Finally, synaptic weights may be limited to a finite range, $s_{\min}A(x,\alpha) \leq S^{\mu}(x,\alpha) \leq s_{\max}A(x,\alpha)$. Typically, $s_{\min} = 0$ and $s_{\max}$ is some positive constant.

2.2.3 Semilinear Models

In semilinear models, the f's in Eqs. (2.1) and (2.2) are chosen to be linear. Then, after substituting for $a(x)$ from Eq. (2.1) and averaging over input patterns (assuming that all inputs have identical mean activity, and that changes in synaptic weights are negligibly small over the averaging time), Eq. (2.2) becomes

$$\Delta S^{\mu}(x,\alpha) = \lambda A(x,\alpha) \left[\sum_{y,\beta,\kappa} I(x-y) \left[C^{\mu\kappa}(\alpha-\beta) - k_2 \right] S^{\kappa}(y,\beta) + k_1 \right]. \tag{2.5}$$

Here, $I(x-y)$ is an element of the intracortical interaction matrix

$$\mathbf{I} \equiv (\mathbf{1}-\mathbf{B})^{-1} = \mathbf{1} + \mathbf{B} + \mathbf{B}^2 + \cdots,$$

where the matrix $\mathbf{B}$ is defined in Eq. (2.1). This summarizes intracortical synaptic influences including contributions via $0, 1, 2, \ldots$ synapses. The covariance matrix

$$C^{\mu\kappa}(\alpha-\beta) = \langle (a^{\mu}(\alpha) - \bar{a})\,(a^{\kappa}(\beta) - \bar{a}) \rangle$$

expresses the covariation of input activities. The factors λ, k_1, and k_2 are constants. Translation invariance has been assumed in both cortex and LGN.

When there are two competing input populations, Eq. (2.5) can be simplified further by transforming to sum and difference variables: $S^S \equiv S^1 + S^2$, $S^D \equiv S^1 - S^2$. Assuming equivalence of the two populations (so that $C^{11} = C^{22}$, $C^{12} = C^{21}$), Eq. (2.5) becomes

$$\Delta S^S(x,\alpha) = \lambda A(x,\alpha) \left\{ \sum_{y,\beta} I(x-y) \left[C^S(\alpha-\beta) - 2k_2 \right] S^S(y,\beta) + 2k_1 \right\} \tag{2.6}$$

$$\Delta S^D(x,\alpha) = \lambda A(x,\alpha) \sum_{y,\beta} I(x-y) C^D(\alpha-\beta) S^D(y,\beta). \tag{2.7}$$

Here, $C^S \equiv C^{11} + C^{12}$, $C^D \equiv C^{11} - C^{12}$. Subtractive renormalization [Eq. (2.3)] alters only Eq. (2.6) for S^S, by subtraction of $2\epsilon(x)A(x-\alpha)$, while leaving Eq. (2.7) for S^D unaltered. Multiplicative renormalization [Eq. (2.4)] alters both Eqs. (2.6) and (2.7), by subtraction of $\gamma(x)S^S(x,\alpha)$ and $\gamma(x)S^D(x,\alpha)$, respectively.

2.2.4 How Semilinear Models Behave

Linear equations like (2.6) and (2.7) can be understood by finding the eigenvectors or "modes" of the operators on the right side of the equations. The eigenvectors are the synaptic weight patterns that grow independently and exponentially, each at its own rate. The fastest growing eigenvectors typically dominate development and determine basic features of the final pattern, although the final pattern ultimately is stabilized by nonlinearities such as the limits on the range of synaptic weights or the nonlinearity involved in multiplicative renormalization [Eq. (2.4)].

We will focus on the behavior of Eq. (2.7) for S^D (for analysis of Eq. (2.6), see [34, 35]). S^D describes the difference in the strength of two competing input populations. Thus, it is the key variable describing the development of ocular dominance segregation, or development under an ON-center/OFF-center competition. In many circumstances, Eq. (2.7) can be derived directly from Eqs. (2.1) and (2.2) by linearization about $S^D \equiv 0$ [40] without need to assume a semilinear model. The condition $S^D \approx 0$ corresponds to an initial condition in which the projections of the two input types are approximately equal. Thus, study of Eq. (2.7) can lend insight into early pattern formation in more general, nonlinear correlation-based models.

Equation (2.7) can be solved simply in the case of full connectivity from the LGN to the cortex, when $A(x,\alpha) \equiv 1$ for all x and α. Then modes of $S^D(x,\alpha)$ of the form $e^{ikx}e^{il\alpha}$ grow exponentially and independently, with rates proportional to $\tilde{I}(k)\tilde{C}^D(l)$, where $\tilde{I}$ and $\tilde{C}^D$ denote the Fourier transforms of I and C^D, respectively (for a description of the modes as real rather than complex functions, see [44]). The wavenumber k determines the wavelength $2\pi/|k|$ of an oscillation of S^D across cortical cells, while the wavenumber l determines the wavelength $2\pi/|l|$ of an oscillation of S^D across geniculate cells. The fastest growing modes, which will dominate early development, are determined by the k and l that maximize $\tilde{I}(k)$ and $\tilde{C}^D(l)$, respectively. The peak of a function's Fourier transform corresponds to the cosine wave that best matches the function, and thus represents the "principal oscillation" in the function.

To understand these modes (Fig. 2.4), consider first the set of inputs received by a single cortical cell, that is, the shape of the mode for a fixed cortical position x. This can be regarded as the *receptive field* of the cortical cell. Each receptive field oscillates with wavenumber l. This oscillation of $S^D \equiv S^1 - S^2$ is an oscillation between receptive field subregions dominated by S^1 inputs and subregions dominated by S^2 inputs. Thus, in ocular dominance competition, monocular cells (cells whose entire receptive fields are dominated by a single eye) are formed only by modes with $l = 0$ (no oscillation). Monocular cells thus dominate development if the peak of the Fourier transform of the C^D governing left/right competition is at $l = 0$. Now, instead, consider an ON/OFF competition: S^1 and S^2 represent ON- and OFF-center inputs from a single eye. Then the receptive fields of modes with nonzero l resemble simple cells: they receive predominantly ON-center and predominantly OFF-center inputs from successive, alternating subregions of the visual world. Thus, simple cells can form if the C^D governing ON/OFF competition has its peak at a nonzero l.

Now consider the arborizations or *projective fields* projecting from a single geniculate point, that is, the shape of the mode for a fixed geniculate position α. These oscillate with wavenumber k. In ocular dominance competition, this means that left- and right-eye cells from α project to alternating patches of the cortex. When monocular cells form ($l = 0$), these alternating patches of the cortex are the ocular dominance columns: alternating patches of the cortex receiving exclusively left-eye or exclusively right-eye input, respectively. Thus, the width of ocular dominance columns — the wavelength of alternation between right-eye– and left-eye–dominated cortical cells — is determined by the peak of the Fourier transform of the intracortical interaction function I. In ON/OFF competition, with $l \neq 0$, the identity of the cortical cells receiving the ON-center or OFF-center part of the projection varies as α varies, so individual cortical cells receive both ON- and OFF-center inputs, but from distinct subregions of the receptive field.

In summary, there is an oscillation within receptive fields, with wavenumber l determined by the peak of $\tilde{C}^D$; and an oscillation within arbors, with wavenumber k determined by the peak of $\tilde{I}$ (Fig. 2.4). These two oscillations are "knit together" to determine the overall pattern of synaptic connectivity. The receptive field oscillation, which matches the receptive field to the correlations, quantitatively describes von der Malsburg's finding that individual receptive fields become selective for a correlated pattern of inputs. Similarly, the arbor oscillation matches projective fields to the intracortical interactions, and thus to the patterns of cortical activity clusters. This quantitatively describes the relationship between activity clusters and maps. Note that the factor e^{ikx} can be regarded as inducing a phase shift, for varying x, in the structure of receptive fields. Thus, cortical cells that are nearby on the scale of the arbor oscillation have similar receptive fields, while cells 1/2 wavelength apart have opposite receptive fields.

An alternative viewpoint on the same pattern is obtained by rewriting the modes as $e^{i(k+l)x}e^{-il(x-\alpha)}$. The argument $l(x - \alpha)$ represents the oscillation with wavenumber l within the receptive field, now expressed in coordinates relative to the center of the receptive field rather than in an absolute position across the geniculate. The argument $(k + l)x$ represents

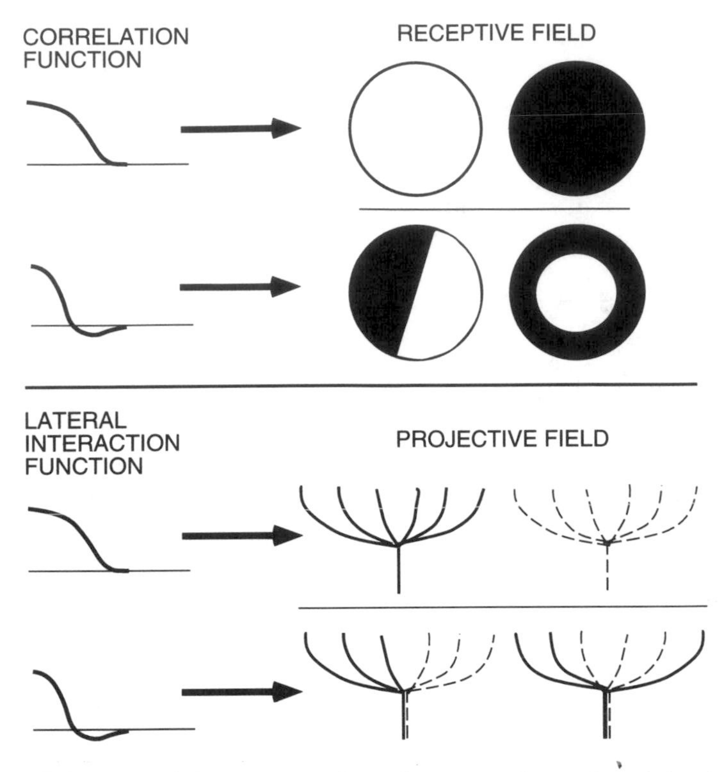

Fig. 2.4. Schematic of the outcome of semilinear correlation-based development. **Top:** The correlation function (C^D) determines the structure of receptive fields (RFs). White RF subregions indicate positive values of S^D; dark subregions, negative values. When C^D does not oscillate, individual cortical cells receive only a single type of input, as in ocular dominance segregation. If C^D oscillates, there is a corresponding oscillation in the type of input received by the individual cortical cells, as in simple-cell RFs. Alternative RF structures could form, as in the center-surround structure shown; but oriented simple-cell–like outcomes predominate for reasonable parameters [41]. Simple cells then develop with various numbers of subregions and various spatial phases; only a single example, of a cell with two subregions and odd spatial symmetry, is pictured. **Bottom:** The intracortical interactions (I) similarly determine the structure of projective fields. Here, solid lines indicate positive values of S^D, while dotted lines indicate negative values. Adapted from [43].

a shift, for varying x, in the phase of the receptive field relative to the receptive field center. For the case of ocular dominance, with $l = 0$, this is just the shift, with wavenumber k, between left-eye dominance and right-eye dominance of the cortical cells. For ON/OFF competition with $l \neq 0$, this represents a periodic shifting, with movement across the cortex, as to which subregions of the receptive field are dominated by ON-center inputs and which subregions are dominated by OFF-center inputs. Thus, we can view the results as an oscillation within receptive fields, with wavenumber l, combined with a shift with cortical position in the spatial phase of receptive fields, this shift occurring with wavenumber $k + l$, the vector sum of the projective field or arbor oscillation and the receptive field oscillation.

The competitive, renormalizing terms [Eqs. (2.3) and (2.4)] do not substantially alter these pictures, except that multiplicative renormalization can suppress ocular dominance development in some circumstances [43].[2] These results hold also for localized connectivity (finite arbors), and thus generally characterize the behavior of semilinear models [39, 44]. The major difference in the case of localized connectivity is that, if k or l corresponds to a wavelength larger than the diameter of connectivity from or to a single cell, then it is equivalent to $k = 0$ or $l = 0$, respectively. A good approximation to the leading eigenvectors in the case of finite connectivity is given simply by $A(x - \alpha)e^{ikx}e^{il\alpha}$, where k and l are determined as above by the peaks of $\tilde{I}(k)$ and $\tilde{C}^D(l)$ (unpublished results).

2.2.5 Understanding Ocular Dominance and Orientation Columns with Semilinear Models

This understanding of semilinear models leads to simple models for the development of both ocular dominance columns [42] and orientation columns [41] as follows (Fig. 2.4).

Monocular cells develop through a competition of left- and right-eye inputs in a regime in which $\tilde{C}^D(l)$ is peaked at $l = 0$. The wavelength of ocular dominance column alternation then is determined by the peak of $\tilde{I}(k)$.

[2] Subtractive renormalization [Eq. (2.3)] has no effect on the development of S^D. Multiplicative renormalization [Eq. (2.4)] lowers the growth rates of all modes of both S^D and S^S by the factor $\gamma(x)$, which depends only on S^S. The result is that, in order for S^D to grow at all, its modes must have larger unconstrained growth rates than those of S^S; that is, the peak of the Fourier transform of C^D must be larger than that of C^S. In practice, this condition is met only if there are anticorrelations between S^1 and S^2, that is, if C^{12} is significantly negative. When this condition is met, then the modes that dominate S^D are just as described above; they are not altered by the constraint term in Eq. (2.4). These and other effects of renormalizing terms are discussed in detail in [43].

Orientation-selective simple cells develop through a competition of ON-center and OFF-center inputs in a regime in which $\tilde{C}^D(l)$ is peaked at $l \neq 0$. The mean wavelength of alternation of ON-center and OFF-center subregions in the simple cells' receptive fields is determined by the peak of $\tilde{C}^D(l)$. This wavelength corresponds to a cell's preferred spatial frequency under stimulation by sinusoidal luminance gratings. In individual modes, all cortical cells have the same preferred orientation, but their spatial phase varies periodically with cortical position. The mixing of such modes of all orientations leads to a periodic variation of preferred orientation across cortex. The period with which preferred orientations change across cortex is more complex to determine [41].

This model of ocular dominance column formation is similar to that of von der Malsburg [59]. The latter model assumed anticorrelation between the two eyes; this was required due to the use of multiplicative renormalization [Eq. (2.4)]. With subtractive renormalization [Eq. (2.4)], ocular dominance column formation can occur even with partial correlation of the two eyes [43]. The model can be compared to experiment, particularly through the prediction of the relation between intracortical connectivity and ocular dominance column width.

The model of orientation-selective cell development is quite different from that of von der Malsburg [57]. Von der Malsburg postulated that oriented input patterns lead to the development of orientation-selective cells. The ON/OFF model instead postulates that ON/OFF competition results in oriented receptive fields in the absence of oriented input patterns; the circular symmetry of the input patterns is spontaneously broken. This symmetry-breaking potential of Hebbian development was first discovered by Linsker [28]. In all of these models, the continuity and periodic alternation of preferred orientation is due to the intracortical connectivity. The ON/OFF model can be compared to experiment most simply by the measurement of C^D, to determine whether it has the predicted oscillation.

2.2.6 Related Semilinear Models

Linsker [27, 28, 29] proposed a model that was highly influential in two respects. First, he pointed out the potential of Hebbian rules to spontaneously break symmetry, yielding orientation-selective cells given approximately circularly symmetric input patterns. Second, he demonstrated that Hebbian rules could lead to segregation *within* receptive fields, so that a cell came to receive purely excitatory or purely inhibitory input in alternating subregions of the receptive field. This model was thoroughly analyzed in [34, 35].

Linsker used a semilinear model with a single input type that could have positive or negative synaptic strengths ($s_{\text{min}} = -s_{\text{max}}$). He largely restricted study to the case of a single postsynaptic cell. Because the equation for a single input type and a single postsynaptic cell [Eq. (2.5), with

$I(x - y) = \delta(x - y)]$ is circularly symmetric,[3] its eigenfunctions also are eigenfunctions of the rotation operator. Thus, the eigenfunctions can be written in polar coordinates (r, θ) as $\cos(n\theta) f_{nj}(r)$ and $\sin(n\theta) f_{nj}(r)$, where $f_{nj}(r)$ is a radial function and n and j are integers indexing the eigenfunctions. In quantum mechanics, atomic orbitals are named Nx, where N is a number representing one plus the total number of angular and radial nodes, and x is a letter denoting the number of angular nodes (s,p,d,f,g,... corresponding to n=0,1,2,3,4,... angular nodes). Thus, 1s is a function with zero nodes, 2s has one node that is radial, 2p has one node that is angular, 3p has two nodes (one radial, one angular), etc. This naming scheme can be applied to any rotationally symmetric system, and in particular can be applied to the eigenfunctions of Linsker's system [34, 35], a fact which physicists have found amusing. The nature of these eigenfunctions, their dependence on parameters, and their role in determining the outcomes Linsker observed in simulations are described in [34, 35].

For our present purposes, the essential results of this analysis are as follows. Two factors underlay Linsker's results. One factor was that oscillations in a correlation function can induce oscillations in a receptive field, as was described above. The other factor was a constraint in the model fixing the percentage of positive or negative synapses received by a cell; this forced an alternation of positive and negative subregions even when the correlation function did not oscillate. These two causes were not disentangled in Linsker's simulations, but only the first appears likely to be of biological relevance.

Tanaka [45, 56] has independently formulated models of ocular dominance and orientation columns that are similar to those described in Sec. 2.2.5. The major difference is that he works in a regime in which each cortical cell comes to receive only a single LGN input. Tanaka defines cortical receptive fields as the convolution of the input arrangement with the intracortical interaction function. This means that a cortical cell's receptive field is due to its single input from the LGN plus its input from all other cortical cells within reach of the intracortical interaction function. Thus, orientation selectivity in this model arises from the breaking of circular symmetry in the pattern of inputs to different cortical cells, rather than to individual cortical cells.

2.3 The Problem of Map Structure

The above models account well for the basic features of the primary visual cortex. However, many details of real cortical maps are not replicated by

[3]The assumption is made that the arbor and correlation functions depend only on distance.

these models [9, 12, 63]. One reason may be the simplicity of the model of the cortex: the real cortex is three-dimensional rather than two; it has cell-specific connectivity rather than connectivity that depends only on distance; and it has plastic rather than fixed intracortical connections. Another reason is that the details of the map structure inherently involve nonlinearities, by which the fastest growing modes interact and compete; whereas the semilinear framework only focuses on early pattern formation, in which the fastest growing modes emerge and mix randomly without interacting.

Some simple models that focus on map development rather than receptive field development strikingly match the map structures observed in monkeys [9]. One such model [46] uses the self-organizing feature map (SOFM) of Kohonen [24, 48], in which only a single cluster of cortical cells is activated in response to a given input pattern. This is an abstraction of the idea that the cortex responds in localized activity clusters. The single activated cluster is centered on the cell whose weight vector best matches the direction of the input activation vector. Hebbian learning then takes place on the activated cells, bringing their weight vector closer to the input activation vector. The size of an activity cluster is gradually decreased as the mapping develops; this is akin to annealing, helping to ensure a final mapping that is optimal on both coarse and fine scales.

Except for the restriction to a single activity cluster and the gradual decrease in cluster size, the SOFM is much like the correlation-based models. However, an abstract representation of the input is generally used. In correlation-based models, the input space may have thousands of dimensions, one for each input cell. In the SOFM model of the visual cortex, the input space instead has five dimensions: two represent retinotopic position, and one represents each of ocular dominance, orientation selectivity, and preferred orientation. Each cortical cell receives five "synapses," corresponding to these five "inputs." Assumptions are made as to the relative "size" of, or variance of the input ensemble along, each dimension. There is no obvious biological interpretation for this comparison between dimensions. Under the assumptions that the ocular dominance and orientation dimensions are "short" compared to the retinotopic dimensions, and that only one input point is activated at a time, Hebbian learning can lead to maps of orientation and ocular dominance that are, in detail, remarkably like those seen in macaque monkeys [9, 46].

The SOFM, and other models based on the "elastic net" algorithm [8, 13], lead to locally continuous mappings in which a constant distance across the cortex corresponds to a roughly constant distance in the reduced "input space." This means that, when one input feature is changing rapidly across the cortex, the others are changing slowly. Thus, the models predict that orientation changes rapidly where ocular dominance changes slowly, and vice versa. It may be this feature that is key to replicating the details of macaque orientation and ocular dominance maps. A model that forces such a relationship to develop between ocular dominance and orientation, while assuring periodic representations of each, also gives a good match to primate visual maps [55].

The SOFM also replicates aspects of the retinotopic maps seen in higher areas of the cat visual cortex [62]. For these studies, the input and output spaces are each taken to be two-dimensional, representing retinotopic positions. The input space is taken to be a half-circle, representing a hemiretina, and the shape of the output space is varied. When this shape is long and narrow, as in cat cortical areas 18 and 19, the retinotopic map developed by the SOFM has a characteristic pattern of discontinuities closely resembling those observed experimentally in those areas [62]. The SOFM achieves maps in which nearby points in the output space correspond to nearby points in the input space, while each area of the input space receives approximately equal representation provided each is equally activated ([48]; see further discussion of the SOFM below). The success of the SOFM models of retinotopic maps suggests that these are constraints that should be satisfied by any model of cortical maps. One would like to determine more precisely the constraints on a retinotopic mapping, embodied by the SOFM, that are sufficient to replicate these results.

It recently has been reported that input correlations can alter the spacing of ocular dominance columns in the cat visual cortex by perhaps 20–30% [32]. A smaller ocular dominance column spacing develops when the activities of the two eyes are correlated by normal vision than when the two eyes' activities are decorrelated (decorrelation is achieved by inducing divergent strabismus, which causes the two eyes to see different parts of the visual world). This effect was anticipated theoretically by Goodhill [12], who argued essentially that correlation of the activities of the two eyes brings them "closer together," and so the two eyes should be brought closer together in their cortical representation by a reduction of the column size. This effect also could have been anticipated by the SOFM models of ocular dominance, because decorrelation corresponds to an increase in the variance of ocular dominance and thus an increase in the "size" of the ocular dominance dimension, which results in increased column size [48]. In semilinear models, in contrast, the column width does not appear to be significantly affected by between-eye correlations. Rather, as the degree of between-eye correlation is increased, more binocular cells form at the column borders, until at some critical level of correlation ocular dominance segregation no longer occurs (unpublished results). That is, the two eyes are brought "closer together" through alteration of the receptive fields rather than through alteration of the map. One can anticipate several biological mechanisms that might be added to instead yield a reduction in the column size, such as nonlinearities that discourage formation of binocular cells, or nonlinearities in cortical activation that cause the size of activity clusters to depend on the correlations of the inputs.

Finally, it recently has been noted that cat orientation maps are significantly smoother than could be achieved by simple linear considerations [63].

The analysis in [63] suggests that these maps could result, mathematically, from a local "diffusion" of preferred orientations. It will be interesting to develop a biologically interpretable model of such a process.

2.4 The Computational Significance of Correlation-Based Rules

2.4.1 Efficient Representation of Information

A simple correlation-based rule for a single postsynaptic cell can, if properly designed, lead to the development of a receptive field that corresponds to the principal component of the input data (that is, to the principal eigenvector of the covariance matrix of the inputs to the cell) [30, 43, 47]. This receptive field in turn maximizes the variance of the postsynaptic cell's activity, given the ensemble of input patterns. It has been argued that correlation-based rules thus maximize the information carried in the postsynaptic cell's activity about the input patterns [30]. Intuitively, by varying as much as possible in its response to different inputs, the postsynaptic cell draws the greatest possible distinction between the different input patterns.

More generally, a number of closely related (and in many circumstances identical) computational functions have been proposed for brain areas near the sensory periphery. These include maximization of information about the inputs [30], minimization of redundancy or correlation in the activities of output cells [3], statistical independence of the output activities [3], or encoding of the input information as compactly as possible (for example, requiring as little dynamic range as possible per neuron) [2]. These functions all involve representing the input information in an efficient way, in the sense of information theory. These measures of efficiency take into account the statistics of the input ensemble but disregard the "semantics," the meaning or survival value to the animal, of the inputs.

The interpretation that the function of a correlation-based rule is to yield such an efficient representation is inviting, but it carries two major problems. First, the principal component representation achieved by correlation-based rules is optimally efficient only for a Gaussian distribution of input patterns, or, in other words, it reflects only the second-order or two-point statistics (the covariance) of the input data. It is possible that a great deal of information may reside in higher order statistics, but a correlation-based rule as conceived above will ignore this information. Intrator has suggested that a variant of standard Hebbian rules can instead maximize a third-order statistic of the output activity, and argues that this may be a better statistic for distinguishing among the elements of real-world ensembles [22, 23]. While one statistic or the other may be best for characterizing a given set of data, both approaches can suffer from the limitation that they are maximizing one particular statistic rather than maximizing some measure of efficiency.

Second, this interpretation applies only to a single, isolated postsynaptic cell. Multiple cells viewing the same input ensemble will extract the same information from it under a given correlation-based rule. This does not add new information about the input, but only redundantly repeats the same information. Thus, although a single cell may have a receptive field that maximizes the information it could carry about the input ensemble, a group of such cells generally will not improve much on the performance of a single cell and will not carry the maximal possible information about the input ensemble.[4]

One way out of this dilemma is to introduce couplings between the postsynaptic cells that force them to learn independent parts of the input ensemble. Unfortunately, excitatory couplings tend to produce correlated cells, while inhibitory couplings produce anticorrelated cells. The ostensible goal, however, is to produce uncorrelated cells, cells whose activities carry independent information about the input ensemble. Thus, biological couplings will not work. A theoretical way out involves using connections between the postsynaptic cells that are modified by anti-Hebbian rules: If two cells have correlated activities, the connection between them becomes more negative; if two cells have anticorrelated activity, the connection between them becomes more positive. The result is that the cells become uncorrelated. Many authors have independently proposed rules that involve such anti-Hebbian learning on lateral connections (e.g., [10, 31, 49]) or related ideas [50]. However, no biological sign of anti-Hebbian synaptic modification thus far has been observed.

An alternative way out of this dilemna stems from the observation that biological receptive fields are localized. Thus, nearby cells see overlapping but not identical sets of inputs. Consider two extreme cases. First, when each input cell is connected to a single output cell, receptive fields are completely localized. In the limit of low noise, the output layer replicates the activity of the input layer, so all information is preserved. However, when noise is significant, some information is lost by this identity mapping, and alternative connectivity schemes may yield greater information about the inputs. Second, when there is global connectivity, so that all input cells are connected to all output cells, receptive fields are completely delocalized. Under a correlation-based rule, each output cell learns the same receptive field. Then, in the low-noise limit, most information is being thrown

[4]For simplicity, in this discussion we will ignore noise. Depending on the signal-to-noise ratio, one will wish to strike a particular balance between variety (carrying more independent components of the input ensemble) and redundancy (e.g., see [2, 30]). However, except in the extreme case of high noise, where complete redundancy is called for, multiple components always will be needed to maximize the information, given multiple output cells.

away — only one dimension of the input pattern is being distinguished. However, suppose that this dimension is the most informative dimension about the input ensemble. Then, in the high-noise limit, this redundant representation of the most information-rich dimension will maximize the information carried about the input ensemble.

Thus, given a correlation-based learning rule, a completely localized representation can maximize information in the low-noise limit, while a completely delocalized representation can maximize information in the high-noise limit. Intermediate levels of localization should be appropriate for intermediate signal-to-noise ratios (this has recently been demonstrated quantitatively [21]). It seems likely that biology, rather than designing an anti-Hebbian learning rule, has used its own correlation-based rules and has made use of its natural tendency to form partially localized receptive fields in order to ensure efficiency of representation.

2.4.2 Self-Organizing Maps and Associative Memories

The above ideas about efficiency consider only the summed information in the responses of the postsynaptic cells, without regard for location or connectivity. Alternative ideas about the computational significance of correlation-based rules focus on the spatial arrangement of postsynaptic response features and the connectivity between the postsynaptic cells.

One such set of ideas stem from the study of the self-organizing feature map (SOFM) of Kohonen [24, 48] and of related *dimension-reducing mappings* [8]. As was previously described, the SOFM corresponds to a Hebbian rule with a nonlinear lateral intracortical interaction, such that each input pattern leads to a single, localized cluster of cortical activity. The SOFM and related algorithms lead to a mapping that matches the topology and geometry of the output space to that of the input space, despite a possible dimensional and/or shape mismatch between the two [8, 24, 48]. That is, nearby points in the output space correspond via the mapping to nearby points in the input space, and input patterns that occur more often develop a larger representation than those that occur less often.

A number of possible functions have been assigned to such mappings. One is the minimization of wiring length, assuming that cortical points representing "nearby" input patterns need to be connected to one another [8]. Another is to represent the input data in a compressed form while minimizing reconstruction error [33, 48]. A specific form of the latter idea is as follows. Suppose that there is noise in the output layer that is distance-dependent, so that the probability of a response being centered at a given output point falls off with its distance from the point that is "correct" for that input. Suppose also that there is a metric on the input space, and that the error in mistaking one input pattern for another is assigned as the distance between the two patterns. Then the SOFM can be interpreted, approximately, as achieving the input–output mapping that minimizes the average error in reconstructing the input pattern from the output responses [33].

The major problem in applying these ideas to biology is the difficulty in assigning biological meaning to the topology and geometry of the non-retinotopic dimensions of the input space. Given an ensemble of visual input patterns on the retina, for example, how large is the corresponding ocular dominance or orientation dimension relative to the retinotopic dimensions? Without a clear prescription for answering this question, it is difficult to make biological predictions from these ideas. Nonetheless, the computational functions of self-organizing maps, their close connection to correlation-based models, and their ability to replicate many features of cortical maps are intriguing.

Another well-known set of ideas concerns the role of correlation-based rules in establishing an associative memory. Suppose one wishes to learn a set of N input–output pairs, $(\mathbf{u}^a, \mathbf{v}^a)$, where $\mathbf{u}^a$ and $\mathbf{v}^a$ are the ath input and output vectors, respectively. Let $\mathbf{v}^a = \mathbf{M}\mathbf{u}^a$ for some synaptic matrix $\mathbf{M}$. If the input patterns are orthonormal, $\mathbf{u}^a \cdot \mathbf{u}^b = \delta_{ab}$, then the input–output association is achieved by setting $\mathbf{M} = \sum_a \mathbf{v}^a(\mathbf{u}^a)^{\mathrm{T}}$ (e.g., [24]). This relation will be learned by a Hebbian rule, $(d/dt)M_{ij} = -M_{ij}/N + v_i u_j$, provided there is a "teacher" to clamp the output to $\mathbf{v}^a$ whenever $\mathbf{u}^a$ is presented. A fully connected network with activity states $\mathbf{v}$ similarly will develop the activity states, or "memories," $\mathbf{v}^a$, as stable attracting states if the connection matrix between the cells is determined by the Hebbian prescription $\mathbf{M} = \sum_a \mathbf{v}^a(\mathbf{v}^a)^{\mathrm{T}}$ (e.g., [18, 19]). Again, to learn a specific set of memories, a "teacher" is required to clamp the network into the appropriate activity states during learning. Given simple nonlinearities in neuronal activation, the stored memories need not be orthogonal to one another, provided the memories are randomly chosen (uncorrelated) and their number is sufficiently small relative to the number of cells (e.g., [17]). It is of biological interest to explore how associative properties can develop through correlation-based rules in the absence of a teacher as well as in the presence of correlated input patterns (for which, see [17]).

2.5 Open Questions

This brief review can only point to a small sample of the rich literature on this topic. Among the many open questions in the field are: How can biologically interpretable models replicate the details of cortical maps? Might orientation selectivity arise from early oriented wave patterns of retinal activity [38, 64] or other mechanisms, rather than through ON/OFF competition? Might the initial development of orientation selectivity occur through the patterning of intracortical connections, rather than through the patterning of LGN connections to the cortex?[5] How might intracortical plasticity affect receptive field and map development [53]? How might input correlations affect column size [12]? How will development be altered by the incorporation of more realistic cortical connectivity, and more realistic, nonlinear learning rules? For example, might input correlations help determine the self-organization of plastic intracortical connections or the size of nonlinearly determined cortical activity clusters, each of which in turn would shape the pattern of input synapses including column size? How can we characterize the computational function of the correlation-based rules used biologically? These and other questions are likely to be answered in the coming years.

Acknowledgments. K.D. Miller is supported by grants from the National Eye Institute, the Searle Scholars' Program, and the Lucille P. Markey Charitable Trust. The author thanks Ed Erwin, Sergei Rebrik, and Todd Troyer for helpful comments on the manuscript.

References

[1] K. Albus, W. Wolf (1984) Early post-natal development of neuronal function in the kitten's visual cortex: A laminar analysis. *J. Physiol.* **348**:153–185

[2] J.J. Atick (1992) Could information theory provide an ecological theory of sensory processing? In: *Princeton Lectures on Biophysics*, W. Bialek (Ed.) (World Scientific, Singapore), pp. 223–289

[3] H.B. Barlow (1989) Unsupervised learning. *Neural Comp.* **1**:295–311

[4] B.O. Braastad, P. Heggelund (1985) Development of spatial receptive-field organization and orientation selectivity in kitten striate cortex. *J. Neurophysiol.* **53**:1158–1178

[5] E.M. Callaway, L.C. Katz (1990) Emergence and refinement of clustered horizontal connections in cat striate cortex. *J. Neurosci.* **10**:1134–1153

[6] B. Chapman, M.P. Stryker (1993) Development of orientation selectivity in ferret visual cortex and effects of deprivation. *J. Neurosci.* **13**:5251–5262

[7] M. Constantine-Paton, H.T. Cline, E. Debski (1990) Patterned activity, synaptic convergence and the NMDA receptor in developing visual pathways. *Ann. Rev. Neurosci.* **13**:129–154

---

[5]See [41] for arguments that the early oriented waves of retinal activity are too large to drive the development of simple cells, i.e., their wavelength is much wider than the set of LGN inputs to a single simple cell; but see [58] for an argument that the waves nonetheless might drive the development of orientation selectivity by determining the patterning of intracortical connections rather than of connections from the LGN to the cortex. The patterning of horizontal connections may take place slightly later than the development of orientation selectivity [1, 5], but both occur sufficiently early that their order remains unclear.

[8] R. Durbin, G. Mitchison (1990) A dimension reduction framework for understanding cortical maps. *Nature* **343**:644–647

[9] E. Erwin, K. Obermayer, K. Schulten (1995) Models of orientation and ocular dominance columns in the visual cortex: A critical comparison. *Neural Comp.* **7**:425–468

[10] P. Foldiak (1989) Adaptive network for optimal linear feature extraction. In: *Proceedings, IEEE/INNS International Joint Conference on Neural Networks*, Vol. 1 (IEEE Press, New York), pp. 401–405

[11] Y. Fregnac, M. Imbert (1984) Development of neuronal selectivity in the primary visual cortex of the cat. *Physiol. Rev.* **64**:325–434

[12] G.J. Goodhill (1993) Topography and ocular dominance: A model exploring positive correlations. *Biol. Cybern.* **69**:109–118

[13] G.J. Goodhill, D.J. Willshaw (1990) Application of the elastic net algorithm to the formation of ocular dominance stripes. *Network* **1**:41–59

[14] C.S. Goodman, C.J. Shatz (1993) Developmental mechanisms that generate precise patterns of neuronal connectivity. *Cell* **72**(Suppl):77–98

[15] R.W. Guillery (1972) Binocular competition in the control of geniculate cell growth. *J. Comp. Neurol.* **144**:117–130

[16] R.W. Guillery, D.J. Stelzner (1970) The differential effects of unilateral lid closure upon the monocular and binocular segments of the dorsal lateral geniculate nucleus in the cat. *J. Comp. Neurol.* **139**:413–422

[17] J.A. Hertz, A.S. Krogh, R.G. Palmer (1991) *Introduction to the Theory of Neural Computation* (Addison-Wesley, Reading, MA)

[18] J.J. Hopfield (1982) Neural networks and physical systems with emergent collective computational abilities. *Proc. Natl. Acad. Sci. USA* **79**

[19] J.J. Hopfield (1984) Neurons with graded responses have collective computational properties like those of two-state neurons. *Proc. Natl. Acad. Sci. USA* **81**

[20] D.H. Hubel, T.N. Wiesel (1963) Receptive fields of cells in striate cortex of very young, visually inexperienced kittens. *J. Neurophysiol.* **26**:994–1002

[21] M. Idiart, B. Berk, L.F. Abbott (1995) Reduced representation by neural networks with restricted receptive fields. *Neural Comp.* **7**:507–517

[22] N. Intrator (1992) Feature extraction using an unsupervised neural network. *Neural Computation* **4**:98–107

[23] N. Intrator, L.N. Cooper (1992) Objective function formulation of the BCM theory of visual cortical plasticity: Statistical connections, stability conditions. *Neural Networks* **5**:3–17

[24] T. Kohonen (1989) *Self-Organization and Associative Memory*, 3rd ed. (Springer-Verlag, Berlin)

[25] S. LeVay, T.N. Wiesel, D.H. Hubel (1980) The development of ocular dominance columns in normal and visually deprived monkeys. *J. Comp. Neurol.* **191**:1–51

[26] Z. Li, J.J. Atick (1994) Efficient stereo coding in the multiscale representation. *Network* **5**:157–174

[27] R. Linsker (1986) From basic network principles to neural architecture: Emergence of spatial-opponent cells. *Proc. Natl. Acad. Sci. USA* **83**:7508–7512

[28] R. Linsker (1986) From basic network principles to neural architecture: Emergence of orientation-selective cells. *Proc. Natl. Acad. Sci. USA* **83**:8390–8394

[29] R. Linsker (1986) From basic network principles to neural architecture: Emergence of orientation columns. *Proc. Natl. Acad. Sci. USA* **83**:8779–8783

[30] R. Linsker (1988) Self-organization in a perceptual network. *Computer* **21**:105–117

[31] R. Linsker (1992) Local synaptic learning rules suffice to maximize mutual information in a linear network. *Neural Comput.* **4**:691–702

[32] S. Löwel, W. Singer (1993) Strabismus changes the spacing of ocular dominance columns in the visual cortex of cats. *Soc. Neuro. Abs.* **19**:867

[33] S. Luttrell (1994) A Bayesian analysis of self-organizing maps. *Neural Comp.* **6**:767–794

[34] D.J.C. MacKay, K.D. Miller (1990) Analysis of Linsker's applications of Hebbian rules to linear networks. *Network* **1**:257–298

[35] D.J.C. MacKay, K.D. Miller (1990) Analysis of Linsker's simulations of Hebbian rules. *Neural Comput.* **2**:173–187

[36] L. Maffei, L. Galli-Resta (1990) Correlation in the discharges of neighboring rat retinal ganglion cells during prenatal life. *Proc. Nat. Acad. Sci. USA* **87**:2861–2864

[37] D.N. Mastronarde (1989) Correlated firing of retinal ganglion cells. *Trends Neurosci.* **12**:75–80

[38] M. Meister, R.O.L. Wong, D.A. Baylor, C.J. Shatz (1991) Synchronous bursts of action-potentials in ganglion cells of the developing mammalian retina. *Science* **252**:939–943

[39] K.D. Miller (1990) Correlation-based models of neural development. In: *Neuroscience and Connectionist Theory*, M.A. Gluck, D.E. Rumelhart, (Eds.) (Lawrence Erlbaum, Hillsdale, NJ), pp. 267–353

[40] K.D. Miller (1990) Derivation of linear Hebbian equations from a nonlinear Hebbian model of synaptic plasticity. *Neural Comput.* **2**:321–333

[41] K.D. Miller (1994) A model for the development of simple cell receptive fields and the ordered arrangement of orientation columns through activity-dependent competition between ON- and OFF-center inputs. *J. Neurosci.* **14**:409–441

[42] K.D. Miller, J.B. Keller, M.P. Stryker (1989) Ocular dominance column development: Analysis and simulation. *Science* **245**:605–615

[43] K.D. Miller, D.J.C. MacKay (1994) The role of constraints in Hebbian learning. *Neural Comput.* **6**:100–126

[44] K.D. Miller, M.P. Stryker (1990) The development of ocular dominance columns: Mechanisms and models. In: *Connectionist Modeling and Brain Function: The Developing Interface*, S.J. Hanson, C.R. Olson (Eds.) (MIT Press/Bradford, Cambridge, MA), pp. 255–350

[45] M. Miyashita, S. Tanaka (1992) A mathematical model for the self-organization of orientation columns in visual cortex. *NeuroReport* **3**:69–72

[46] K. Obermayer, G.G. Blasdel, K. Schulten (1992) A statistical mechanical analysis of self-organization and pattern formation during the development of visual maps. *Phys. Rev. A* **45**:7568–7589

[47] E. Oja (1982) A simplified neuron model as a principal component analyzer. *J. Math. Biol.* **15**:267–273

[48] H. Ritter, T. Martinetz, K. Schulten (1992) *Neural Computation and Self-Organizing Maps: An Introduction* (Addison-Wesley, Reading, MA)

[49] J. Rubner, K. Schulten (1990) Development of feature detectors by self-organization. *Biol. Cybern.* **62**:193–199

[50] T.D. Sanger (1989) An optimality principle for unsupervised learning. In: *Advances in Neural Information Processing Systems*, Vol. 1, D. Touretzky (Ed.) (Morgan Kaufmann, San Mateo, CA), pp. 11–19

[51] C.J. Shatz (1992) The developing brain. *Scientific Am.* **267**:60–67

[52] C.J. Shatz, M.P. Stryker (1978) Ocular dominance in layer IV of the cat's visual cortex and the effects of monocular deprivation. *J. Physiol.* **281**:267–283

[53] J. Sirosh, R. Mikkulainen (1995) A unified neural network model for the self-organization of topographic receptive fields and lateral interactions. *Neural Comput.* (to appear)

[54] M.P. Stryker, S.L. Strickland (1984) Physiological segregation of ocular dominance columns depends on the pattern of afferent electrical activity. *Inv. Opthal. Supp.* **25**:278

[55] N.V. Swindale (1992) A model for the coordinated development of columnar systems in primate striate cortex. *Biol. Cyb.* **66**:217–230

[56] S. Tanaka (1991) Theory of ocular dominance column formation: Mathematical basis and computer simulation. *Biol. Cybern.* **64**:263–272

[57] C. von der Malsburg (1973) Self-organization of orientation selective cells in the striate cortex. *Kybernetik* **14**:85–100

[58] C. von der Malsburg (1993) Network self-organization in the ontogenesis of the mammalian visual system. Internal Report 93-06, Ruhr-Universität Bochum, Institut für Neuroinformatik, 44780 Bochum, Germany

[59] C. von der Malsburg, D.J. Willshaw (1976) A mechanism for producing continuous neural mappings: ocularity dominance stripes and ordered retino-tectal projections. *Exp. Brain Res.* (Supp.) **1**:463–469

[60] T.N. Wiesel, D.H. Hubel (1965) Comparison of the effects of unilateral and bilateral eye closure on cortical unit responses in kittens. *J. Neurophysiol.* **28**:1029–1040

[61] T.N. Wiesel, D.H. Hubel (1974) Ordered arrangement of orientation columns in monkeys lacking visual experience. *J. Comp. Neurol.* **158**:307–318

[62] F. Wolf, H-U. Bauer, T. Geisel (1994) Formation of field discontinuities and islands in visual cortical maps. *Biol. Cyb.* **70**:525–531

[63] F. Wolf, K. Pawelzik, T. Geisel, D.S. Kim, T. Bonhoeffer (1994) Optimal smoothness of orientation preference maps. In: *Computation in Neurons and Neural Systems* (Kluwer, Boston), pp. 97–102

[64] R.O. Wong, M. Meister, C.J. Shatz (1993) Transient period of correlated bursting activity during development of the mammalian retina. *Neuron* **11**:923–938

3

Associative Data Storage and Retrieval in Neural Networks

Günther Palm[1] and Friedrich T. Sommer[2]

with 9 figures

Synopsis. Associative storage and retrieval of binary random patterns in various neural net models with one-step threshold-detection retrieval and local learning rules are the subject of this chapter. For different heteroassociation and autoassociation memory tasks specified by the properties of the pattern sets to be stored and upper bounds on the retrieval errors, we compare the performance of various models of finite as well as asymptotically infinite sizes. In infinite models, we consider the case of asymptotically sparse patterns, where the mean activity in a pattern vanishes, and study two asymptotic fidelity requirements: constant error probabilities and vanishing error probabilities. A signal-to-noise ratio analysis is carried out for one retrieval step where the calculations are comparatively straightforward and easy. As performance measures we propose and evaluate information capacities in bits/synapse which also take into account the important property of fault tolerance. For autoassociation we compare one-step and fixed-point retrieval that is analyzed in the literature by methods of statistical mechanics.

3.1 Introduction and Overview

With growing experimental insight into the anatomy of the nervous system as well as the first electrophysiological recordings of nerve cells in the first half of this century, a new theoretical field was opened, namely, the modeling of the experimental findings at one or a few nerve cells, leading to very detailed models of biological neurons [1]. But, different from most biological phenomena, where the macroscopic function can be understood by revealing the cellular mechanism, the function of the nervous system as

---

[1]Abteilung Neuroinformatik, Fakultät für Informatik, Universität Ulm, Oberer Eselsberg, D-89081 Ulm, Germany.

[2]Institut für Medizinische Psychologic und Verhaltensneurobiologic der Universität Tübingen, Gartenstr. 29, D-72074 Tübingen, Germany.

a whole turned out to be constituted by the *collective* behavior of a very large number of nerve cells, and the activity of a large fraction of cells, a whole activity pattern, had to be considered instead.

The modeling had to drop the biological faithfulness at two points: on the cellular level, the models had to be simplified such that a large number of nerve cells could be described; and on the macroscopic level, the function had to be reduced to simple activity pattern processing like pattern completion, pattern recognition, or pattern classification, allowing a theoretical description and quantification.

McCulloch and Pitts [2] argued that, due to the *all-or-none* character of nervous activity, the neurophysiological findings can be reproduced in models with simple two-state neurons, in particular, in *associative memory models* which exhibit binary activity patterns.

In the 1950s and 1960s small feedforward neural nets were suggested for simple control tasks, among them the associative memory [3], [4] and the simple perceptron [5]. All of these models employ *one-step retrieval*, which means that in one parallel update step the initial or input pattern is transformed to the output pattern. Such models which contain no feedback loops will be the main subject of this chapter.

Little, who introduced the Ising-spin analogy of the neural states[3] [6], opened the door to analyzing the feedback retrieval process in neural nets with methods of statistical mechanics. The analysis that was developed during the 1970s [7] for lattices of coupled spins with randomly distributed interactions to describe spin glasses could be applied successfully to *fixed-point retrieval* in an associative memory [8].[4] In fixed-point retrieval, the retrieval process is iterated until a stable state is reached. This method has been described in several recent books, e.g., van Hemmen and Kühn [9], Amit [10], and Hertz, Krogh, and Palmer [11].

This chapter takes as its starting point a larger class of simple processing tasks: the association between members of binary pattern sets. Depending on the properties of the randomly generated pattern sets, we will characterize different memory tasks (Sec. 3.1) and concentrate on the question of how a neural model has to be designed to yield optimal performance.

We consider feedforward neural associative memory models with one-step retrieval (Sec. 3.2). To keep our model as variable as possible, Ising-spin symmetry of the neural states is not assumed, and arbitrary local learning

[3]The two states of a binary neuron are identified with up and down states of a spin particle in the Ising model; the synaptic couplings correspond to the spin–spin interactions.

[4]Pattern completion with fixed-point retrieval in a neural net can be treated like relaxation in a solid, once the storage process has determined the dynamics. The macroscopic observables of the system (corresponding to specific heat, conductivity, or magnetization in solids) are then the overlaps to stored patterns or, equivalently, the recall errors.

rules are admitted to form the synaptic connections. One-step retrieval can be analyzed by elementary probability theory, and it is compatible with a larger class of memory tasks, not only pattern completion. On the other hand, as we will discuss, in cases of pattern completion, a feedback retrieval model is preferable. Section 3.3 contains the detailed signal-to-noise ratio analysis, where we have included most of the calculations because the intention of this work is to provide not only results, but also the methods.

Another important question concerns the judgement of the performance of different memory models. Unfortunately, in the literature, many different measures are used. Instead of staying with the mean retrieval errors obtained from the analysis, we apply elementary information theory to the memory process, leading us to the definition of information capacities, which allow us to compare models with different memory tasks (Sec. 3.4).

In Sec. 3.5 we evaluate these performance measures for the various models. The last section resumes the previous sections and points out the relations to the literature. It compares one-step and fixed-point retrieval, taking advantage of the works based on methods of statistical mechanics. The results of the different approaches, which seem to be quite incoherent at first sight, turn out to be not only comparable but also consistent.

3.1.1 Memory and Representation

A memory process often can be considered as a mapping from one set of events into another set of events; as a trivial example, one may think of the problem as how to establish a phone line to a friend. To solve the problem, one has to map the friend's name to his phone number. For the construction of a memory device like a phonebook, which helps you with this problem, one first has to map or to code the events "the friend's name" and "his phone number" into symbols, in this case strings of letters and numbers, which can be written and read by a user. This mapping will be called the *representation* of the events. The memory device has to store these pairs of strings in some way. It can solve the problem if the representation maps the events into unique data strings. Thus, a given set of patterns specifies the memory task that a memory device has to solve.

Without loss of generality, we focus on binary patterns as data strings. A binary pattern is a string containing only two types of elements, for instance, "B" and "W" (for black and white pixels). We restrict ourselves to such pattern sets where every member has approximately the same *ratio p between the number of "B" and "W" digits.* We call a pattern *distributed* if both fractions of pixels have more than one member. Throughout this chapter we distinguish between three different *patterns types*:

1. A *singular pattern* with m digits has only a single "B" digit and $m-1$ "W" digits. A singular pattern is, by definition, not distributed.

2. A *sparse pattern* is distributed, but the ratio p between the number of "B" and "W" digits satisfies $p \ll 0.5$. In the infinite model $m \to \infty$, we will consider the *sparse limit* $p \to 0$ with $mp \to \infty$, which leads to nontrivial distributed patterns.

3. In a *nonsparse pattern*, the fraction p between the number of "B" and "W" digits has to be away from 0. In the infinite model, $p = \text{const}$ as $m \to \infty$.

3.1.2 Retrieval from the Memory

The memory device has to store a set of patterns in such a way that a desired pattern can be selectively recalled at the output port. In *memory retrieval* a desired output pattern is selected by applying a pattern at the input port of the device. We denote the set of output patterns as the *content patterns* $\mathcal{S}^C$. An input pattern that selects a content pattern is called its *address pattern*, or simply its *address*. The set of address patterns is denoted by $\mathcal{S}^A$. Thus, in the retrieval, the memory device has to map from an address pattern to its corresponding content pattern. This map is defined by the set of pairs consisting of address and content patterns:

$$\left\{(x^1, y^1), \ldots, (x^M, y^M) : x^k \in \mathcal{S}^A, y^k \in \mathcal{S}^C\right\}.$$

3.1.3 Fault Tolerance in Addressing

Between two patterns x and $\hat{x}$, the number of different bits $h(x, \hat{x})$ defines a natural distance relation called the *Hamming distance*. Via this distance a whole set of input patterns may specify one desired content pattern uniquely: all patterns $\hat{x}$ with the property $h(\hat{x}, x) < h(\hat{x}, x^k)$ for all $x^k \neq x$ and $x, x^k \in \mathcal{S}^A$. We call a memory retrieval *fault-tolerant* if it allows input noise in the sense that many input patterns which have a unique closest address are mapped on the content pattern belonging to this address.

For a set of singular address patterns, normally no $\hat{x} \notin \mathcal{S}^A$ has a unique closest address and, therefore, fault tolerant retrieval is impossible. Thus, fault-tolerant retrieval can only be expected if the address patterns are distributed.

3.1.4 Various Memory Tasks

We call *heteroassociation* the general memory task where the set of address patterns $\mathcal{S}^A$ and the set of content patterns $\mathcal{S}^C$ can be chosen arbitrarily.

The following special cases of heteroassociation will be considered:

- If the address patterns are singular patterns, the memory task is called the *look-up–table task*. Then the singular pixel of an address pattern points into a table of content patterns like the usual access in a look-up table.

- For singular content patterns, we identify each bit of the content pattern with a class in the set of address patterns. This memory task can be interpreted as *pattern classification*, which separates the set of address patterns in disjunct classes. This task (with one-bit content patterns) has been executed by the classical simple perceptron models; see [5].
- *Autoassociation* is the case of heteroassociation where the address and content patterns are identical; therefore, it also may be denoted as content addressability. Only for fault-tolerant retrieval does the autoassociation task make sense; then, the memory performs *pattern completion* from a distorted version $\hat{x}^k$ as an input pattern to the error-free content pattern x^k; see also Forrest and Wallace in [9].

3.1.5 Retrieval Errors

A memory that allows errors in the addressing perhaps also will recall erroneously the wrong content pattern or put at least some errors in the output.

In the retrieval of binary patterns there may occur two types of flip errors in a digit of the output pattern $\tilde{y}^k$: a "W" of the content pattern y^k may be turned into a "B", and a "B" in the content pattern y^k may be turned into a "W". Of course, with increasing addressing noise these errors also will increase. But again via the distance relation it is possible that a memory output containing errors in some digits still will specify the event coded by the original content pattern. A given memory task together with the sets $\mathcal{S}^A$ and $\mathcal{S}^C$ will fix the maximal mean errors that can be tolerated in the retrieval. These upper bounds, which have to be satisfied by the error probabilities, will be called the *fidelity requirement.*

3.2 Neural Associative Memory Models

The typical ingredients of an artificial neural network model are a large number of similar processor units called *neurons*, which obtain signals through adjustable connections from a large number of input fibers and/or other neurons. In this model the adjustable connections, the *synapses*, connect an input port to each neuron (see Fig. 3.1).

The two different types of calculations in the model, the processing of the neural input signal in the retrieval, on the one hand, and the synaptic adjustment according to the data in the storage phase, on the other, are separated in time in this model; we distinguish the *storage process* and the *retrieval process.*

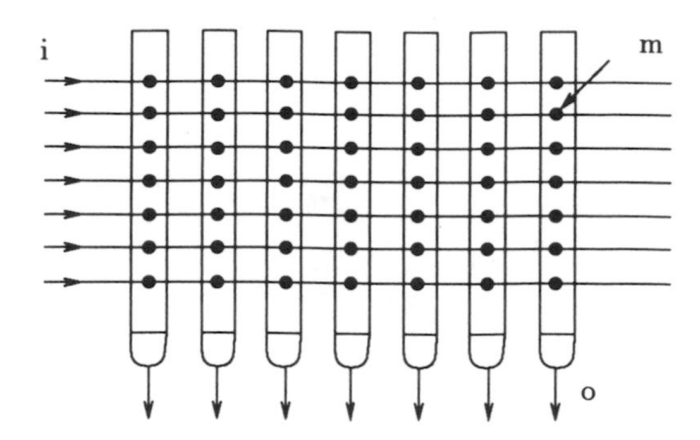

Fig. 3.1. Schematic view of a neural associative memory: i — retrieval input fibers, o — retrieval output fibers (axons), m — modifiable synaptic connection between neuron and input fiber. The horizontal lines are wires that propagate the input signals to the synapses. Each column represents one neuron. The larger upper section where the synaptic connections access corresponds to the dendritic tree, and the lower section the cell body. The arrow pointing below from the cell body corresponds to the axon.

To perform the calculations the pixel types "B" and "W" in the input patterns have to be translated into signals that can propagate through the network. Two different values, 1 and $a \in [-1, 0]$, will be assigned to the pixel types "B" and "W", respectively. Each pattern is identified with an n-vector $x \in \{a, 1\}^n$, and we will use synonymously the expressions pattern and $\{a, 1\}$-vector. Of course, we are free to exchange "W" and "B" in the assignment; the flip transformation $\mathcal{F}$ applied to all components in the data will not change the memory problem. Here, $\mathcal{F}(x_i = W) := B$ and $\mathcal{F}(x_i = B) := W$. Therefore, we can always assign the value 1 to the smaller pixel fraction so that

$$p = \#\{i : x_i = 1\}/(n - \#\{i : x_i = 1\}) \leq 0.5.$$

Such models already have been proposed and analyzed many years ago, e.g., Uttley [12], Steinbuch [3], Rosenblatt [5], Longuett-Higgins et al. [13], Amari [14], Gardner-Medwin [15], and Kohonen [16].

3.2.1 Retrieval Process

In the retrieval phase an address pattern is applied to the input port of the memory. The input signals are propagated via a synaptic connection strength matrix $\mathcal{M}_{ij}$ to all neurons. In *one-step retrieval* every neuron j actualizes its state, the *axonal activity* $\tilde{y}_j$, according to this input, and the vector $\tilde{y}$ is the retrieval output pattern.

Each neuron has to form the *dendritic potential* d_j, the sum over all its incoming activities,

$$d_j := \sum_i \mathcal{M}_{ij} x_i, \tag{3.1}$$

and then to determine the new activity value in the neural update equation

$$\tilde{y}_j = f(d_j - \Theta). \tag{3.2}$$

The output signal of a biological neuron is a train of short electric pulses, the neural spikes. It is the *spike rate* and not the amplitude or the duration of a spike that grows with increasing dendritic potential. These properties have been modeled in the so-called *spike coding models*; cf [17, 18, 19, 20]. Here we focus on *rate coding models*, where the *neural transferfunction* $f(x)$ describes only the spike rate. In almost all of these models, $f(x)$ is a monotonously increasing function and Θ is the *threshold value*, which can be adjusted globally for all neurons in each retrieval step.

Models with linear transfer functions, as, for instance, those proposed in Kohonen [16] or Anderson [21, 22], lead for large networks to quasi-continuous–valued output patterns.

Binary output patterns are obtained if the neural transfer function is a two-valued stepfunction: $f(x) = 1$ for $x \geq 0, f(x) = a$ otherwise. The neural state $\tilde{y}_j = 1$ is called *firing* or *active*, $\tilde{y}_j = 0$ *silent* or *passive*. The *retrieval error probabilities* for *on errors* and *off errors*, respectively, are expressed by the conditioned probabilities

$$e_1 := \text{Prob}\left[\tilde{y}_j^k = a | y_j^k = 1\right], \qquad e_a := \text{Prob}\left[\tilde{y}_j^k = 1 | y_j^k = a\right]. \tag{3.3}$$

Such models have been treated in Willshaw et al. [4], Palm [23], and Nadal and Toulouse [24]. In one-step retrieval the output pattern is evaluated from the input pattern after one synchronious parallel calculation of all neurons.

Step-shaped neural transfer functions also have been used in the spin-glass literature on autoassociation, e.g., in [25, 8, 26, 27]. These works consider an *iterative retrieval procedure* where, via a feedback loop, the signal flow through the system is iterated until a stationary state, a fixed point, is reached. Such *fixed-point retrieval* has been considered for two different ways of performing the iteration. In models with parallel update, the complete one-step retrieval process is iterated in the manner that the output is fed back as new input; see, for instance, [6, 15, 28, 29, 30, 31]. In models with sequential random update, only one neuron, randomly selected, is updated [Eq. (3.2)] in one iteration step, leading to the new input, which only deviates in one component from the preceding one; see again [25, 8, 26, 27]. The improvement due to iterated retrieval for the pattern completion task obtained in simulations can be observed in Fig. 3.9.

3.2.2 Storage Process

In this process, which is also called the *learning process*, the synaptic matrix, or the storage medium, is formed from the set of patterns to be stored.

During the storage process, each pair (x^l, y^l) of patterns to be learned is applied at the in- and output ports of the memory. This provides pre- and postsynaptic values for every synapse $\mathcal{M}_{ij}$.

Learning Rules

For a given pair (x, y) of pre- and postsynaptic activity values, the *local synaptic rule* $R(x, y)$ determines explicitly the amount of synaptic connectivity change. For binary patterns, there are only four different constellations possible for pre- and postsynaptic activities, viz., (a, a), $(1; a)$, $(a, 1)$, and $(1, 1)$. Thus, a synaptic rule is determined by four numbers:

$$R = (r_1, r_2, r_3, r_4). \tag{3.4}$$

The following two famous local learning rules will be focused on in the subsequent analysis:

- The *Hebb rule*, or asymmetrical coincidence rule, $H := (0, 0, 0, 1)$ increases the synaptic matrix element for coinciding pre- and postsynaptic firings only. In his *neurophysiological postulate* Hebb [32] proposed this type of synaptic modification between pairs of firing nervous cells.

- The *agreement rule*, or Hopfield rule or symmetrical coincidence rule, $A := (1, -1, -1, 1)$ increases the synaptic matrix element for agreeing pre- and postsynaptic states and decreases the synaptic weight for disagreeing states. This rule was used in the original Hopfield model [25].

The above rules are both product rules: $R(x, y) = xy$. For $a = 0$ we obtain the Hebb rule, and for $a = -1$ the agreement rule, and, sometimes, for instance in [33], both are considered as Hebbian learning. We retained the distinction because in the original formulation of his postulate Hebb clearly talks of the influence of synchronously firing neurons on their interconnecting synapses. The psychologist Hebb claimed this postulate to be inspired by physiological and psychological findings, while the symmetry between firing and silence in the agreement rule is biologically very implausible.

Storage Procedures

We consider *one-step learning*, which means that, after one single presentation of every pair, the formation of the *synaptic matrix* is finished. Two different types of storage procedures will be examined:

- The *incremental storing procedure*, where the synaptic matrix is given by

$$\mathcal{M} = (\mathcal{M}_{ij}) := \sum_{k=1}^{M} R(x_i^k, y_j^k). \tag{3.5}$$

- The *binary storage procedure*, where the synaptic matrix $\bar{\mathcal{M}}$ is obtained from $\mathcal{M}$ by another highly nonlinear operation:

$$\bar{\mathcal{M}}_{ij} := \text{sgn}(\mathcal{M}_{ij}) \tag{3.6}$$

  with $\text{sgn}(0) := 0$.

Storage procedures can be strictly local (as in most of the papers cited here) or nonlocal (as, for example, in Personnaz et al. [34, 35]). Depending on the sign of the average connectivity change, they can be productive, destructive, or balancing for the total network connectivity (cf. [36, 37]). Local storage procedures can make use of two (probably the majority), three (supervised learning with additional teacher signal, e.g., Barto et al. [38]), or more terms to compute a synaptic change (compare Palm [36] again). In this chapter we concentrate on storage procedures employing strictly local two-term learning rules.

The most common synaptic arrangement in biological neural nets as in the cerebral cortex (and the hippocampus) is the simple dyadic synapse. It connects just two neurons: the presynaptic and the postsynaptic; therefore, there are just two natural, locally available activity signals: the presynaptic and the postsynaptic.

3.2.3 Distributed Storage

One reason for the big comeback of systems with neural architecture in the last decade is the fact that, in computer science, distributed processing turned out more and more to be an indispensable goal. How do the simple memory models introduced in this section display the properties of distributed storage?

For heteroassociation, local rules store second-order correlations between address and content pattern activities; for instance, with the Hebb rule, each pair of active neurons (x_i^k, y_j^k) affects one synapse $\mathcal{M}_{ij}$.

The storage is called *distributed* if the storage of one single pattern pair causes nonlocal changes in the storage medium. More than one element of the synaptic matrix is affected if at least one pattern in the pair is nonsingular, that is, if either set of address or content patterns contains nonsingular patterns.

Here we define distributed storage in a stricter sense: we require that many matrix elements carry information about more than one pattern pair.

In this sense distributed information storage for arbitrary local rules is provided only if both pattern sets, address and content patterns, contain nonsingular and overlapping patterns. Then, the storage of several pattern pairs will affect the same synapses, so that each entry in the synaptic connectivity matrix $\mathcal{M}$ may contain the superposition of several memory traces; i.e., for most index pairs (i, j) the sum $\sum_k R(x_i^k, y_j^k)$ should have more than one nonzero contribution. Like in holography, an accessible content segment (a pattern pair) is written widely spread in the storage medium and different content segments will overlap.

In the case of autoassociation, local rules store the second-order autocorrelation of the pattern activity; with the Hebb rule, each pair of active neurons in a learning pattern causes a change in one synapse. Distributed storage requires the patterns to be nonsingular and overlapping.

3.3 Analysis of the Retrieval Process

The aim of the present section is the analysis of one-step retrieval in the associative memory after learning, i.e., after the storage process has formed the memory matrix for a given memory task $(\mathcal{S}^A, \mathcal{S}^C)$. In Sec. 3.1.5 and by Eq. (3.3) we have introduced the quantities of interest in the analysis of this feedforward system, viz., the mean retrieval error probabilities in an output pattern for a given input pattern.

We already mentioned in the introduction that different spatial scales can be distinguished in the treatment of neural nets, the microscopic scale of synapses and model neurons, and the macroscopic scale of the collective behavior of all neurons. What we presume about the model is on the microscopic scale (neuron model, learning rules, etc.); what we would like to know from a theory is on the macroscopic scale, the collective behavior of the *whole* set of neurons (retrieval errors). In physics it is quite usual to deal with separable scales, for instance, in thermodynamics the molecular versus the macroscopic scale. Physical mean-field theories that originally have been developed for spin glasses[5] yield asymptotic results for the retrieval errors[6] in the limit of infinite system size: $m, n \to \infty$, which often

---

[5]Spin glasses are magnetic solids with two different competing fractions of spin couplings. One fraction favors parallel and the other antiparallel spin alignment, which causes irregular (glasslike) stable spin configurations. The mean-field theory provides values for the mean magnetization as macroscopic order parameters.

[6]The order parameters of a mean-field theory treating neural networks are the M *overlaps* $\{m_l, l = 1, \dots, M\}$, where each overlap m_l counts the number of common pixels between the retrieval output and the content pattern y^l. If we apply a (distorted) address pattern $\tilde{x}^k$ as an input pattern, particularly, one overlap is important for the retrieval quality, namely, the overlap m_k corresponding to the input pattern. The theory provides a mean value $< m_k >$, averaged over a large number of retrieval events, which is equivalent to the retrieval error probabilities in Sec. 3.5.

is called the *thermodynamic limit* of fixed-point retrieval in the associative memory after learning.

We will consider memory tasks with different mean ratios p between the elements 1 and a in the pattern sets in the finite model and in the thermodynamic limit, i.e., $m \to \infty$. Curiously, memory tasks with sparse patterns, as defined in Sec. 3.1.1., will turn out to yield optimal asymptotic performance.

3.3.1 Random Pattern Generation

To apply probability theory for the estimation of mean retrieval error probabilities, we have to assume the following properties of the memory data and of the distortion of the input patterns.

Content and Address Patterns

In the memory tasks we assume the simplest model of the data to be stored, namely, sets of randomly generated patterns. The value of each of the n digits in a pattern $x^k \in \mathcal{S}$ is chosen independently with the probability $p := \text{Prob}[x_i^k = 1]$. A set of randomly generated patterns is fixed by three parameters: the probability p, the dimensionality of a pattern n, and the number of patterns M. We will use the following notation for address and content patterns: $\mathcal{S}^A := \mathcal{S}(p, m, M), \mathcal{S}^C := \mathcal{S}(q, n, M)$. For heteroassociation, the sets $\mathcal{S}^A$ and $\mathcal{S}^C$ will be generated mutually independently.

Input Patterns

The signal detection problem will be treated in three different cases of addressing:

1. A *perfect address pattern* as an input pattern x^k, with $n_1 := \#\{i : x_i^k = 1\}$ being the number of 1 components.

2. An *ensemble of perfect input patterns*, where now the number of ones in the input pattern n_1 also becomes a random variable. It is a binomially distributed variable and, for large m, the fraction n_1/m will be close to its expectation value p because of the strong law of large numbers [39]. In the analysis, the *average input activity* μ of the ensemble will become an important quantity which, for large m, equals

$$\mu := [n_1 + (m - n_1)a]/m = p + (1 - p)a. \tag{3.7}$$

3. An *ensemble of noisy input patterns* $\hat{\mathcal{S}}^A$, which is generated by a second random generation process from the set of address patterns $\mathcal{S}^A$ used for learning. Here we concentrate on noisy input patterns, where $\hat{x}^k \in \hat{\mathcal{S}}^A$ is a "part" of an address pattern x^k in the following sense: Prob $[\hat{x}_i^k = 0 | x_i^k = 0] = 1$ and Prob $[\hat{x}_i^k = 1 | x_i^k = 1] =: p'$. As

for the faultless ensemble, we describe the input activity for large m by the average input activity of the address ensemble

$$\mu' := pp' + (1 - pp')a. \tag{3.8}$$

In the analysis that follows we will use the prime to indicate the results for the noisy input ensemble.

3.3.2 Site Averaging and Threshold Setting

Depending on its dendritic potential [Eq. (3.1)] and the threshold value Θ_j, each neuron j "decides" in the update process [Eq. (3.2)] whether it should be active or silent. This can be regarded as a signal detection problem on the random variable d_j that every neuron has to solve.

To find the probabilities for on and off errors in Eq. [3.3] we have to consider the neurons separated in two fractions: the *on-neurons*, which should be active in the original content pattern y^k, and the *off-neurons*, which should not be active. In our model, the threshold of each neuron is set to the same value depending only on the total activity of the input pattern. Therefore, it is sufficient to analyze the averaged dendritic potentials in each of the fractions. We will use the notation $d^1 =< d_j >_{j\in\{j:y_j^k=1\}}$ and $d^a =< d_j >_{j\in\{j:y_j^k=a\}}$. With the assumptions of the last subsection these averaged quantities can be treated as random variables.

Of course, the synapses — randomly generated in the storage process — are "quenched" in the retrieval so that dendritic potentials at different on-sites or off-sites will behave differently. This suggests a memory model where the threshold is adjusted separately for each neuron, which has been treated in [49] and will be discussed in Sec. 3.6.3.

3.3.3 Binary Storage Procedure

For binary storage, the dendritic potential at neuron j is $d_j = \sum_i x_i^k \bar{\mathcal{M}}_{ij}$, where the values of the binary Hebb matrix $\bar{\mathcal{M}}$ are distributed on $\{0, 1\}$. The probability that a matrix element is 0 can be easily calculated:

$$p_0 := \text{Prob}[\mathcal{M}_{ij} = 0] = (1 - pq)^M. \tag{3.9}$$

We discuss the three cases of addressing in Sec. 3.3.1 separately.

1. Given x^k as an input pattern, the expectation $E(d^1 - d^a) = n_1(1 - p_0)$ is independent of the value a but the variance $\sigma^2(d_j)$ is minimal for $a = 0$. So, optimally we choose $a = 0$. Then we obtain for the dendritic potential at an on-neuron $d^1 = n_1$. Thus we maximally can put $\Theta = n_1$ to obtain $e_1 = 0$.

The second error probability is determined from the dendritic potential at an off neuron:

$$e_a = \text{Prob}[d^a > \Theta] = \text{Prob}\left[\prod_{i \in \{i:x_i^k=1\}} \mathcal{M}_{ij} = 1 \middle| y_j^k = 0\right] \simeq (1-p_0)^{n_1}. \tag{3.10}$$

2. If we average over an ensemble of perfect patterns, where we adjust the threshold individually for each input to $\Theta = n_1$, then the threshold also becomes a random variable. Now consider the fixed threshold setting $\Theta = En_1$ for all input patterns. For this threshold choice we simply have to insert the expectation of n_1 into Eq. (3.10):

$$e_a \simeq (1 - p_0)^{mp}. \tag{3.11}$$

   This fixed threshold setting leads to $e_1(E\Theta) > 0$ because of patterns with $n_1 < En_1$ and to $e_a(E\Theta) < Ee_a(\Theta)$ because of the concavity of the function $e_a(\Theta)$. We will use Eq. (3.11) as approximation for the retrieval error e_a with the individual threshold adjustment.

3. Finally, for noisy addressing we obtain for the same fixed threshold setting $\Theta = p' E(n_1)$

$$e'_{a1} = (e_a)^{p'}. \tag{3.12}$$

Strictly speaking, the above calculation requires independence of the entries $\bar{\mathcal{M}}_{ij}$. Although this is not the case, it is shown Appendix 3.1 that at least for sparse address patterns with $m^{2/3}p \to 0$ the entries $\bar{\mathcal{M}}_{ij}$ become asymptotically independent for large m.

3.3.4 Incremental Storage Procedure

In incremental storage, the contribution of each pattern pair is simply summed up in the synaptic weights; and we can divide the dendritic potential into two parts: the signal part s, which is the partial sum coming from the storage of the pattern pair (x^k, y^k), and the noise part N, the remaining partial sum that contains no information about y_j^k. From Eqs. (3.1) and (3.5) we obtain

$$\begin{aligned} d_j &= N + s := \sum_i x_i^k \mathcal{M}_{ij} = \sum_i \sum_l x_i^k R(x_i^l, y_j^l) \\ &= \sum_i \sum_l x_i^k R(x_i^l, y_j^l) + \sum_i x_i^k R(x_i^k, y_j^k). \end{aligned}$$

The dendritic potential and its signal part have to be regarded separately at an on-neuron ($y_j^k = 1$) and at an off-neuron ($y_j^k = a$):

$$s_1 := \sum_i x_i^k R(x_i^k, 1), \qquad s_a := \sum_i x_i^k R(x_i^k, a).$$

We now assume that, for the noise parts, $E(N_1) = E(N_a)$ holds and that it is the *variance of the noise* $\sigma(N)$, which determines the mean facility to solve the neural detection problem. Inspired by engineering methods we introduce the *signal-to-noise ratio* as a threshold setting independent retrieval quality measure:

$$r := E(s_1 - s_a)/\sigma(N). \tag{3.13}$$

The motivation to do so is quite intuitive: the threshold detection problem can be solved for many neurons for the same value Θ if $E(s_1 - s_a)$ is large and $\sigma(N)$ is low.

The fidelity requirement that e_a and e_1 should be small is equivalent to the corresponding requirement that the signal-to-noise ratio r should be large. How the retrieval errors are balanced between the two possible types of retrieval errors is governed by the threshold setting. If both retrieval error probabilities have to be below 0.5, the threshold has to satisfy $Ed^a \leq \Theta \leq Ed^1$, Ed^a being the expectation of the dendritic potential at an off-site. Thus we put $\Theta = Ed^a + \vartheta\sigma(N)r = Ed^1 - (1-\vartheta)\sigma(N)r$ with $\vartheta \in [0.1]$.

For large m the noise term N can be considered as sum of a large number of independent random variables and the central limit theorem holds. Then we can estimate the error probabilities using a normal distribution and get

$$\begin{aligned} e_1 &= \text{Prob}[d^1 - \Theta < 0] \simeq G[-E(d^1 - \Theta)/\sigma(N)] = G[-(1-\vartheta)r] && (3.14)\\ e_a &= \text{Prob}[d^a - \Theta > 0] \simeq G[-\vartheta r] && (3.15)\end{aligned}$$

with the normal or Gaussian distribution $G[x] := (1/\sqrt{2\pi}) \int_{-\infty}^{x} e^{-x^2/2} dx$.

To obtain explicit values for the error probabilities we now have to analyze the signal and noise term in Eq. (3.13) for the different ensembles of input patterns and different learning rules (Sec. 3.2).

For input ensembles we are interested in the mean retrieval errors where, for every input, the threshold has been set in the optimal way according to the number of active input digits n_1. We insert the signal-to noise ratio averaged over an input ensemble into Eq. (3.14) and consider a fixed threshold setting that is equal for all input patterns. For binary storage, we take this result as an approximation for the individual threshold adjustment, which is equivalent to an exchange of the expectations of the pattern average and the input average in the calculation.

Signal-to-Noise Calculation

Again we discern the three cases of addressing described in Sec. 3.3.1.

1. For the faultless address x^k as input the signal is sharply determined as
$$s_1 - s_a = n_1(r_4 - r_3) - (m - n_1)a(r_2 - r_1).$$
The noise decouples into a sum of $(M-1)$ independent contributions corresponding to the storage of the pattern pairs (x^l, y^l) with

$l \neq k$. For every pair the input x^k generates a sum of n_1 random variables $R(x,y)$ and of $(m-n_1)$ random variables $aR(x,y)$ at a neuron j. The variable $R(x,y) = R(x_i^l, y_j^l)$ is the four-valued discrete random variable [Eq. (3.4)] with the distribution $(1-p)(1-q), p(1-q), (1-p)q, pq$.

With $E(R)$ and $\sigma^2(R)$ denoting expectation and variance of $R(x,y)$, a simple [but for $\sigma^2(N)$ tedious] calculation yields

$$E(N) = (M-1)[n_1 + (m-n_1)a]E(R) \tag{3.16}$$

$$\sigma^2(N) = (M-1)\{Q_1\sigma^2(R) + Q_2\mathrm{Cov}[R_iR_h]\}, \tag{3.17}$$

where we have used the abrevations

$$\begin{aligned} Q_1 &:= n_1 + (m-n_1)a^2 \\ Q_2 &:= n_1(n_1-1) + 2an_1(m-n_1) + a^2(m-n_1)(m-n_1-1) \\ \mathrm{Cov}[R_iR_h] &= q(1-q)[p(r_4-r_3) + (1-p)(r_2-r_1)]^2. \end{aligned}$$

The covariance term $\mathrm{Cov}[R_iR_h] := \mathrm{Cov}[R(x_i^l, y_j^l)R(x_h^l, y_j^l)]$ measures the dependency between two contributions in the ith and hth places of the column j on the synaptic matrix.

2. If we average over the ensemble of perfect input patterns, we can use again for large m the approximations $n_1/m \simeq (n_1-1)/m \simeq (n_1+1)/m \simeq p$ and $(M-1)/m \simeq M/m$ and obtain

$$E(s_1 - s_a) = m[p(r_4-r_3) - (1-p)a(r_2-r_1)] \tag{3.18}$$

$$E(N) = (M-1)m\mu E(R)$$

In Eq. (3.17) we have to insert

$$Q_1 = m[p + (1-p)a^2], \qquad Q_2 = m^2\mu^2. \tag{3.19}$$

3. Finally, we consider the ensemble of noisy address patterns. In this case,

$$E(s_1' - s_a') = m[p(p' + (1-p')a)(r_4-r_3) - (1-p)a(r_2-r_1)]. \tag{3.20}$$

In the description of the noise we only to replace p by pp' and μ by μ' in (3.18) and (3.19).

Signal-to-Noise Ratios for Explicit Learning Rules

Regarding Eqs. (3.17) and (3.18), we observe that the signal-to-noise ratio is the same for the rules R and $bR+c$, where c is an arbitrary number and b is a positive number. Two rules that differ only in this way will be called *essentially identical.* Thus we may denote any rule R as

$$R = (0, r_2, r_3, r_4). \tag{3.21}$$

The following formulas are written more concisely if we introduce instead of r_2, r_3, r_4 the mutually dependent parameters

$$\gamma := r_4 - r_3 - r_2, \qquad \kappa := r_2 + \gamma p, \qquad \eta := r_3 + \gamma q.$$

In this notation, the variance of the rule becomes

$$\begin{aligned} \sigma^2(R) :&= E(R^2) - (E(R))^2 \\ &= \eta^2 p(1-p) + \kappa^2 q(1-q) + \gamma^2 p(1-p)q(1-q). \end{aligned}$$

In the description of the input ensemble we transform from the parameters p, a to the quantities p, μ, see Eq. (3.7).

The signal-to-noise ratio averaged over perfect address patterns (2) is then obtained from Eq. (3.13) as

$$r^2 = (m/M)\frac{[\mu\kappa + (1-\mu)p\gamma]^2}{[p + (\mu - p)^2/(1-p)]\sigma^2(R) + mq(1-q)\mu^2\kappa^2}. \tag{3.22}$$

Averaged over noisy address patterns (c) we obtain equivalently

$$r'^2 = (m/M)\frac{[\mu'\kappa + (1-\mu)pp'\gamma]^2}{[pp' + (\mu' - pp')^2/(1-pp')]\sigma^2(R) + mq(1-q)\mu'^2\kappa^2} \tag{3.23}$$

with the definition for μ' taken from Eq. (3.8).

Optimal Learning Rule

The expression (3.22) invites optimization of the signal-to-noise ratio in terms of the three parameters γ, κ, and η so as to yield the *optimal learning rule* R_0.

The parameter η appears only in $\sigma^2(R)$ in the denominator. We first minimize $\sigma^2(R)$ with $\eta = 0$ and obtain

$$r^2 = \left(\frac{m}{M}\right)\frac{[\mu\kappa + (1-\mu)p\gamma]^2}{q(1-q)\{[p + (\mu-p)^2/(1-p)][\kappa^2 + \gamma^2 p(1-p)] + m\mu^2\kappa^2\}}. \tag{3.24}$$

The (large) factor m in the second term of the denominator in Eq. (3.24) makes this term dominating unless at least one of the other factors κ or μ vanishes.

At first sight we have two distinct cases that differ with respect to the average activity μ of the input patterns:

1. Either μ stays away from 0, and then it is optimal to choose $\kappa = 0$ (case 1);

2. or $\mu \to 0$ fast enough to make the second term negligible in the sum of the denominator in Eq. (3.24) (case 2). However, if we insert $\mu = 0$ in Eq. (3.24), again $\kappa = 0$ turns out to be the optimal choice.

Thus, both cases leave us with $\kappa = 0$ and $\eta = 0$ and yield the *covariance rule* as general optimal rule:

$$R_0 = (pq, -p(1-q), -q(1-p), (1-p)(1-q)). \tag{3.25}$$

The condition $\mu = 0$ will occur several times in the sequel, and will be referred to as the condition of *zero-average input* activity. In particular, for $p = 0.5$ it implies $a = -1$, and for $p \to 0$ this implies $a \to 0$. This condition, which is equivalent to $a = -p/(1-p)$ or to $p = -a/(1-a)$, fixes the optimal combination between input activity and the model parameter a.

For arbitrary p and a in the input patterns, and for arbitrary μ, the optimal signal-to noise ratio is evaluated by inserting R_0 into Eq. (3.24),

$$r_0^2 = (m/M)\frac{(1-\mu)^2 p}{q(1-q)[p + (\mu - p)^2/(1-p)](1-p)}. \tag{3.26}$$

Transforming back from μ to a, we obtain

$$r_0^2 = (m/M)\frac{p(1-p)(1-a)^2}{[p + (1-p)a^2]q(1-q)}. \tag{3.27}$$

Insertion of the zero-average input condition $\mu = 0$ into Eq. (3.26) yields the optimal signal-to-noise ratio,

$$r_0^2 \simeq \frac{m}{Mq(1-q)}. \tag{3.28}$$

Optimizing the signal-to-noise ratio for noisy addresses 3, Eq. (3.23) leads to the same optimal rule [Eq. (3.25)]. Then the signal-to-noise ratio value for perfect addressing is reduced from the noise in the input patterns. For the optimal rule R_0 with $\mu = 0$, it is given by

$$r_0'^2 \simeq \frac{(1-p)p'^2}{p' - 2pp' + p} r_0^2. \tag{3.29}$$

For learning rules with $\kappa \neq 0$, which have a nonzero covariance term only, $\mu = 0$ can suppress the m^2 term in the variance of the noise. Therefore, $\kappa \neq 0$ and $\mu \neq 0$ lead to vanishing r as $m \to \infty$. A little algebra shows that learning rules with $\mu \neq 0$ and finite γ also yield a vanishing r. In conclusion, all suboptimal rules need $\mu = 0$ to achieve a nonvanishing r.

Table 3.1. Squared signal-to-noise ratios $r^2(m, M, p, q)$ for $\mu = 0$.

	Optimal Rule R_0	Hebb Rule H	Agreement Rule C
$r^2 =$	$\frac{m}{Mq(1-q)}$	$\frac{m(1-p)}{Mq(1-pq)}$	$\frac{8mp(1-p)}{M[p(1-q)+(1-p)q]}$

Hebb and Agreement Rule

If we compare the Hebb rule and the agreement rule to the optimal learning rule R_0, we realize that, in general, both rules are suboptimal. But nevertheless, for $p = q = 0.5$ the optimal rule becomes equal to the agreement rule, $R_0 = (0.25, -0.25, -0.25, 0.25)$, and for $p, q \to 0$ the Hebb rule is approximated by the optimal rule, $R_0 \to H$.

By Eq. (3.22) one can compute the signal-to-noise ratio for these rules, the results of which for $\mu = 0$ may be found in Table 3.1.

As expected, the Hebb rule becomes essentially identical to R_0 for $p, q \to 0$. In the $a = 0$ model, where the parameter a is not adjusted to guarantee $\mu = 0$, we need a stricter sparseness in the address patterns, $mp^2 \to 0$, to provide $\mu \to 0$ fast enough to preserve the essential identity between H and R_0.

By comparing the r^2-values corresponding to the different rules in Table 3.1, we will derive the performance analysis of the Hebb and agreement rules (see Secs. 3.5.2 and 3.5.4) from the analysis of R_0 carried out in this section.

Summary

With incremental storage procedures the signal-to-noise ratio analysis of one-step threshold-detection retrieval led to the following results:

- If a rule R yields the signal-to-noise ratio r, then any rule $bR + c$, with b positive, yields the same signal-to-noise ratio. We call these rules *essentially identical.*

- For any rule R, the best combination of the parameters p and a is given by the *zero-average input condition* $\mu = p + (1-p)a = 0$.

- The maximal signal-to-noise ratio r_0 is always achieved for the covariance rule R_0 [Eq. (3.25)]. For increasing μ, the value r_0 continuously decreases and reaches $r_0 = 0$ at $\mu = 1$.

- Every rule essentially different from R_0 has a 0 asymptotic signal-to-noise ratio, if the condition $\mu = 0$ is violated.

- The Hebb rule becomes essentially identical to R_0 for memory tasks with $q \to 0$ and $p \to 0$, i.e., for sparse address and content patterns.
- The agreement rule is equal to R_0 for $p = q = 0.5$.
- *Storage of extensively many patterns, i.e.,* $M/m > 0$ as $m \to \infty$: In this case, R_0 and H achieve asymptotically vanishing errors ($r \to \infty$) for memory tasks with sparse content patterns: $q \to 0$ as $m \to \infty$. The agreement rule A only achieves $r =$ const as $m \to \infty$.

3.4 Information Theory of the Memory Process

How can the performance of an associative memory model be measured? In our notation, a given memory task specifies the parameters p, q, M, p', e_a, e_1. From the signal-to noise ratio analysis we can determine for randomly generated patterns the maximal number of pattern pairs M^* for which the required error bounds e_a, e_1 are still satisfied. Then the first idea is to compare the M^* to the number of neurons used in the memory model. This quotient of patterns per neuron $\alpha = M^*/n$ is used in many works, but this measure disregards the parameter q used in the random generation of the content patterns as well as the whole process of addressing.

In the following we use the description of elementary information theory to find performance measures for the memory task and compare them with the size of the storage medium, viz., the number of synaptic connections $n \times m$.

3.4.1 Mean Information Content of Data

Every combination of a memory problem and a coding algorithm will lead to a set of content patterns that exhibit in general very complicated statistical correlations.

For a set of *randomly generated patterns* $\mathcal{S}$, which we have used to carry out the signal-to-noise ratio analysis, each digit was chosen independently. The mean information contained in one digit of a pattern is then simply given by the Shannon information [40] for the two alternatives with the probabilities p and $1 - p$,

$$i(p) := -p \log_2 p - (1-p) \log_2 (1-p),$$

and the *mean information content* in the set of randomly generated content patterns $\mathcal{S}^C$ is $I(\mathcal{S}^C) = Mn\, i(q)$, where q is the ratio between 1- and a-components in each content pattern. The *pattern capacity* compares the mean information content of the content patterns with the actual size $m \times n$ of the storage medium and is defined as

$$P(m,n) := \max_M \{I(\mathcal{S}^C)\}/nm = M^* i(q)/m. \tag{3.30}$$

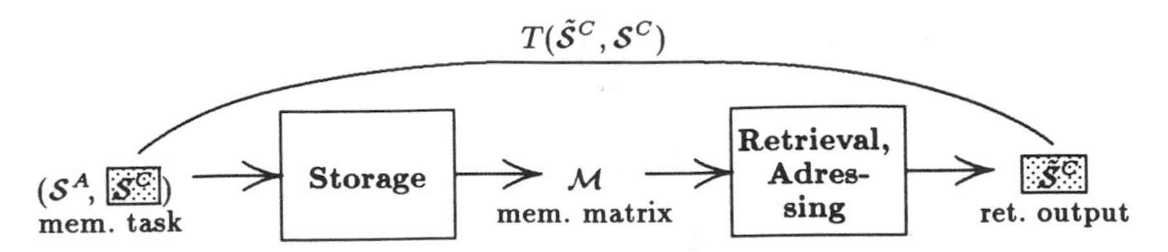

Fig. 3.2. Output capacity: Information channel of storage and retrieval; (mem. = memory, ret. = retrieval).

Here, M^* equals the maximum number of stored patterns under a given retrieval quality criterion. The definition (3.30) is an adequate measure of how much information can be put in the memory, but not at all of how much can be *extracted* during the retrieval. A performance measure should also consider the information loss due to the retrieval errors.

3.4.2 Association Capacity

The memory can be regarded as a noisy information channel consisting of two components (see Fig. 3.2): The channel input is the set of content patterns $\mathcal{S}^C$, and the channel output is the set of recalled content patterns $\tilde{\mathcal{S}}^C$ afflicted with the retrieval errors. The two components correspond to the *storage process*, where the sets $\mathcal{S}^A$ and $\mathcal{S}^C$ are transformed into the synaptic matrix and to the *retrieval process* where the matrix is transformed into a set of memory output patterns $\tilde{\mathcal{S}}^C$. The retrieval error probabilities specify the deviation of $\tilde{\mathcal{S}}^C$ from $\mathcal{S}^C$ and thus the channel capacity.

The capacity of an information channel is defined as the transinformation that is contained in the output of the channel about the channel's input. The transinformation between $\tilde{\mathcal{S}}^C$ and $\mathcal{S}^C$ can be written as

$$T(\tilde{\mathcal{S}}^C, \mathcal{S}^C) = I(\mathcal{S}^C) - I(\mathcal{S}^C|\tilde{\mathcal{S}}^C), \tag{3.31}$$

where the *conditional information* $I(\mathcal{S}^C|\tilde{\mathcal{S}}^C)$ is subtracted from the information content in $\mathcal{S}^C$. It describes the information necessary to restore the set of perfect content patterns $\mathcal{S}^C$ from the set $\tilde{\mathcal{S}}^C$. For random generation of the data we obtain

$$I(\mathcal{S}^C|\tilde{\mathcal{S}}^C)/nm = \frac{M}{m} I(y_i^k|\tilde{y}_i^k) \tag{3.32}$$

with the contribution of one digit

$$\begin{aligned} I(y_i^k \mid \tilde{y}_i^k &= \text{Prob}[\tilde{y}_i^k = 1] i(\text{Prob}[y_i^k = 0 \mid \tilde{y}_i^k = 1]) \\ &+ \text{Prob}[\tilde{y}_i^k = 0] i(\text{Prob}[y_i^k = 1 \mid \tilde{y}_i^k = 0]) \\ &= [q(1-e_1) + (1-q)e_a] i \left(\frac{(1-q)e_a}{q(1-e_1) + (1-q)e_a} \right) \end{aligned}$$

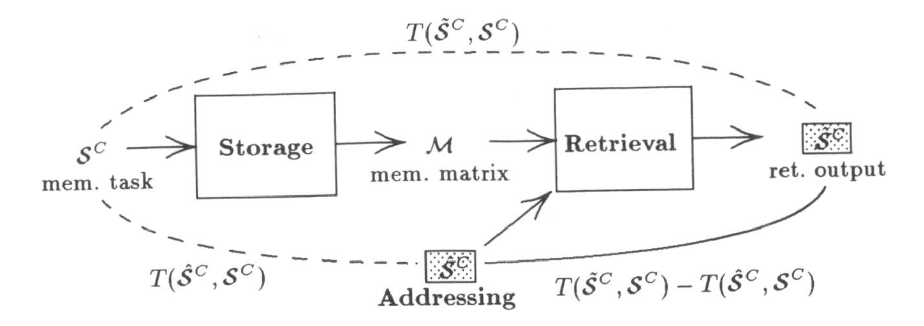

Fig. 3.3. Completion capacity: Information balance for autoassociation. (mem. = memory, ret. = retrieval).

$$+ \; [qe_1 + (1-q)(1-e_a)]i\left(\frac{qe_1}{qe_1 + (1-q)(1-e_a)}\right). \quad (3.33)$$

Now we define the *association capacity* as the maximal channel capacity per synapse:

$$A(m,n) := \max_{M} T(\tilde{S}^C, S^C)/mn = P(m,n) - \frac{M^*}{m} I(y_i^k \mid \tilde{y}_i^k). \qquad (3.34)$$

The capacity of one component of the channel is an upper bound for the capacity of the whole channel: The capacity of the first box in Fig. 3.2 will be called *storage capacity* (discussed in [41]). The maximal memory capacity that can be achieved for a fixed retrieval procedure (i.e., fixing only the last box in Fig. 3.2) will be called the *retrieval capacity.*

3.4.3 Including the Addressing Process

The defined association capacity is a quality measure of the retrieved content patterns, but the retrieval quality depends on the properties of the input patterns and on the addressing process. Of course, maximal association capacity is obtained for faultless addressing; and with growing addressing faults (decreasing probability p') the association capacity A decreases because the number of patterns has to be reduced to satisfy the same retrieval error bounds. To include judgement of addressing fault tolerance for heteroassociation, we have to observe the dependency $A(p')$.

For autoassociation where $S^A = S^C$, we will consider the information balance between the information already put into the memories input and the association capacity (see Fig. 3.3).

This difference gives the amount of information that is really gained during the retrieval process. We define the *completion capacity* for autoassociation as the maximal difference of the transinformation about S^C contained in the output patterns and contained in the noisy input patterns $\tilde{\mathcal{S}}^A$,

$$C(n) := \max_{\hat{\mathcal{S}}^C} \left\{ T(\mathcal{S}^C \mid \tilde{\mathcal{S}}^C) - T(\mathcal{S}^C \mid \hat{\mathcal{S}}^C) \right\} / n^2. \tag{3.35}$$

From Eq. (3.31) we obtain

$$\begin{aligned} C(n) &= \max_{\hat{\mathcal{S}}^C} \left\{ I(\mathcal{S}^C \mid \hat{\mathcal{S}}^C) - I(\mathcal{S}^C \mid \tilde{\mathcal{S}}^C) \right\} / n^2 \\ &= \max_{p'} \left\{ M^* [I(y_i^k \mid \hat{y}_i^k) - I(y_i^k \mid \tilde{y}_i^k)] \right\} / n. \end{aligned} \tag{3.36}$$

In Eq. (3.36) we have to insert again the maximum number of stored patterns M^* and the conditioned information to correct the retrieval errors; cf. Eq. (3.33). In addition, the one-digit contribution of the conditioned information necessary to restore the faultless address patterns $\mathcal{S}^A$ from the noisy input patterns $\hat{\mathcal{S}}^A$ is required. It is given by

$$I(y_i^k \mid \hat{y}_i^k) = (1 - pp') i \left(\frac{p(1-p')}{1-pp'} \right). \tag{3.37}$$

Note that, for randomly generated content patterns, i.e., with complete independence of all of the pattern components y_i^k, one usually reaches the optimal transinformation rates and thus the formal capacity.

3.4.4 Asymptotic Memory Capacities

In Sec. 3.3 we analyzed the model in the thermodynamic limit, the limit of diverging memory size. For asymptotic values of the capacities in this limit we not only will examine memory tasks where the fidelity requirement remains constant; we also will examine the following *asymptotic fidelity requirements* on the retrieval which distinguish asymptotically different ranges of the behavior of the quantities e_a and e_1 with respect to $q \to 0$ as $m, n \to \infty$:

- The high-fidelity or *hi-fi* requirement: $e_1 \to 0$ and $e_a/q \to 0$. Note that for $q \to 0$ the hi-fi requirement demands for both error types the same behavior of the ratio between the number of erroneous and correct digits in the output: $d_a \simeq d_1 \to 0$ with the *error ratios* defined by $d_a := e_a/q$ and $d_1 := e_1/(1-q)$.

- The low-fidelity or *lo-fi* requirement: e_1 and e_a stay constant (but small) for $n \to \infty$.

With one of these asymptotic retrieval quality criteria the *asymptotic capacities* P, A, and C are defined as the limits for $n, m \to \infty$ and $n \to \infty$, respectively.

3.5 Model Performance

3.5.1 Binary Storage

Output Capacity

In this memory model the probability $p_0 = \text{Prob}(\bar{\mathcal{M}}_{ij} = 0)$ is decreased if the number of stored patterns is increased. Since obviously no information can be drawn from a memory matrix with uniform matrix elements, we will exclude the cases $p_0 = 1$ and $p_0 = 0$ in the following.

For faultless addressing, the maximal number M^* of patterns that can be stored for a given limit on the error probabilities can be calculated by Eqs. (3.9) and (3.10):

$$M^* = \frac{\ln[p_0]}{\ln[1-pq]} = \frac{\ln[1-(e_a)^{1/mp}]}{\ln[1-pq]}. \tag{3.38}$$

From Eq. (3.34) we obtain for $e_1 = 0$ and $e := e_a \ll q$ the association capacity

$$A(m,n) \simeq (M^*/m)\{i(q) - (1-q)e\log_2[e(1-q)/q]\}. \tag{3.39}$$

In Fig. 3.4 we have plotted a) $\alpha = M^*/m$ from Eq. (3.38), and b) the association capacity from Eq. (3.39) against p for $q = p$ and the constant error ratio $d = e_a/p = 0.01$ for three finite memory sizes. Figure 3.5 shows simulation results for the error ratio d with the parameters as in Fig. 3.4. For p-values near the information optima in Fig. 3.4b, the experimental value d_{exp} is close to the value d used in Fig. 3.4a. For lower and higher p-values, there are deviations between theory and experiment; see the caption for Fig. 3.5.

Nonvanishing asymptotic association capacity requires $M^*/m > 0$ as $m \to \infty$. In Eq. (3.38) this can be obtained either for $p_0 \to 0$, which we have already excluded, or for $pq \to 0$. In this case, we obtain

$$M^* \simeq \frac{\ln[p_0]}{-pq}. \tag{3.40}$$

The hi-fi requirement leads with Eq. (3.11) to the following condition on p and q:

$$e_a/q = \exp(mp\ln[1-p_0] - \ln[q]) \to 0. \tag{3.41}$$

In the case $q \to 0$, the requirement (3.41) is satisfied if we put

$$p = u\frac{-\ln[q]}{m} \tag{3.42}$$

with the positive number $u > -(\ln[1-p_0])^{-1}$. Inserting Eq. (3.42) into (3.40), we obtain the inequality

$$M^* < \frac{m\ln[p_0]\ln[1-p_0]}{-q\ln[q]}, \tag{3.43}$$

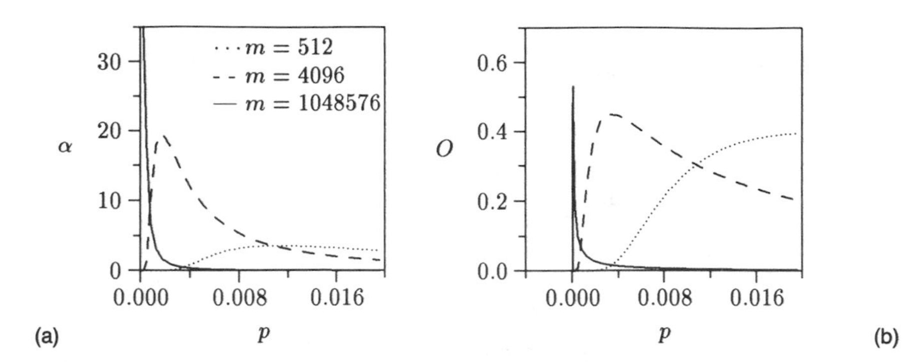

Fig. 3.4. Binary storage in finite memory sizes: Number of stored patterns α and output capacity A in bits/syn with the lo-fi requirement $d = 0.01$ for $p = q$ and $n = m$.

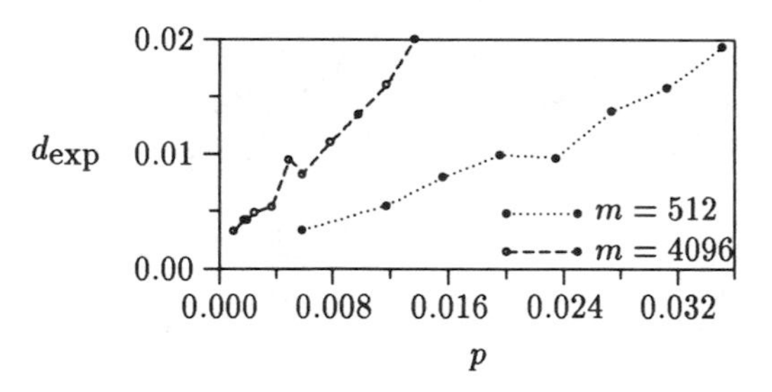

Fig. 3.5. Retrieval error ratio $d = e_a/k$ of simulations along the α-p curves of Fig. 3.4 for $d_{\text{theor}} = 0.01$. For low p-values, the experimental error is even lower than predicted because we used learning patterns with a nonfluctuating activity in the simulations. For higher p-values, the theoretic values are too small because, in this range, the effects of statistical dependence between different matrix elements should not be neglected.

which can be put into Eq. (3.39), yielding, for $p_0 = 0.5$ and $m \to \infty$, the maximal association capacity $A \simeq 0.69$ bits/syn.

Note that for autoassociation and heteroassociation with $p = q, m = n$, Eq. (3.42) implies that

$$p \propto \ln[n]/n \tag{3.44}$$

and

$$M^* \propto \left(\frac{n}{\ln[n]}\right)^2. \tag{3.45}$$

The relation (3.45) already has been obtained in [42, 43] for sparse memory patterns with arbitrary learning rules by regarding the space of all possible synaptic interactions; cf. Sec. 3.6.3.

For singular address patterns and arbitrary $q = \text{const}$, however, error-free retrieval is possible for $M^* \leq m$, which is the combinatorial restriction for nonoverlapping singular patterns. In this case, with Eq. (3.39), as association capacity of $A = i(q) \leq 1$ bits/synapse is obtained. For constant p, Eq. (3.42) demands asymptotically empty content patterns, $q \propto \exp(-mp/u)$, leading to vanishing association capacity. For singular content patterns, the combinatorial restriction $M^* \leq m$ also yields vanishing association capacity.

Fault Tolerance and Completion Capacity

In the case of noisy input patterns [Eq. (3.12)], the hi-fi condition becomes $e_a/q = \exp(mpp' \ln[1 - p_0] - \ln[q]) \to 0$. As in the preceding subsection, we obtain the maximal number of patterns by $M'^* = p'M^*$, where M^* is the value for faultless addressing [Eq. (3.43)]. Thus, for heteroassociation, the association capacity exhibits a linear decrease with increasing addressing fault, $A(p') = p'A$.

For autoassociation with the hi-fi requirement, the retrieval error term in the completion capacity [Eq. (3.36)] can be neglected as in the association capacity, and we obtain for $p \to 0$

$$\begin{aligned} C &= \max_{p'} \left\{ (M'^*/n)(1 - pp')i\left(\frac{p(1-p')}{1-pp'}\right) \right\} \\ &= \max_{p'} \left\{ \frac{\ln[p_0]\ln[1-p_0]p'(1-p')}{\ln[2]} \right\} = 0.17 \text{ bits/syn} \end{aligned} \tag{3.46}$$

for $p_0 = 0.5$ and $p' = 0.5$. In Fig. 3.6, the completion capacity is plotted against p for three finite memory sizes and for the constant error ratios a) $d = e_a/p = 0.01$, and b) $d = 0.05$. The optimum is always obtained for $p' = 0.5$.

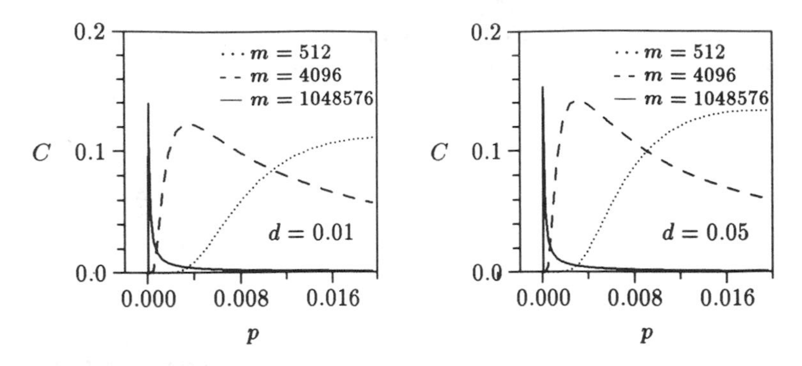

Fig. 3.6. Binary storage in finite memory sizes: Completion capacity C in bits/syn for two lo fi values; the maximum has always been achieved for addressation with $p' = 0.5$.

3.5.2 Incremental Storage

Output Capacity

For faultless addressing, zero-average input, and the optimal rule R_0, the maximal number of stored patterns for a given signal-to-noise ratio value r is obtained from Eq. (3.28):

$$M^* = m/(r^2 q(1-q)). \tag{3.47}$$

If the threshold setting provides $e_a/q = e_1/(1-q) =: d$, the association capacity can be computed for small fixed values of the error ratio d from Eqs. (3.34) and (3.47):

$$A \simeq \frac{i(q) + q(1-q)d\{\log_2[qd] + \log_2[(1-q)d]\}}{r^2 q(1-q)}. \tag{3.48}$$

With substitution of $r = G^{-1}[qd] + G^{-1}[(1-q)d]$ in Eq. (3.48) we obtain the association capacity for the rule R_0 for a constant d error ratio, the lo fi requirement. ($G^{-1}[x]$ is the inverse Gaussian distribution.) In Fig. 3.7 we display the association capacity values for the optimal, Hebb, and agreement rules, the latter two obtained by comparison of the signal-to-noise ratios in Table 3.1.

The hi-fi requirement only can be obtained for $r \to \infty$ as $m \to \infty$ in Eq. (3.47), which is possible either for $M^*/m \to 0$, leading to vanishing association capacity, or for $q \to 0$, the case of *sparse* content patterns, which we focus on in the following.

We now choose a diverging signal-to-noise ratio by

$$r = \sqrt{-2\ln[q]}/\vartheta. \tag{3.49}$$

The threshold has to be set asymmetrically, $\vartheta \to 1$, because for sparse patterns $e_a/e_1 \to 0$ is demanded. (This implies $q = \exp[-(\vartheta r)^2/2]$, yielding,

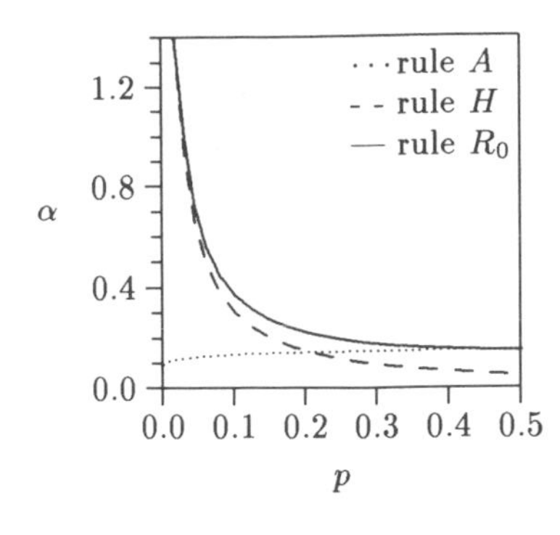

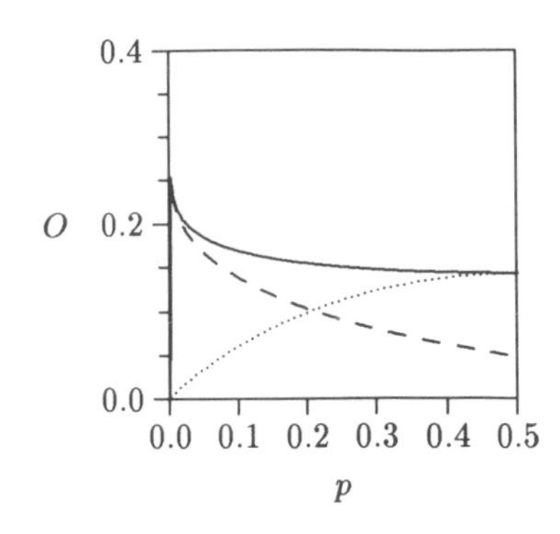

Fig. 3.7. Model with incremental storage, fulfilled condition of zero-average input, and $m, n \to \infty$: Number of stored patterns α (left) and asymptotic output capacity A in bits/synapse (right) for $p = q$ with the lo-fi requirement $d = 0.01$. The optimal rule R_0 is approached by the agreement rule A for $p = 0.5$ and by the Hebb rule for $p \to 0$. For $p \to 0$, the lo-fi output capacity values of the optimal and Hebb rules reach but do not exceed the hi-fi value of $A = 0.72$ bits/synapse (this only can be observed if the p-scale is double logarithmic; see Fig. 5 in [51]).

with Appendix 3.2, $e_a/q \simeq (\pi r^2/2)^{-1/2} \to 0$. If the threshold ϑ approaches 1 slowly enough that $(1-\vartheta)r \to \infty$ still holds, then $e_1 \to 0$ also is true and the hi-fi requirement is fulfilled.)

With vanishing e/q, Eq. (3.48) simplifies asymptotically to

$$A \geq P + \frac{2e\log_2[e]}{r^2} \simeq P.$$

Again, the information loss due to retrieval errors can be neglected due to the high-fidelity requirement.

Inserting Eq. (3.49) into (3.47) we obtain for zero-average input and the optimal rule R_0,

$$M^* = m/(-2q(1-q)\ln[q]), \tag{3.50}$$

which, like our result (3.49), can be calculated alternatively with the Gardner method [42, 43]; cf. Sec. 3.6.3.

With Eqs. (3.50) and (3.30) we obtain as asymptotic association capacity with the hi-fi requirement, $A = 0.72$ bits/syn.

In contrast to the model with binary storage — where a positive association capacity only for sparse content *and* address patterns has been obtained — with incremental storage, an association capacity $A = 0.72$ bits/syn is achieved even for memory tasks with nonsparse address patterns. However, for $\{0, 1\}$-neurons we again are restricted to sparse address patterns because, for nonsparse address patterns, the zero-average input condition cannot be satisfied.

With singular address or content patterns that are not interesting cases for associative memory, as we will discuss in Sec. 3.6.1, incremental and

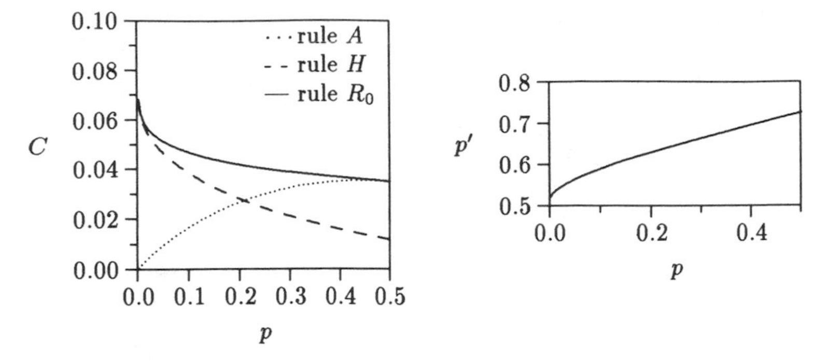

Fig. 3.8. Incremental storage for $n \to \infty$: Completion capacity in bits/syn with the lo-fi requirement $d = 0.01$ (left diagram). The optimal p' in the addressing has been determined numerically (right diagram).

binary storage form the same memory matrix and achieve exactly the same performance; see the last part of Sec. 3.5.1.

Fault Tolerance and Completion Capacity

For heteroassociation with noisy addressing we obtain the association capacity for zero-average input and R_0 by using Eq. (3.29) (remember that $r^2 \propto m/M$):

$$A(p') = \frac{(1-p)p'^2}{p' - 2pp' + p} A. \tag{3.51}$$

For $p = 0.5$ this implies $A(p') = p'^2\, A$, and for $p \to 0$, as in the binary case, $A(p') = p'\, A$. For autoassociation with the hi-fi requirement we obtain in a way similar to Eq. (3.46)

$$\begin{aligned} C(n) &= \max_{p'} \left\{ \frac{\vartheta^2 p'(1-p')\log_2[p(1-p')]}{2\ln[p]} \right\} \\ &\simeq \max_{p'} \left\{ \frac{\vartheta^2 p'(1-p')}{2\ln[2]} \right\} = 0.18 \text{ bits/syn.} \end{aligned}$$

Again, the maximum is reached for $p' = 0.5$ and $\vartheta \to 1$.

A similar optimization in p' can be carried out for fixed values of p and the lo-fi requirement; see Fig. 3.8. In this case, the optimum is reached for p' larger than 0.5.

3.6 Discussion

3.6.1 HETEROASSOCIATION

In applications of associative memory, the coding of address and content patterns plays an important role. In Sec. 3.1 we distinguished three types of patterns leading to the memory tasks defined in Sec. 3.4: singular patterns with only a single 1-component, sparse patterns with a low ratio between the numbers of 1- and a-components, and nonsparse patterns. To get a general idea, Table 3.2 shows those memory models which achieve association capacity values $A > 0$ under the hi-fi requirement. Note that only the Hebb and the optimal learning rules in memory tasks with sparse or singular patterns yield nonvanishing hi-fi association capacities. In the following, we consider the different types of content patterns subsequently.

Nonsparse Content Patterns

Only in combination with singular address patterns do nonsparse patterns achieve high association capacity. In this case, qualified in Sec. 3.4 as the look-up–table task, all rules achieve $A = 1$. The associative memory works like a RAM device, where each of the m content patterns is written into one row of the memory matrix $\overline{\mathcal{M}}$ and, therefore, trivially $A = i(q)$. However, this is not an interesting case for associative storage because the storage is not distributed, and in the recall no fault tolerance can be obtained: $A(p') = 0$ for $p' < 1$.

Table 3.2. Models that yield $A > 0$ for the hi-fi requirement in different memory tasks (incr. = incremental storage, bin. = binary storage, incr.R_0, H, for instance, denotes the incremental storage model with either optimal rule or Hebb rule).

	Nonsparse Content	Sparse Content	Singular Content
Nonsparse address	—	incr. R_0	—
Sparse address	—	incr. R_0, H bin. H	—
Singular address	incr. R_0, H bin. H	—	—

Table 3.3. Hi-fi association capacity values of the different models for sparse content patterns. As a measure of addressing fault tolerance (cf. Sec. 3.3), in the second line of each cell the reduction factor for faulty addressing is displayed. For instance, with sparse address and content patterns the Hebb rule in the incremental storage yields $A = 0.36$ bits/syn if, in the addressing, $p' = 0.5$ is chosen.

	Binary	Incremental	
	H	H	R_0
Nonsparse address	— —	— —	$A = 0.72$ p'^2
Sparse address	$A = 0.69$ p'	$A = 0.72$ p'	$A = 0.72$ p'

Sparse Content Patterns

Combined with sparse or nonsparse address patterns, sparse content patterns represent the most important memory task for neural memory models with Hebb or optimal learning rules, where high capacity together with associative recall properties are obtained. For optimal association capacity, many patterns in the set of sparse learning patterns will overlap. Therefore, in the learning process, several pattern pairs affect the same synapse, and distributed storage takes place. In Table 3.3, the hi-fi association capacity values can be compared. For sparse address patterns, the Hebb and optimal rules achieve exactly the same performances because, with the zero-average input condition, both rules are essentially identical. Even the binary Hebb rule shows almost the same performance. At first sight it is striking that binary storage, using only one-bit synapses, yields almost the same performance as incremental storage, which uses synapses that can take many discrete values. This fact becomes understandable if we consider the mean contributions of all of the patterns at one synapse by incremental and by binary storage: $E\mathcal{M} = 0.69$ for incremental compared with $E\bar{\mathcal{M}} = 0.5$ for binary storage. In both cases, the sparseness requirement prevents the matrix elements from extensive growth; also, in incremental storage the vast majority of synapses take only the values 0, 1, and 2.

For nonsparse address patterns, only the optimal setup, namely, the rule R_0 in the incremental storage, achieves nonvanishing association capacity. This case is of less importance for applications since implementation is much more difficult (higher computation effort for $a \neq 0$, and the determination of the value of a requires the parameter p of the patterns).

Relaxing the quality criterion does not enhance the association capacity value in the sparse limit. The lo-fi association capacity values plotted in Figs. 3.4 and 3.7 do not exceed the hi-fi values in Table 3.3. With the agreement rule, finite lo-fi association capacity values can be achieved (see Fig. 3.7), whereas the hi-fi association capacity always vanishes.

Singular Content Patterns

The neural pattern classifier that responds to a nonsingular input pattern with a single active neuron often is called the *grandmother model* or *perceptron*. Here, the information contained in the content patterns is asymptotically vanishing compared to the size of the network: $A = 0$. Again, no distributed storage takes place.

3.6.2 AUTOASSOCIATION

If content and address patterns are identical in order to accomplish pattern completion in the retrieval, we have only to regard the cases of sparse and nonsparse learning patterns.

Asymptotic Results

The amount of information that really can be extracted by pattern completion with high quality is given by the asymptotic hi-fi completion capacity. It always vanishes in cases of nonsparse patterns. For one-step retrieval with sparse patterns, we have determined $C = 0.18$ and $C = 0.17$ bits/syn for the Hebb rule in incremental and binary storage, respectively (Secs. 3.5.1 and 3.5.2).

Using a practically unrealistic fixed-point readout scheme[7] and the Hebb rule, we have found completion capacity values of $C = 0.36$ bits/syn for incremental and $C = 0.35$ bits/syn for binary storage [30, 23]. Thus, one would expect the performance of one-step retrieval to be improved by fixed-point retrieval, i.e., starting from a single address pattern and *iterating* the retrieval process until the fixed point is reached. Asymptotically, however, fixed-point retrieval does not improve the one-step capacity results [44, 45, 46]. It is a consequence of the fulfilled hi-fi condition that already after only the first step we get asymptotically vanishing errors for diverging system size.

Finite-Size Systems

Although Fig. 3.6 illustrates that the asymptotic capacity bounds are only reached for astronomic memory sizes, even for realistic memory sizes sparse

[7]Fixed points are patterns that remain unchanged during a retrieval step, i.e., input and output patterns are identical.

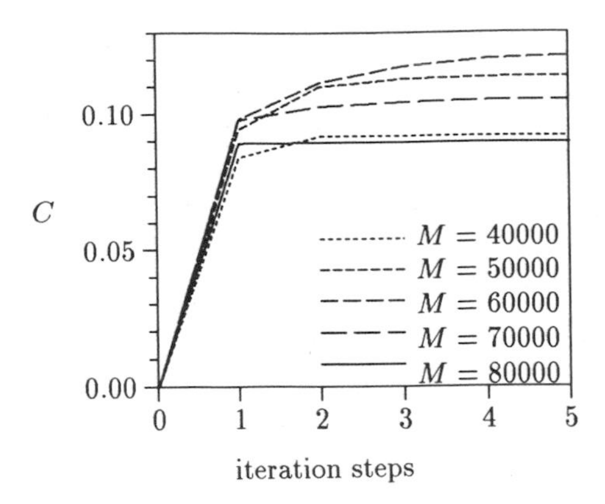

Fig. 3.9. Completion capacity C in bits/syn for iterative retrieval for addressation with $p' = 0.5$ which has been achieved in simulations in binary storage with 4096 neurons. Depending on the number of stored patterns M an improvement up to twenty percent (for $M = 60000$) can be obtained after the first step through iteration.

patterns yield better performance than nonsparse patterns. Simulations and analysis have revealed (again cf. [44, 45]) that iterative retrieval methods with an appropriate threshold-setting scheme (indicating how the threshold should be aligned during the sequence of retrieval steps) yield superior exploitation of the autoassociation storage matrix as compared to one-step retrieval; see Fig. 3.9. For finite systems, fixed-point retrieval even improves the performance and capacity values above the asymptotic value; e.g., for $n = 4096$, about $C = 0.19$ bits/syn can be obtained.

For a certain application and a given finite memory size, however, we cannot reduce the pattern activity ad libitum by modifying the coding algorithm. Thus we sometimes may be faced with $p \gg \ln[n]$; cf. Eq. (3.42). In this case, binary Hebbian storage is ineffective — see Fig. 3.6 — and incremental storage does not work either.

3.6.3 Relations to Other Approaches

Heteroassociation

The zero-average input condition for memory schemes with nonoptimal local synaptic rules was first made explicit by Palm [47] but appeared implicitly in some closely related papers. Horner [48] has used it to define the neural off-value a in his model, and Nadal and Tolouse [24] have exploited it (through their condition of "safely sparse" coding) as a justification for their approximations.

The optimization of the signal-to-noise ratio r carried out by Willshaw and Dayan [37] and independently by Palm [47] already has been suggested

— though not carried out — by Hopfield [25]. Also, Amit et al. [8] have proposed the covariance rule R_0.

The signal-to-noise ratio is a measure of how well threshold detection can be performed in principle, independent of a certain strategy of threshold adjustment. We have examined the model where the threshold assumes the same value Θ for all neurons during one retrieval step and optimized the response behavior depending on the individual input activity. So we could lump together the on- and off-fractions of output neurons and calculate the average signal-to-noise ratio.

In a recent work, Willshaw and Dayan [49] carried out a signal-to-noise analysis using quite similar methods for a different model. In their model, the threshold setting Θ_j was chosen individually for each neuron for the average total activity of input patterns. Thus, the signal-to-noise ratio at a single neuron was optimized for averaged input activity. Due to this difference, the results only agree for zero-average input activity and in the thermodynamic limit; for the same optimal rule, the same signal-to-noise ratio is obtained. In general, their model is not invariant under the addition of an arbitrary constant in the learning rule because, for $E(R) \neq 0$, activity fluctuations in an individual input patterns are not compensated for by threshold control as in our model.

Most of the results for heteroassociation discussed here can be found in Peretto [50], Nadal and Toulouse [24], Willshaw and Dayan [37], and Palm [47, 51]. Some of our results are numerically identical to those of Nadal and Toulouse, who employ different arguments [e.g., approximation of the distribution of the noise term, Eq. (3.13), by a Poisson distribution]. In our framework one also could define a "no fidelity requirement," namely, e_a and $e_1 \to 0.5$, which would correspond to the "error-full regime" of Nadal and Toulouse. This leads to the same numerical result, $A = 0.46$, which, however, is not very interesting from an engineering point of view since it is worse than what can be achieved with high fidelity. The result for binary storage stems from Willshaw et al. [4] for the Hebb rule, and to Hopfield [25] for the agreement rule. A new aspect is the information-theoretical view on the trade-off between association capacity and fault tolerance.

Autoassociation

Autoassociation has been treated extensively in the literature; see, for example, [8, 25, 43, 26, 29]. In two points, our treatment differs from most of the papers on autoassociation:

- Usually, models with fixed-point retrieval (and only with incremental storage) have been considered.
- As the appropriate performance measure for pattern completion, we evaluate and compare the completion capacity which takes into account the entire information balance during the retrieval.

With one exception [48, 52], other authors regard (in our terms) the pattern capacity, i.e., the retrieval starts from the perfect pattern as address.[8] Hence, to compare the existing fixed-point results with our one-step retrieval for autoassociation, we should take the association capacity or pattern capacity results calculated in Sec. 3.5.2 for heteroassociation in the case $p = q$.

For nonsparse patterns with $p = 0.5$, fixed-point retrieval with the lo-fi requirement stays below one-step retrieval: For the same fidelity of $d = 0.002$, the one-step result for the agreement rule (Fig. 3.4) is higher than the Hopfield bound for the fixed-point retrieval in [10, p. 296]. Here, one-step retrieval behaves more smoothly with respect to increasing memory load because the finite retrieval errors after the first step are not increased further by iterated retrieval. If the lo-fi fidelity requirement is successively weakened, a smooth increase of the one-step association capacity can be observed, and no sharp overload breakdown of the capacity (the Hopfield catastrophy) takes place, as would be the case for fixed-point retrieval at the Hopfield bound α_c [25, 8, 29].

The pattern capacity for the binary agreement rule has been estimated by a comparison of the signal-to-noise ratios for binary and nonbinary matrices in [25] and has been exactly determined in [26] as $A^b = (2/\pi)A$. For nonsparse learning patterns, binary storage is really worse than incremental storage.

Again, as for heteroassociation, only for sparse patterns can nonzero values for the asymptotic hi-fi capacities can be achieved. For one-step retrieval with $a = 0$, we have found a hi-fi pattern capacity of $P = 0.72$ bits/syn. For fixed-point retrieval, it has been possible to apply the statistical mechanics method to sparse memory patterns; cf. for instance [53, 27]. In [27] just the same value $P = 0.72$ bits/syn has been obtained. By a combinatorial calculation we also have obtained this pattern capacity value for fixed-point retrieval [30]. One-step and fixed-point retrievals yield the same pattern capacities because, for sparse patterns, the hi-fi condition is satisfied. It guarantees that almost all learned patterns are preserved in the first retrieval step and hence are fixed points.

Quite a different way to analyze the storage of sparse and nonsparse patterns through statistical mechanics has been developed by Gardner [42, 43]. In the space of synaptic interactions, she has determined the subspace in which every memory pattern is a stable fixed point. For sparse patterns this method yields the same pattern capacity value.

---

[8] To obtain the pattern capacity, it is sufficient to study the properties of the fixed points as a static problem. In evaluating the completion capacity, one has to study how the system state evolves from a noisy input pattern in order to determine the properties of the output pattern with a given address. This is a dynamic problem which is in fact very difficult.

3.6.4 Summary

The main concerns of this chapter can be summarized as follows:

- The statistical analysis of a simple feedforward model with one-step retrieval provides the most elementary treatment of the phenomena of distributed memory and associative storage in neural architecture.
- The asymptotic analytical results are consistent with the literature. For autoassociation, most of the cited works consider fixed-point retrieval, which allows us to compare one-step with fixed-point retrieval.
- Our information-theoretic approach introduces the capacity definitions as the appropriate performance measures for evaluating for the different memory tasks the *information per synapse* which can be stored and recalled. Note that nonvanishing capacity values imply that the information content is proportional to the number of synapses in the model.
- For *local* learning rules, *sparse* content patterns turn out to be the best possible case, cf. [54]. High-capacity values and distributed storage with fault-tolerant retrieval are provided by the Hebb rule and $\{0, 1\}$ neurons. Here, the number of stored patterns is much higher than the number of neurons constituting the network. The binary Hebb rule — much easier to implement — yields almost the same performance as the incremental Hebb rule. For autoassociation, one-step retrieval achieves the same asymptotic capacity values as fixed-point retrieval (for the finite-size model, fixed-point retrieval yields higher capacity values). The hi-fi condition can always be fulfilled by sparse content patterns and only by these.

Acknowledgment. We are indebted to F. Schwenker for Fig. 3.9 and for many helpful discussions. We thank J. L. van Hemmen for a critical reading of the manuscript. This work was partially supported by the Bundesministerium für Forschung und Technologie.

Appendix 3.1

In this section we show, for the Hebb rule in binary storage, the independence of two different matrix elements. This is required in Sec. 3.3.2.

Proposition 1 *For the binary storage matrix $\mathcal{M}$ we have, as $n \to \infty$,*

$$\frac{Prob[\mathcal{M}_{1j} = 1 \text{ and } \mathcal{M}_{2j} = 1]}{Prob[\mathcal{M}_{1j} = 1]Prob[\mathcal{M}_{2j} = 1]} \to 1 \qquad \text{and}$$

$$\frac{Prob[\mathcal{M}_{j1} = 1 \text{ and } \mathcal{M}_{j2} = 1]}{Prob[\mathcal{M}_{j1} = 1]Prob[\mathcal{M}_{j2} = 1]} \to 1,$$

provided p and $q \to 0$ and $x := Mpq$ stays away from 0 for $n \to \infty$.

Proof. $\text{Prob}[\mathcal{M}_{ij} = 1] = 1 - (1-pq)^M$:

$$\begin{aligned}\text{Prob}[\mathcal{M}_{1j} = 1 \text{ and } \mathcal{M}_{2j} = 1] &= \text{Prob}[(\exists k : x_1^k = x_2^k = 1 \text{ and } y_j^k = 1) \text{ or} \\ &\quad (\exists l, m : x_1^l, x_2^l = 0, x_1^m = 0, x_2^m, y_j^l \\ &\quad = 1, y_j^m = 1)] \\ &= 1 - (p(E_1) + p(E_2) - p(E_1 \cap E_2)),\end{aligned}$$

where

$$E_1 = [\forall k : \text{not } (x_1^k = x_2^k = 1 \text{ and } y_j^k = 1) \text{ and not } (x_1^k = 1, x_2^k = 0, y_j^k = 1)]$$

and

$$E_2 = [\forall k : \text{not } (x_1^k = x_2^k = 1 \text{ and } y_j^k = 1) \text{ and not } (x_1^k = 0, x_2^k = 1, y_j^k = 1)].$$

Thus, $\text{Prob}(E_1) = \text{Prob}(E_2) = (1-pq)^M$ and $\text{Prob}(E_1 \cap E_2) = (1 - q(2p - p^2))^M$. Therefore, we obtain

$$\begin{aligned}&\text{Prob}[\mathcal{M}_{1j} = 1 \text{ and } \mathcal{M}_{2j} = 1] - \text{Prob}[\mathcal{M}_{1j} = 1] \cdot \text{Prob}[\mathcal{M}_{2j} = 1] \\ &\quad = (1 - 2qp + qp^2)^M - (1-pq)^{2M} = (1 - 2qp + qp^2)^M \\ &\quad\quad -(1 - 2pq + p^2q^2)^M \\ &\quad = e^{-M(2pq - p^2 q)} - e^{-M(2pq - p^2q^2)} = e^{-2pqM}(e^{Mp^2q} - e^{Mp^2q^2}).\end{aligned}$$

Thus we find

$$\begin{aligned}&\frac{\text{Prob}[\mathcal{M}_{1j} = 1 \text{ and } \mathcal{M}_{2j} = 1] - \text{Prob}[\mathcal{M}_{1j} = 1] \cdot \text{Prob}[\mathcal{M}_{2j} = 1]}{\text{Prob}[\mathcal{M}_{1j} = 1] \cdot \text{Prob}[\mathcal{M}_{2j} = 1]} \\ &\quad = \frac{e^{-2x}(e^{px} - e^{qpx})}{(1 - e^{-x})^2} \to 0,\end{aligned}$$

since $px \to 0$ and $pqx \to 0$.

This proposition shows the asymptotic pairwise independence of the entries $\mathcal{M}_{ij}$ in the memory matrix $\mathcal{M}$, since entries which are not in the same row or column of the matrix are independent anyway.

In order to show complete independence, one would have to consider arbitrary sets of entries $\mathcal{M}_{ij}$. In this strict sense, the entries cannot be independent asymptotically. For example, if one considers all entries in one column of the matrix, then $\text{Prob}[\mathcal{M}_{ij} = 0 \text{ for all } i] = (1-q)^M \approx e^{-Mq}$, which is with Eq. (3.9) in general not equal to $p_0^m = (1-pq)^{Mn} \approx e^{-Mmpq}$.

Thus independence can at best be shown for sets of entries of the matrix $\mathcal{M}$ up to a limited cardinality $L(n)$. The worst case, which is also important for our calculations of storage capacity, is again when all entries are in the same column (or row) of the matrix. This case is treated in the next proposition, which gives only a rough estimate.

Proposition 2

$$\frac{\mathit{Prob}[\mathcal{M}_{ij} = 1 \ \mathit{for}\ i = 1, \ldots, l]}{\mathit{Prob}[\mathcal{M}_{ij} = 1]^l} \to 1 \qquad \mathit{for}\ n \to \infty$$

as long as $pl^2 \to 0$ *and* $x = Mpq$ *stays away from* 0 *for* $n \to \infty$.

Proof.

$$\text{Prob}[\mathcal{M}_{ij} = 1] \leq \text{Prob}[\mathcal{M}_{lj} = 1 | \mathcal{M}_{ij} = 1 \text{ for } i = 1, \ldots, l-1]$$

$$\leq \text{Prob}[\mathcal{M}_{lj} = 1 | \text{ there are at least } l-1 \text{ pairs } (x^k, y^k) \text{ with } y_j^k = 1]$$

$$= 1 - (1-p)^{l-1}(1-pq)^{M-l+1}.$$

Therefore,

$$0 \leq \log \frac{p[\mathcal{M}_{ij} = 1 \text{ for } i = 1, \ldots, l]}{p[\mathcal{M}_{ij} = 1]^l} \leq \sum_{i=0}^{l-1} \log \frac{1 - (1-p)^i (1-pq)^{M-i}}{1 - (1-pq)^M}$$

$$= \sum_{i=0}^{l-1} \log \frac{1 - \left(\frac{1-p}{1-pq}\right)^i p_0}{1 - p_0} \leq \sum_{i=0}^{l-1} \log \frac{1 - (1 - ip)p_0}{1 - p_0},$$

since

$$\left(\frac{1-p}{1-pq}\right)^i \geq (1-p)^i \geq 1 - ip,$$

$$\leq \sum_{i=0}^{l-1} ip \frac{p_0}{1-p_0},$$

since

$$\log(1+x) \leq x,$$

$$\leq \frac{p \cdot p_0}{1-p_0} \cdot \frac{l^2}{2} \to 0 \text{ for } p \cdot l^2 \to 0,$$

and if $p_0 = (1-pq)^M \approx e^{-Mpq} = e^{-x} \not\to 1$. For Eq. (3.10) we need the independency of $l = mp$ matrix elements; thus, for sparse address patterns with $m^{2/3}p \to 0$, the requirement of Proposition 2 is fulfilled and the independence can be assumed.

Appendix 3.2

The following estimation of the Gauss integral $G(t)$ is used in Sec. 3.5.2.

Proposition 3

$$(2\pi t^2)^{-1/2}e^{-t^2/2}(1-t^2) \leq G(-t) = 1 - G(t) \leq (2\pi t^2)^{-1/2}e^{-t^2/2}$$

Proof. Since $x^2 = t^2 + (x-t)^2 + 2t(x-t)$, we have

$$\int_t^\infty e^{-x^2/2}dx = e^{-t^2/2}\int_0^\infty e^{-x^2/2}e^{-xt}dx.$$

From this and with $e^{-x^2/2} \leq 1$, we obtain the second inequality directly since $\int_0^\infty e^{-xt}dx = 1/t$ and the first one after partial integration because $\int_0^\infty xe^{-xt}dx = 1/t$.

References

[1] Hodgkin, A.L., Huxley, A.F. (1952) A quantitative description of membrane current and its application to conduction and excitation in nerve. *J. Physiol. (Lond.)* **117**:500–544

[2] McCulloch, W.S., Pitts, W. (1943) A logical calculus of the ideas immanent in neural activity. *Bull. Math. Biophys.* **5**

[3] Steinbuch, K (1936) Die Lernmatrix. *Kybernetik* **1**:36

[4] Willshaw, D.J., Buneman, O.P., Longuet-Higgans, H.C. (1969) Nonholographic associative memory. *Nature (London)* **222**:960–962

[5] Rosenblatt, F. (1962) *Principle of Neurodynamics* (Spartan Books, New York)

[6] Little, W.A. (1974) The existence of persistent states in the brain. *Math. Biosci.* **19**:101–120

[7] Kirkpatrick, S., Sherrington, D. (1978) Infinite-ranged models of spin-glasses. *Phys. Rev. B* **17**:4384–4403

[8] Amit, D.J., Gutfreund, H., Sompolinsky, H. (1987) Statistical mechanics of neural networks near saturation. *Ann. Phys.* **173**:30–67

[9] Domany, E., van Hemmen, J.L., Schulten, K. (1991) *Models of Neural Networks* (Springer-Verlag, Berlin)

[10] Amit, D. J. (1989) *Modelling Brain Function* (Cambridge University Press, Cambridge)

[11] Hertz, J., Krogh, A., Palmer, R. G. (1991) *Introduction to the Theory of Neural Computation* (Addison Wesley, Redwood City, CA)

[12] Uttley, A.M. (1956) Conditional probability machines and conditional reflexes. In: *An. Math. Studies* **34**, Shannon, C.E., McCarthy, J. (Eds.) (Princeton Univ. Press, Princeton, NJ), pp. 237–252

[13] Longuett-Higgins, H.C., Willshaw, D.J., Buneman, O.P. (1970) Theories of associative recall. *Q. Rev. Biophys.* **3**:223–244

[14] Amari, S.I. (1971) Characteristics of randomly connected threshold-element networks and network systems. *Proc. IEEE* **59**:35–47

[15] Gardner-Medwin, A.R. (1976) The recall of events through the learning of associations between their parts. *Proc. R. Soc. Lond. B.* **194**:375–402

[16] Kohonen, T. (1977) *Associative Memory* (Springer-Verlag, Berlin)

[17] Caianiello, E.R. (1961) Outline of a theory of thought processes and thinking machines. *J. Theor. Biol.* **1**:204–225

[18] Holden, A.V. (1976) *Models of the Stochastic Activity of Neurons* (Springer-Verlag, Berlin)

[19] Abeles, M. (1982) *Local Cortical Circuits* (Springer-Verlag, Berlin)

[20] Buhmann, J., Schulten, K. (1986) Associative recognition and storage in a model network of physiological neurons. *Biol. Cybern.* **54**:319–335

[21] Anderson, J.A. (1968) A memory storage model utilizing spatial correlation functions. *Kybernetik* **5**:113–119

[22] Anderson, J.A. (1972) A simple neural network generating an interactive memory. *Math. Biosci.* **14**:197–220

[23] Palm, G. (1980) On associative memory. *Biol. Cybern.* **36**:19–31

[24] Nadal, J.-P., Toulouse, G. (1990) Information storage in sparsely coded memory nets. *Network* **1**:61–74

[25] Hopfield, J.J. (1982) Neural networks and physical systems with emergent collective computational abilities. *Proc. Natl. Sci.* **79**:2554–2558

[26] van Hemmen, J.L. (1987) Nonlinear networks near saturation. *Phys. Rev. A: Math. Gen.* **36**:1959–1962

[27] Tsodyks, M.V., Feigelman, M.V. (1988) The enhanced storage capacity in neural networks with low activity level. *Europhys. Lett.* **6**:101–105

[28] Amari, S.I. (1989) Statistical neurodynamics of associative memory. *Neural Networks* **1**:63–73

[29] Fontanari, J.F., Köberle, R. (1988) Information processing in synchronous neural networks. *J. Phys. France* **49**:13–23

[30] Palm, G., Sommer, F. T. (1992) Information capacity in recurrent McCulloch-Pitts networks with sparsely coded memory states. *Network* **3**:1–10

[31] Gibson, W.G., Robinson, J. (1992) Statistical analysis of the dynamics of a sparse associative memory. *Neural Networks* **5**:645–662

[32] Hebb, D.O. (1949) *The Organization of Behavior* (Wiley, New York)

[33] Herz, A., Sulzer, B., Kühn, R., van Hemmen, J.L. (1988) The Hebb rule: Storing static and dynamic objects in an associative neural network. *Europhys. Lett.* **7**:663–669; (1989) Hebbian learning reconsidered: Representation of static and dynamic objects in associative neural nets. *Biol. Cybern.* **60**:457–467

[34] Personnaz, L., Dreyfus, G., Toulouse, G. (1986) A biologically constrained learning mechanism in networks of formal neurons. *J. Stat. Phys.* **43**:411–422

[35] Personnaz, L., Guyon, I., Dreyfus, G. (1986) Collective computational properties of neural networks: New learning mechanisms. *Phys. Rev. A: Math. Gen.* **34**:4217–4228

[36] Palm, G. (1982) *Neural Assemblies* (Springer-Verlag, Berlin)

[37] Willshaw, D.J., Dayan, P. (1990) Optimal plasticity from matrix memories: What goes up must come down. *Neural Comp.* **2**:85–93

[38] Barto, A.G., Sutton, R.S., Brouwer, P.S. (1981) Associative search network: A reinforcement learning associative memory. *Biol. Cybern.* **40**:201–211

[39] Lamperti, J. (1966) *Probability* (Benjamin, New York)

[40] Shannon, C., Weaver, W. (1949) *The Mathematical Theory of Communication* (University of Illinois Press, Urbana, IL)

[41] Palm, G. (1992) On the information storage capacity of local learning rules. *Neural Comp.* **4**:703–711

[42] Gardner, E. (1987) Maximum storage capacity in neural networks. *Europhys. Lett.* **4**:481–485

[43] Gardner, E. (1988) The space of interactions in neural network models. *J. Phys. A: Math. Gen.* **21**:257–270

[44] Schwenker, F., Sommer, F.T., Palm, G. (1993) Iterative retrieval of sparsely coded patterns in associative memory. *Neuronet'93 Prague*

[45] Sommer, F.T. (1993) Theorie neuronaler Assoziativspeicher; Lokales Lernen und iteratives Retrieval von Information. Ph.D. thesis, Düsseldorf

[46] Palm, G., Schwenker, F., Sommer, F.T. (1993) Associative memory and sparse similarity perserving codes. In: *From Statistics to Neural Networks: Theory and Pattern Recognition Applications*, Cherkassky, V. (Ed.) (Springer NATO ASI Series F) (Springer-Verlag, New York)

[47] Palm, G. (1990) Local learning rules and sparse coding in neural networks. In: *Advanced Neural Computers*, Eckmiller, R. (Ed.) (Elsevier, Amsterdam), pp. 145–150

[48] Horner, H. (1989) Neural networks with low levels of activity: Ising vs. McCulloch-Pitts neurons. *Z. Phys. B* **75**:133–136

[49] Willshaw, D.J., Dayan, P. (1991) Optimizing synaptic learning rules in linear associative memories. *Biol. Cybern.* **50**:253–265

[50] Peretto, P. (1988) On learning rules and memory storage abilities. *J. Phys. France* **49**:711–726

[51] Palm, G. (1991) Memory capacities of local rules for synaptic modification. *Concepts in Neuroscience* **2**:97–128

[52] Horner, H., Bormann, D., Frick, M., Kinzelbach, H., Schmidt, A. (1989) Transients and basins of attraction in neural network models. *Z. Phys. B* **76**:381–398

[53] Buhmann, J., Divko, R., Schulten, K. (1989) Associative memory with high information content. *Phys. Rev. A* **39**:2689–2692

[54] Palm, G. (1987) Computing with neural networks. *Science* **235**:1227–1228

4

Inferences Modeled with Neural Networks

H.-O. Carmesin[1]

with 8 figures

Synopsis. We study changes of synaptic couplings as a consequence of received inputs *and* of an internal mechanism. We adopt three approaches. First, we study the relation between formal logic and networks using the *McCulloch–Pitts mapping* from formulas to networks. We observe that transformations of logical formulas correspond to internal changes in a network, which in turn correspond to deductive inferences. In contrast, inductive inferences correspond to learning in networks and to the "guessing of axioms." Thus, formal logic does not address learning. This deficit is reflected in Wittgenstein's paradox (unique learning of counting by children), which can be "solved in terms of networks." Second, under appropriate conditions, the Hebb rule causes the minimization of complexity (number of couplings) during learning, and this makes the learning of counting unique. The minimization also supports the view that, in psychological experiments, test persons solve transitive and more complicated inferences in a parallel rather than a sequential fashion. Third, a mechanism for internal changes in networks is studied that achieves both proofs by complete induction and an axiom system for any given consistent task.

4.1 Introduction

You want to catch a cat. It runs into a small room. You follow, and when you enter the door, the cat has hidden. You know that there are only two places to hide: behind the chest or on the cupboard. If you approach the wrong place, the cat will escape through the door. You remember that the cat has played this game with you quite often, and it always hid behind the chest. So you infer that the cat is behind the chest. But before you approach the chest, you consider additionally: Most likely, my brother forgot his suitcase behind the chest. Thus, there is insufficient space left for the cat. Hence,

[1]Institut für Theoretische Physik, Universität Bremen, D-28334 Bremen, Germany.

you infer that the cat is on top of the cupboard. After you have caught your cat, you sit in your armchair and wonder how your nervous system, which presumably is organized according to the Hebb rule [1], provided you with such useful inferences. Traditionally, inferences have been studied mainly by logicians [2–5], computer scientists [6], cognitive psychologists [7], and philosophers [8, 9]. Here, we will model inferences with neural networks and work out essential relations to the traditional approaches.

As is illustrated in the above example, the inference is caused by inputs that are taken at different times and in different contexts. From all of the inputs taken, relatively few relevant inputs are selected and coordinated to an appropriate inference. Accordingly, we will propose a framework in which a network takes inputs in a first phase, reorganizes internal states in a second phase, and performs an action in a third phase.

For the sake of a clear understanding of inferences, we concentrate our attention on three efficient approaches, each of which is possible in the proposed framework. First, we use mappings [10] from logical formulas to networks. Second, we model the counting ability [8]. Although this ability may appear trivial, it provides the basis for most infinite procedures[2] and allows the study of learning. Third, we establish a cognitive system that generates to a given task a corresponding axiom system in terms of networks. Thus, we model the formation of axioms from experience. Now that we have characterized these three approaches, we begin our investigation with definitions.

4.1.1 Useful Definitions

By *inference* we mean the combination of inputs by a neural network. In our example, the nervous system combines remembered and actually perceived inputs. The problem with generating such combinations of inputs is the *binding problem in its full generality*, because here the combined inputs are taken at different times and in different contexts. What are these combinations or coordinations of inputs? Combinations occur during the performance of the network. The performance includes changes of neural activity *and* of couplings. Consequently, combinations occur either directly through neural activities, or indirectly through changes of (synaptic) couplings. Such changes are described by differences between *full network states*[3] $N_{full}(t)$, which are characterized by the couplings *and*

[2]The counting ability is the guideline along which *intuitionistic logic* was built [11–13]. To support an orientation in the literature, we note that the functions that exist in *intuitionistic logic* are all *general recursive*. The *general recursive functions* are the same [13] as those studied by Turing (*computable functions*), and Church (λ-*definable functions*).

[3]Geometrically, the full network states are elements (of a subset) of the $N+N^2$-space, which has as subspaces the N-space of the neuronal states and the N^2-space of the synaptic states.

the neurons. In order to study changes of couplings, we call two network states $N(t_a)$ and $N(t_b)$ at times t_a and t_b *synaptically equal* if they have the same couplings. A network is permanently changing its network state, or $N(t) \rightarrow N(t+1)$ for short. A network is in fact a sequence of *network states*, $N(t_i)$, or N_i for short. By a *master mechanism* we mean any rule that determines the changes of couplings. For instance, the Hebb rule is a master mechanism.

4.1.2 Proposed framework

We separate the combinations of inputs into the following three phases.

Learning: First, the network receives inputs and achieves its first network state, N_1. We describe this first network state in terms of synapses, basins of attraction, rules, etc.

Internal change: Second, the network state N_1 may be active without receiving inputs, whereby it *changes internally* to become N_2. For simplicity, in this second phase we allow only such changes that leave *invariant* the output generated to a given input in the third phase, but which possibly will speed up (or slow down) the third phase. That is, N_1 and N_2 combine the same inputs to the same outputs. We call such internal changes *conservative.* Two network states that differ only by a conservative internal change are called *cognitively equivalent.* If N_2 is faster than N_1, then N_2 can *predict* the behavior of N_1. The study of nonconservative internal changes is beyond the scope of this chapter. For instance, internal changes might have been involved in the above example of recalling the suitcase.

Action: Third, the network state N_2 receives other inputs and combines them. The retrieval of a pattern [14] can be such an action; if inputs during the learning phase define the couplings through the Hebb rule, then these training inputs are in effect combined with those inputs that are received during retrieval. For simplicity, we neglect the change of couplings in this phase. In the following, it is clear from the context which phase we are discussing and which network state we are considering.

By *inductive inference* we denote a coordination of the first phase (learning phase), while by *deductive inference* we denote one of the second phase (internal change). The third phase (retrieval) finishes inductive *and* deductive inferences and leaves the full network state synaptically equal. Altogether, we expect this framework to be especially appropriate for the modeling of inferences, because it contains inductive inference in the first phase *and* deductive inference in the second phase. In full generality, the second phase of internal change includes changes of neuronic values. However, it is expected that the changes of couplings are more important, because there are far more couplings than neurons.

$$\begin{array}{ccccc} (t/f,\ldots,t/f) & & & & (\pm,\ldots,\pm) \\ \downarrow & \mathrm{T} & & \tau & \downarrow \\ M\{p\} & \longleftarrow & p & \longrightarrow & N\{p\} \\ \downarrow & & & & \downarrow \\ t/f & & & & \pm \end{array}$$

Fig. 4.1. *Middle*: A formula p is mapped via T and τ. *Left*: A mapping $M\{p\}$ maps a tupel of t/f to one t/f. *Right*: A network $N\{p\}$ maps a tupel of $+/-$ to one $+/-$.

4.1.3 How Far Can We Go with the Formal-Logic Approach?

McCulloch and Pitts [10] studied this question by an ingenously simple and effective mapping:

1. *The calculus of propositions* [2–5] is the (ancient) starting point[4]:

2. *Model*: So far, the calculus contains meaningless sequences. This is changed by the original *"interpretation"* [5]: We define [10] a mapping T, which maps each p to its *Boolean function* $M\{p\}$: That is, each variable q takes one of the values t/f, *"true"* or *"false."* The formula p determines the number d of input variables q. Each $M\{p\}$ maps d such q to one r. This "interpretation" is called a *"model"* (according to [5]), since $p_1 \equiv p_2$ if and only if $M\{p_1\} = M\{p_2\}$, $M\{\neg p\} = t$ if and only if $M\{p\} = f$, and $M\{p_1 \vee p_2\} = t$ if and only if $M\{p_1\} = t$ or $M\{p_2\} = t$; see Fig. 4.1.

3. *The mapping* τ (McCulloch–Pitts mapping) maps each formula p onto a feedforward network (dynamics defined in Sec. 4.2) $N\{p\}$, which performs as $M\{p\}$; see Fig. 4.1, whereby a unique $N\{p\}$ is achieved by some convention.

4. *The mapping* τ^- maps each feedforward network N to a formula p, such that $M\{p\}$ performs as N; cf. Fig. 4.2.

5. *Transformation* T_N: To a given N we form the corresponding p via τ^-.

---

[4]Primitive connections are $\neg$ (negation) and $\vee$ (disjunction); they combine variables or formulas; the formulas are the possible combinations. Popular abbreviations are $p \rightarrow q$ for $\neg p \vee q$ (implication), $p \wedge q$ for $\neg(\neg p \vee \neg q)$ (conjunction), and $p \equiv q$ for $(p \rightarrow q) \wedge (q \rightarrow p)$ (equivalence). The axioms are (1) $p \vee p \rightarrow p$, (2) $p \rightarrow p \vee q$, (3) $p \vee q \rightarrow q \vee p$, and (4) $(p \rightarrow q) \rightarrow (r \vee p \rightarrow r \vee q)$, where p, q, and r can be variables, or formulas. A formula r is called an *immediate consequence* of p and q if p is the formula $q \rightarrow r$. The class of derivable formulas is defined to be the class of formulas that contains the axioms and all immediate consequences of derivable formulas.

$$\begin{array}{ccc} & \tau^{-} & \\ p & \longleftarrow & N \\ \downarrow \text{logic} & & \downarrow T_N(p,q) \\ q & \longrightarrow & N\{q\} \\ & \tau & \end{array}$$

Fig. 4.2. *Upper part:* N is mapped to p. *Middle left part:* p is transformed to the equivalent q. *Lower part:* q is mapped to $N\{q\}$. *Middle right part:* Altogether, N is mapped to $N\{q\}$.

We transform p to an equivalent q through the application of axioms and the immediate consequence. We map q to the respective $N\{q\}$ via τ (Fig. 4.2). We observe that T_N is a candidate for a conservative internal change.

6. *Consistency problem:* If the axiom system (see footnote 3) of the calculus of propositions were inconsistent, then $\neg p \equiv p$ would be derivable. Then, the corresponding induced transformation T_N would transform a network N_1 into a network N_2 that maps to the output $+$ if N_1 maps to the output $-$. We conclude that, through τ the consistency problem is mapped to networks (i.e., the induced transformations T_N are conservative if and only if the axiom system is consistent).

7. *Logical operations* $\vee$ and $\neg$: The logical operations by which formulas are connected are $p \vee q$ and $\neg p$; corresponding operations are possible for networks.

8. *Networks as models*: To each network we define the *class of equally deciding networks*, i.e., of networks that map identically. These classes of networks are another model for the calculus.

9. *Discussion*: The above items characterize the relation between the calculus of propositions and feedforward neural networks[5]. In particular, the axioms of the logical calculus describe "generally valid" relations. Specific knowledge is expressed in additional axioms. For instance, the knowledge about classical mechanics is contained in Newton's three axioms. However, the process of establishing the axioms (i.e., the above first phase of learning) is not addressed. Newton had to "learn" his axioms, possibly by observing the famous apple falling

[5]By a feedback network we mean a network that contains at least one loop of couplings. Analogous items 1–8 establish a similar relation between the calculus of predicates with natural numbers as individuals and feedback neural networks [15].

from the tree. The formulas provided by logic address the second phase of internal changes. The third phase of action is established through an interpretation of the formulas. Finally, with regard to an application of the above considerations to neural network models, we identify two problems that occur in the second phase of internal change. The first one is to make internal changes conservative, because otherwise they are not reliable, and the second is to search for such sequences of applications of axioms and immediate consequences that speed up the network.

Facts About the Two Problems

The calculus of propositions is consistent [2–5]; thus, we can generate conservative internal changes in feedforward networks through τ. In neurobiology, recursive networks occur as well. In order to generate conservative internal changes in them, we have two alternatives: Either we limit the allowed transformations of formulas [11–13, 16] and, as a consequence, obtain conserved internal changes only, but at the same time the number of internal changes is limited; or we have to make a hypothesis [4] (e.g., *transfinite induction* [17]) (for a detailed analysis of such questions see [18]) from which we can conclude that the induced internal changes are conservative.

For instance, two pupils, Mary and Bob, have learned how to calculate with variables. In the afternoon, they both derive new formulas. The next day they compare their results. Most of the formulas Mary derived do not occur in Bob's derivations, and some have been derived by Bob, too. But for one formula F derived by Mary, Bob derived the negation $\neg F$. Both are puzzled and confirm that they made no mistakes in their derivations. Is this possible? (There are four possibilities: Mary and Bob made an error, only Bob made an error, only Mary made an error, or neither Mary nor Bob made an error. In the latter case, the transformations of formulas are not consistent.) This example also illustrates the goal of deductive inference, namely, to make predictions about the domain of (if the domain contains one element only, then a single activity is predicted) future activities of nervous systems, here about those of Mary and Bob.

The history: At the beginning of the century, logicians were looking for a consistency proof (Hilbert's program [19]) for a system with natural numbers as individuals (*Peano arithmetic*) and a logical calculus like that of the *Principia Mathematica* [2]. A change was initiated by the logician Gödel [4], who argued that within such a calculus there are propositions U that can neither be proven nor disproven. First, this result gave rise to consistency proofs which rely on additional hypotheses [17] (first of the above problems treated with the second of the above alternatives). Second, this result was used pragmatically by Turing, who proposed quite a general class of computing machines, which now are called Turing machines[6] [6], and showed that, for a given proposition U, there is no general procedure from which a Turing machine could decide whether U is provable (second of the above problems). We address these two problems for the particular case of networks (Sec. 4.6): *(1) How do conservative internal changes emerge in neural networks? (2) Which internal changes are especially effective in networks?*

Limitation of the Formal-Logic Approach

Formal logic does not address learning, although learning precedes internal change. This limitation becomes especially apparent when logic generates statements about infinite sequences. How can finite, "mechanically generated" formulas predict anything about possibly infinite processes, like counting or forming sequences of primes? Consequently, it is not satisfactory to neglect the study of learning or of the link between learning and internal change. This link was studied by *intuitionistic logicians* who organized consistency proofs along the idea that counting already has been learned [11]. Wittgenstein [8, 9, 20] went one step further toward basic mechanisms and asked: How can counting be learned? To solve the above problems, we focus our whole study on counting[7] and in particular on Wittgenstein's question. If we explain in some terms how to count, we have to explain these terms through other terms, etc., and we would end up with an infinite regress. Accordingly, we consider pupils who learn counting *from examples*, e.g., 1, 2, 3, ..., 121. A pupil who can count up to 121 (i.e., who adapted this) can usually continue to 122, ... How is this possible? Wittgenstein was not able to answer this question, because the answer requires knowledge about the nervous system [20]. We will give an explanation in terms of a self-organization process that begins with the Hebb rule [1]. So, the used key knowledge is the Hebb rule.

Hebb's Rule

Hebb's neurophysiological postulate says that a synaptic efficiency increases, if the pre- and postsynaptic neurons fire simultaneously, and that this increase is due to some metabolic process. Recently, a roughly similar metabolic process has been observed [21].

---

[6]A Turing machine consists of a head and a tape. The head contains statements that establish its performance. The tape is a linear sequence of sections, called fields. In each field there is a symbol out of a finite set of symbols. At each instant of time, the head is at a field. It reads the respective symbol and maps it to the pair (symbol to be written to the field, move to be performed). The move is either to the left, to the right, no move, or the end of processing.

[7]Together with calculating, counting covers all three phases of combinations, is a possible basis for analysis and geometry with all transformations, and can be studied efficiently.

Synapses from Correlations

By its nature, the Hebb rule transforms correlations among neural activities into synaptic efficiencies. This motivated Hebb to *speculate* that cell assemblies emerge as a consequence of the Hebb rule. The Hopfield rule is highly related to the Hebb rule [14] and transforms (input) patterns into synaptic efficiencies. Legendy [22] explained observed correlations in spike patterns by "unspecified synapse forming mechanisms," which occur according to postulated principles that form synapses from correlations.

Synapses from Successful Correlations

Legendy was fully aware that synapses from correlations are too simple; in his third section, his 14th remark is: "Presumably template formation is, in certain systems, biologically censored when correlations are 'too perfect,' for, the alternative would be the unchecked formation and boundless proliferation of useless templates. One may *speculate* that the notorious difficulties in eliciting plasticity in physiological experiments and the relative scarcity of successes might come from such a censorship mechanism." Thorndike [23] formulated such a censorship mechanism before neural networks were invented: "When a modifiable connection between a situation and a response is made and is accompanied or followed by a satisfying state of affairs, that connection's strength is increased."

The presented mechanism that solves Wittgenstein's paradox is the *Hebb rule with success*, i.e., with some censorship mechanism (see below). As a further result, cell assemblies of *few* synapses emerge. Accordingly, we idealize the postulate: The couplings will be chosen such that a given task is performed *and* the number of couplings (complexity) is minimized [20, 24]. Then, we show that counting is learned with that postulate. We study properties and further consequences of this postulate: *How can inductive inference be performed most effectively? Is the experimental evidence in favor of parallel rather than sequential processing?* Altogether, the mechanism presented here shows under which conditions Hebb's and Legendy's speculations are confirmed.

4.2 Model for Cognitive Systems and for Experiences

4.2.1 Cognitive Systems

All cognitive systems considered here consist of networks, master mechanisms, and peripheral processors. The latter provide a perfect transfer of signals[8] and symbols to and from the cognitive system and are not discussed in detail, while the master mechanism is a rule (see below) for the change of couplings. The neurons s_i of the network take values $s_i = \pm 1$ at discrete time steps. Their dynamics is determined by the neuronic equations [25] $s_i(t+1) = \text{sgn}(\Sigma_j J_{ij} s_j(t) - \lambda_i)$, where sgn is the signum function, λ_i is a threshold parameter, and the J_{ij} are the couplings.

4.2.2 EXPERIENCE

For the case of inductive inference, data or experience are given. Thus, in addition to the model of the cognitive system, we need a model of these experiences. Here, *experiences are modeled in terms of elementary tasks and tasks* as follows.

We use a trainer,[9] like in studies on the committee machine [26]. The trainer generates questions q_i with uniquely determined answers $a_i = M(q_i)$. Both q_i and a_i are sequences of symbols, each of which is taken from a finite set of symbols. The pair (q_i, a_i) is called an *elementary task*. By a *task* we mean a set of elementary tasks. For a *consistent task* we additionally require that to each question q_i there be only one answer a_i. The mapping M can be evaluated by a *finite Turing machine*, i.e., a Turing machine with a finite tape that stores up to α symbols and a finite number of statements in its program. Each statement consists of a finite number of *elementary operations*. Thereby, elementary operations are either *elementary motions*[10] or reading or writing a symbol from or to the actual field of the tape or *elementary mappings*. An elementary mapping is a mapping from a finite set of elements to another finite set of elements; e.g., the logical OR and NOT can be elementary mappings, and the combinations thereof are sufficient to determine any function from configurations of two-valued variables to other configurations of two-valued variables [3]. The set Q of possible questions and the set A of possible answers are the sets of sequences of up to α symbols. So, a mapping that is evaluated by a finite Turing machine is such a mapping M.

The trainer begins a *dialogue* by asking q_1, the cognitive system replies $\tilde{a}_1$, and the trainer answers with V_1 = *yes* if $\tilde{a}_1 = a_1$, otherwise with V_1 = *no but* a_1. The dialogue continues analogously. The triple (q_i, a_i, V_i) is called the ith *training situation*. The cognitive system is *adapted to the dialogue* consisting of i elementary tasks if the cognitive system generates only correct answers $\tilde{a}_j = a_j$ for $j \leq i$. The mapping M is called *induced by the trainer to the network* if, for any q_j, the answer of the cognitive

[8]Most generally, anything that can be transformed to symbols by peripheral processors is included.

[9]We also include the case without a trainer but with experiences in an environment.

[10]Elementary motions are single moves to the right or to the left.

system is correct. The number of nonzero couplings of the network is called the *complexity* $c(N)$. The principle of minimization of complexity is the following postulate.

Postulate: After the ith training, the master mechanism determines the couplings such that the dialogue consisting of i elementary tasks of a consistent task is adapted to the network N *and* $c(N)$ is minimized.

4.2.3 From the Hebb Rule to the Postulate?

1. Basic Considerations

We now study the conditions under which networks of minimal complexity emerge from the Hebb rule. For this purpose we formulate and then analyze an appropriate class of network models [27–30]. A network has S sensor, I inner, and M motor neurons. We define for each elementary task μ

$$\tau^\mu = \begin{cases} 1, & \text{if the network was successful at } \mu; \\ 0, & \text{otherwise.} \end{cases} \tag{4.1}$$

The Hebb rule shall be applied with a learning rate a, a decay rate b, and under the condition of success. So the change of a coupling is

$$\Delta J_{ij}(t) = \tau^\mu \Big(a s_i(t) s_j(t) - b J_{ij}(t) \Big). \tag{4.2}$$

The s_i assume values +1 (firing) and −1 (not firing). For each elementary task, the configuration of sensor neurons is given by the question q_μ. The network generates an answer $\tilde{a}_\mu$ at its motor neurons. By $\{s_i^\mu\}$ we denote a neuronal configuration so that the values of the sensor neurons are given by q_μ. The inner neurons and motor neurons take their values according to a corresponding Boltzmann distribution P^μ. For the change of the couplings only configurations with $\tau^\mu = 1$ are relevant, so that

$$H(\{s_i\}) = -\frac{1}{2} \sum_{ij, i \neq j}^{N} J_{ij} s_i s_j, \qquad P^\mu(\{s_i^\mu\}) = \frac{\tau^\mu \exp(-\beta H)}{\sum_{\{s_i^\mu\}}^{2^{I+M}} \tau^\mu \exp(-\beta H)}. \tag{4.3}$$

The sum over $\{s_i^\mu\}$ is the sum over all 2^{I+M} states of the inner and motor neurons. The network is permanently stimulated by its environment. This is taken into account through an adiabatic approximation as follows. To compute ΔJ_{ij} that occurs after performing all 2^S elementary tasks, we sum over all configurations of the 2^I inner neurons and the 2^M motor neurons taken with their probability,

$$\Delta J_{ij} = \sum_{\mu}^{2^S} \sum_{\{s_q^\mu\}}^{2^{I+M}} P^\mu(\{s_q^\mu\})(a s_i^\mu s_j^\mu - b J_{ij}) = a \sum_{\mu}^{2^S} \sum_{\{s_q^\mu\}}^{2^{I+M}} s_i^\mu s_j^\mu P^\mu(\{s_q^\mu\}) - b 2^S J_{ij}. \tag{4.4}$$

Accordingly, each coupling matrix J_{ij} can be written

$$J_{ij} = \sum_{\mu}^{2^S} \sum_{\{s_q^\mu\}}^{2^{I+M}} \lambda_{ij}^\mu(\{s_q^\mu\}) s_i^\mu s_j^\mu. \tag{4.5}$$

We call the above $\lambda_{ij}^\mu(\{s_q^\mu\})$ *amplitudes* and insert them into Eq. (4.4) so as to obtain

$$\Delta J_{ij} = \sum_{\mu}^{2^S} \sum_{\{s_q^\mu\}}^{2^{I+M}} \left(aP^\mu(\{s_q^\mu\}) - b2^S \lambda_{ij}^\mu(\{s_q^\mu\})\right) s_i^\mu s_j^\mu. \tag{4.6}$$

New stimuli steadily come in through the sensory neurons and, since the set of input patterns is finite (2^S), the network cannot continue learning indefinitely. We therefore look for *stationary* coupling matrices, i.e., $\Delta J_{ij} = 0$. To this end, it suffices that each term in the sum (4.6) vanishes so that

$$\lambda_{ij}^\mu(\{s_q^\mu\}) = \lambda_0 P^\mu(\{s_q^\mu\}) \qquad \text{with} \quad \lambda_0 = \frac{a}{b2^S}. \tag{4.7}$$

This is a fixed-point equation for the amplitudes. As a result, the amplitudes do not differ for different ij, i.e., $\lambda_{ij}^\mu(\{s_q^\mu\}) = \lambda^\mu(\{s_q^\mu\})$.

Fixed-Point theorem: *All solutions of the fixed-point equation are stationary networks [Eq. (4.7)].*

So the fixed-point equations are sufficient for J to be stationary.

Generating Function

We insert Eqs. (4.3) and (4.7) into Eq. (4.5) to get the equivalent fixed-point equation for couplings $0 = J_{ij} - \lambda_0 \sum_\mu^{2^S} (\partial F^\mu / \partial J_{ij})$ with $F^\mu = T \ln(\sum_{\{s_q^\mu\}}^{2^{I+M}} \tau^\mu \exp(-\beta H))$. We express it with a generating function W:

$$0 = \frac{\partial W}{\partial J_{ij}} \qquad \text{with} \quad W = \frac{1}{2} \sum_{kl} J_{kl}^2 - \lambda_0 \sum_{\mu}^{2^S} F^\mu. \tag{4.8}$$

A linear stability analysis shows that each local minimum, maximum, and saddle point is a stable fixed point [31]. In order to obtain networks with minimal complexity, we modify the neural dynamics so that the motor neurons have no noise (zero temperature), which gives the new value of a motor neuron as $s_i = \text{sgn}(\sum_j J_{ij} s_j)$.

Illustrative Example

In order to study the emergence of a small network with inner neurons and minimal complexity, we model one sensor neuron s_1, one motor neuron s_2, and two inner neurons s_3 (necessary) and s_4 (redundant). We consider the

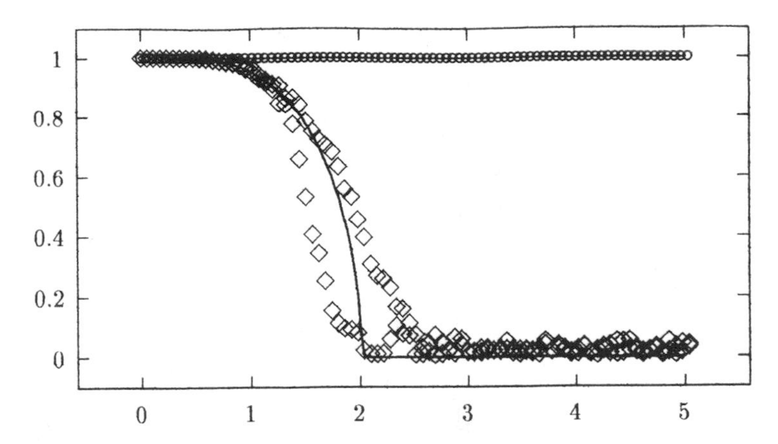

Fig. 4.3. *Network emerging after training the negation task.* x-axis: temperature, y-axis: coupling times b/a, solid lines: solutions of fixed-point equation (4.7). The data have been obtained by computer simulation. The upshot is that above T=2 only necessary couplings (∘; upper branch) are present, near T=2 hysteresis occurs, and below T=2 redundant couplings (⋄; lower branch) appear.

negation task $s_2 = -s_1$. We require the condition $J_{12} = J_{21} = 0$ so that an inner neuron becomes necessary. As a result, above a critical temperature 2, the couplings with the necessary neuron s_3 are 1 while the others are 0. That is, there occurs a spontaneous breaking of the symmetry so that one inner neuron is taken to form a *network of minimal complexity.* Below $T = 2$, the couplings with the unnecessary neuron are nonzero; cf. Fig. 4.3.

In biological terms, the condition $J_{12} = J_{21} = 0$ means that there happen to be no synapses J_{12} and J_{21}, the weight of which could be modified by the Hebb mechanism. Consequently, the task is performed via inner neurons. The used neurons become coupled with large weights; this emerging structure may be regarded as a cell assembly.

2. Analysis of Symmetry Breaking

In the above example the solutions J of the fixed-point equation exhibit a spontaneous breaking of symmetry. As a consequence, there occurs a network of minimal complexity. To *understand* symmetry breaking for three learning procedures (2a–c below), we study fluctuations. For detailed arguments, see [27]. So we consider the fixed-point equation at $\beta = 0$ [see Eq. (4.8)],

$$J_{ij} = \lambda_0 \sum_{\mu}^{2^S} \frac{\sum_{\{s_q^\mu\}}^{2^{I+M}} \tau^\mu s_i^\mu s_j^\mu}{\sum_{\{s_q^\mu\}}^{2^{I+M}} \tau^\mu}. \tag{4.9}$$

(2a) By chance, one of the couplings J_{23} and J_{24} is larger, say it is J_{23}. Then s_4 does not influence s_2, that is, τ^μ does not depend on s_4, i.e.,

s_4 is not necessary for success. Consequently, the couplings with s_4 vanish [see Eq. (4.9)]. This is not so for s_3. So small networks emerge, because *neurons that are necessary for success become coupled.*

(2b) If the correct answer is fixed at s_2 (supervised learning), then no neuron is necessary for success; thus, no neuron becomes coupled at $\beta = 0$.

(2c) If s_2 fluctuates, then τ^μ depends on s_2 only. Then, at $\beta = 0$, success is achieved only randomly; so, no inner neuron becomes necessary for success; thus, no inner neuron becomes coupled.

4.3 Inductive Inference

Under what conditions does inductive inference occur? What is necessary, sufficient, and optimal for inductive inference?

Lemma: For a given mapping M, a network N_M of finite complexity $c(N_M)$ exists that maps each q_i correctly to $a_i = M(q_i)$.

Two proofs will be outlined. The first is a direct construction, the second is an application of [10] and is stated only briefly.

First Proof: By definition, M can be generated by a finite Turing machine. The proof is performed by constructing a finite network that simulates a given finite Turing machine. Without restriction of generality, we assume that, at each field of the finite tape of the Turing machine, either a -1 or a 1 is stored. Each such field can be simulated in the network by a neuron that is coupled to itself by a positive coupling, has zero threshold, and thus stores the value once given to it. There is a network N_c of finite complexity c that counts up to the number of fields of the tape [20,30]. N_c can simulate the actual position of the head of the Turing machine. It also can be modified such that it can count forward or backward selectively [32]. Thus, the elementary motions can be simulated by N_c. To each neuron that simulates a field one can associate a neuron that takes the value 1 if and only if the respective number is represented by N_c. A simple network N_{rw} can be constructed that reads and writes if desired and if the respective associated neuron takes the value 1. Hence, reading and writing can be simulated by N_{rw}. Finally, any elementary mapping can be simulated by a network of finite complexity c since the logical OR and NOT, and combinations thereof, can be simulated by a network. Altogether, the Turing machine can be simulated by the network constructed above.

Idea of Second Proof: Since the Turing machine is finite, its tape is finite; hence, the set of questions Q and answers A is finite and accordingly the number of mappings M is finite. Furthermore, such mappings are realizable in a finite network according to [10].

The first proof is applicable more generally to Turing machines with unlimited tape and networks with unlimited external memory (see Sec. 4.4). Both proofs are applicable to dialogues in which some symbols are hidden.

Straightforward consequences of the lemma demonstrate under which conditions a mapping M is established by a network. Among all networks that map each q_i correctly, there are one or more networks N_0 of smallest complexity $c(N_0)$. By construction, any network generated by the master mechanism has a c smaller than or equal to $c(N_0)$. The number m_0 of dynamically nonequivalent[11] networks of c smaller than or equal to $c(N_0)$ is finite [33]. Thus, the number m_e of errors ($\tilde{a}_j \neq a_j$) that the network can make is $m_e \leq m_0$. After a finite time t_0, the network makes no more errors. Let us call a question q_i to which the network answers incorrectly *instructive* (in a given dialogue). If at time $t_0 + 1$ the mapping M has not yet been induced to the network, then the trainer failed[12] to ask at least one additional instructive question. We define: A trainer who does not fail to ask an instructive question is called *instructive.* By a *rule* we mean a set of $\mid Q \mid$ different questions, each with its answer. We call a rule *reducible,*[13] if $\mid Q \mid > m_0$. As an immediate consequence, we obtain Theorem 1.

Theorem 1:[14] An *instructive trainer* induces a given mapping M to the network in a *finite* dialogue. To a *consistent task* the network incorporates a rule that depends on the task. To each *reducible* rule there is an instructive trainer that provides a dialogue consisting of less than $\mid Q \mid$ elementary tasks.

4.3.1 Optimal Inductive Inference

We now turn to the comparison of alternative master mechanisms μ and networks ν. Now a "generalized" cognitive system consists of peripheral processors, a master mechanism μ, and a network ν that is made up of interconnected elements (e.g., neurons, couplings, wheels, tubes, pipes) and performs according to a dynamics d_ν. The elements belong to K types $E_k, k = 1, ..., K$, the number of elements of type k is n_k (elementary comlexity). The master mechanism provides a coordination of these elements. For each such coordination the cognitive system establishes a mapping from

[11]Dynamically equivalent networks generate the same dynamics.

[12]Even if the trainer was instructive and the cognitive system identified the mapping, it could not be aware of it; thus, an ambiguity remains.

[13]Most rules of practical interest are reducible because they have relatively low complexity.

[14]This theorem holds for recursive networks, feedforward networks, attractor networks, and essentially also for networks made of wheels, tubes, pipes, etc.; see below.

each input (and possibly from initial values) to a corresponding output. A *generalized complexity* is any linear combination $c_g = \Sigma_k a_k n_k$ with positive coefficients a_k. We require that the generalized complexity be bounded, $c_g \leq \hat{c}$.

Because $c_g \leq \hat{c}$, only a finite number of elements is contained in the network. Consequently, only a finite set M_{max} of mappings M can be incorporated by the network. The cardinality of M_{max} is called the *creative capacity* κ_c of the network, because the answers need to be created by the network. By *inductive capacity* κ_i we denote the number of mappings that can be incorporated by a given cognitive system. During the training, the master mechanism provides realizations of mappings $M_j \in M_{max}$. The master mechanism that realizes adaptation of the dialogue and minimization of complexity with the generalized complexity c_g is called μ_1. We call μ_1 *optimal* because $\kappa_i = \kappa_c$ for μ_1. In general, $\kappa_i \leq \kappa_c$[15] (it would be interesting to observe the ratio κ_i/κ_c for various animals). There are other master mechanisms that are optimal as well,[16] e.g., master mechanisms that adapt to any dialogue are optimal.

4.3.2 Unique Inductive Inference

A master mechanism μ provides *unique inductive inference* if it identifies each reducible rule through an appropriate dialogue consisting of less than $\mid Q \mid$ elementary tasks. The minimization of μ_1 is important for the *uniqueness* of inductive inference. According to Theorem 1, μ_1 provides *unique inductive inference.* In contrast, a master mechanism μ' that adapts to any dialogue and gives the first answer of the dialogue in a novel elementary task does not identify each reducible rule through a dialogue with $\mid Q \mid -1$ questions.

4.3.3 Practicability of the Postulate

Typically, the minimization of complexity [35] requires much computing time if a general or random set of elementary tasks is considered [36]. For the special case of a feedforward network, the time required for minimizing the number of neurons of the network grows faster than polynomially with the number of the hidden units, i.e., it is NP-complete. However, this is of little relevance for many important and nonrandom tasks. For example, the minimization of complexity in networks has been successfully applied to the modeling of transitive inference in pigeons [23], learning orthography

[15]For instance, the Hebb rule is a master mechanism which does not provide adaptation if the network contains neurons that do not take inputs. As a consequence, $\kappa_i < \kappa_c$ for the Hebb rule and such a network. (In the human brain, most neurons do not take inputs.)

[16]For a game simulating inductive inference, see [34].

[37], electrostatics [38], geometry [39], counting [20], and calculating [32]. Furthermore, inductive inference works essentially in the same manner if the minimization of complexity either is used in a statistical procedure with finite computing time or emerges from a statistical network model [27–30]. Finally, in certain applications, decoupling into modules is possible [32].

4.3.4 Biological Example

A pigeon in a Skinner box[17] had to choose between two stimuli; this is the elementary task [40, 24]. The stimuli were A, B, C, D, E. To each pair we designate the answer q_i; and the correct answer is rewarded. In the training phase, the dialogue consisted of four elementary tasks (arrow to rewarded stimulus): (A ← B), (B ← C), (C ← D), (D ← E). After the pigeons learned to respond correctly, (B ← D) was given as a novel, fifth elementary task, but without reward. 87.5% of the answers were correct, i.e., the pigeons inferred transitively.

The network model shows that transitive inference is of minimal complexity. However, a Turing machine likewise requires minimal complexity, i.e., program length, for transitive inference. In order to decide whether the pigeon's performance was sequential or parallel, we suggest considering the following dialogue:

$$(A \leftarrow B), (B \leftarrow C), (C \leftarrow D), (D \leftarrow E), (E \leftarrow A),$$
$$(A \leftarrow C), (B \leftarrow D), (C \leftarrow E), (D \leftarrow A), (E \leftarrow B).$$

Altogether, essentially 12 dialogues exist in this framework. Among these, the suggested dialogue is relatively complex for a network, but not for a Turing machine. Meanwhile, experiments with humans have been performed with this dialogue. The suggested task was relatively difficult for humans and pigeons [41]. This supports the assumption that humans dealt with this situation in a parallel fashion, i.e., that they performed "network-like." The point is that the "system of that task" is obvious to the reader, because here the elementary tasks are ordered systematically. However, the test persons received the same elementary tasks in terms of a computer game without useful order, could not reorganize, and hence performed "network-like."

4.3.5 Limitation of Inductive Inference in Terms of Complexity

Complexity measures are likewise used for inductive inference in frameworks (e.g., parameters for fits to data, coding data, approximate representaion of data in relatively low dimension) without networks; see, e.g., [42,

[17] A Skinner box is an experimental device, in which the response of an animal to a stimulus is studied.

43]. In particular, if inductive inference is addressed, then the formation of scientific theories is addressed as well [38]. First, we ask: Is a network with the minimizing master mechanism μ_1 a reasonable tool for the formation of scientific theories from "isolated phenomena"? We consider the following examples: pattern formation in clouds, the crystalline structure of a diamond, and a cobweb. Although these examples exhibit significant geometric structures (which would be detected through μ_1), they are explained differently. The structure in clouds is explained as a result of the mechanical motion of many molecules, the crystalline structure is explained as a result of quantum mechanical interactions, while the coweb is explained by its purpose — a tool for catching insects. Hence, the answer is no. Second, we ask: Is a network with the minimizing master mechanism μ_1 a reasonable tool for the formation of scientific theories from "sufficiently large sets of isolated phenomena"? Because there exist so many phenomena, we cannot even study, let alone answer, this question.

4.3.6 Summary for Inductive Inference

An a priori principle is necessary for inductive inference and is provided by the minimizing master mechanism μ_1. The postulate is an optimal a priori principle. Among all complexities, only c_u is asymptotically relevant and is, therefore, considered in the following, i.e., the complexity is the number of couplings. The master mechanism minimizes c under certain conditions, which we treat as modifications of the model developed so far. Consequently, the results can be interpreted as *solutions of a minimization problem with additional conditions.*

This minimization is specified as follows. If the cognitive system needs a certain amount of complexity, it generates that complexity only for the time it is needed, and it deletes the respective couplings as soon as possible. This final deletion of synapses is in agreement with the above self-organization mechanism, in which couplings to unnecessary neurons are destabilized.

4.4 External Memory

How does a cognitive system with external memory perform its tasks, and what is its relation to a Turing machine? To answer these questions, we consider two modifications of the theory developed so far.

First modification. The cognitive system shall have access to external memory,[18] the elementary units of which are called *locations*. The periph-

[18]In a biological cognitive system, external memory might be realized by neurons or assemblies of neurons. In particular, the *formatio reticularis* performs primarily operational tasks, while other parts of the brain perform primarily memorizing tasks.

eral processor guarantees reading from and writing to locations.[19] Because the complexity c is minimized, the cognitive system stores questions and the corresponding answers on locations without using the network. The dialogue is adapted to the cognitive system and the complexity vanishes, i.e., $c = 0$. In the case of counting, such a cognitive system will be unable to generate new numbers and will perform worse than a cognitive system without locations [20]. If the available locations are unlimited, no inductive inference is performed by the cognitive system.

Second modification. From now on the locations are limited appropriately. (For the sake of simplicity, we will assume that the cognitive system applies locations only after it has incorporated M.) The cognitive system contains several networks N_i. Let us define an *instruction* to be a set of symbols on locations that is readable by a peripheral processor and activates a specific performance of a peripheral processor. More precisely, the instruction specifies under which condition a certain symbol is written on a certain position and at which position the next instruction is to be read. (The condition is obeyed if certain symbols are at certain positions.) From now on it also is assumed that the peripheral processor can read and perform such instructions.[20]

The above modification leads to several interesting consequences. First, by means of inductive inference, a rule in a given set of training situations will be incorporated into a network. In the following, we denote by N_1 the network that incorporates the rule. Second, the cognitive system becomes[21] a Turing machine.[22] The application of the incorporated rule can be performed by a finite set of discrete operations on a finite set of symbols on locations, because the rule has already been incorporated into a finite network. These operations can be handled by the peripheral processor without any network if appropriate instructions are written on locations. Hence, N_1 is unnecessary if the cognitive system writes appropriate instructions on locations. Because the above possibility to reduce c to 0 exists, the master mechanism realizes that possibility, i.e., writes the instructions, and sets c to 0. In that final state, the cognitive system can be understood as a Turing machine, and, for that purpose, all locations have to be interpreted in a linear order by some convention. It remains to specify how the cognitive system generates appropriate instructions.

These instructions need not be guessed; rather, they can be extracted

---

[19]These skills can be learned in the sense of Sec. 4.3.

[20]This can be trained as specified in Sec. 4.3.

[21]This result generally can be applied to automatic programming. Its realization is straightforward, because only minimization procedures need to be implemented.

[22]Also in the first case of unlimited locations the cognitive system can be interpreted as a (very trivial) Turing machine that handles the storage of questions and answers on its tape.

from N_1. For this purpose, the cognitive system specifies one location for each neuron of N_1 and records all values of these (two-valued) neurons while processing the incorporated rule. Then, another network N_2 is "trained" as follows. After every action (reading or writing) of the peripheral processor, network N_2 is asked: "What is the next action of the peripheral processor, and by which instruction is it expressed?" Thereby N_2 can use as inputs only signals that are inputs to the peripheral processor. These signals are transferred by appropriate couplings that are generated by the master mechanism. Due to the first part of the question, the network N_2 will incorporate a rule that allows the prediction of the action of the peripheral processor as a function of input signals to the peripheral processor. Due to the second part of the question, N_2 generates the required instructions.

4.4.1 COUNTING

We specify an elementary counting task as follows. Map a natural number given in its binary representation to its successor. In the final state of $c = 0$ (see first modification), the peripheral processor has to perform an algorithm that finds the successor to a given natural number. In the following, one such algorithm is given. (1) Write the given number on a first line. (2) Write a 1 below with corresponding digits one below the other. (3) Leave a third line free underneath.[23] (4) Start with the rightmost digit and, for each digit, do the following. If there is no 1 in the first three lines, write a 0 on the fourth line at the position of the corresponding digit. If there is one 1 in the first three lines, write a 1 on the fourth line at the corresponding digit. If there are two 1's in the first three lines, write a 0 on the fourth line at the respective digit and a 1 on the third line, one digit to the left.

4.5 Limited Use of External Memory

Is it possible to systematically divide a given task into subtasks? What is the essential subtask of counting? What is its complexity? How can the cognitive system learn from a finite set of elementary tasks an infinite set of elementary tasks, namely, to count numbers, i.e., to generate numbers successively without restriction by a largest number. Once again, we first treat a modification and then indicate its consequences.

The idea is to make certain texts on locations "taboo," namely, the instructions, and thereby to force the cognitive system to incorporate the mapping corresponding to a subtask: For every question, only *empty* locations are given to the cognitive system. These are the only available

[23]The number to be written on this line can be interpreted as carry.

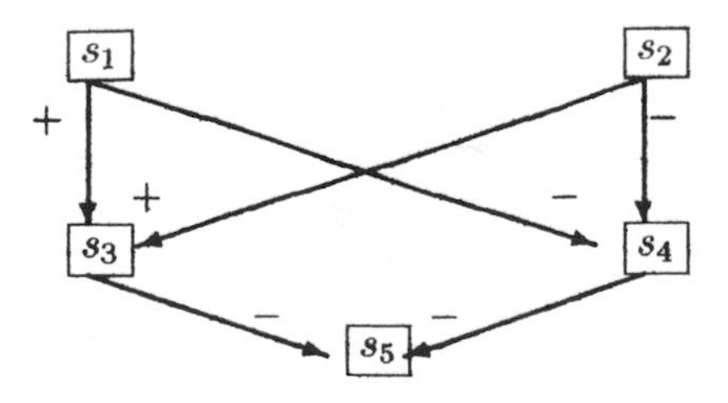

Fig. 4.4. XOR: An arrow denotes $J_{ij} = \pm 1$. In this figure, all thresholds are 1. $s_5 = \mathrm{XOR}(s_1, s_2)$ and $nec = 6$. For counting, $nec = 9$.

locations to find an answer; thus, no instructions are available. Then, a question is written on locations. Finally, the cognitive system is asked to answer. The consequence is that the network will incorporate a rule; see Theorem 1. Thereby, it will incorporate neither what the rule acts on, namely, on questions, nor any (including intermediate) results that the cognitive system generates, because these are on locations. A more general modification is the following: One can construct analogous procedures of presenting locations to the cognitive system with "auxiliary texts" and questions in order to incorporate any desired aspect in the network while keeping all other aspects on the locations. The main point of the above procedure is that a network can be driven selectively. Thus, *complexities of tasks or subtasks can be investigated selectively.*

4.5.1 COUNTING

With this modification, we are prepared to study the incorporation of a rule for counting by a network. For the neuronic equations, we denote the complexity by nec and prove the following.

Lemma: In order to map the pair (s_1, s_2) according to $\mathrm{XOR}(s_1, s_2)$, six couplings are necessary and sufficient. Here, XOR denotes the "exclusive or" operation, and feedforward networks are considered.

Idea of the Proof (as presented in [32]). The pair (s_1, s_2) can take four configurations from which (-1,-1) and (1,1) must be separated. With the sums $(J_{a1}s_1 + J_{a2}s_2 + \lambda_a)$ in the sign, one neuron can separate only one configuration; in Fig. 4.4, s_4 separates (-1,-1) and so does s_3 with (1,1). Both neurons must be connected to s_5; hence, $nec = 6$.

With this lemma, we prove the desired (Sec. 4.1) result about counting.

Proposition: For a network with limited use of locations, and for the task of mapping a given number onto its successor, $nec = 8$. For the task of counting, $nec = 9$.

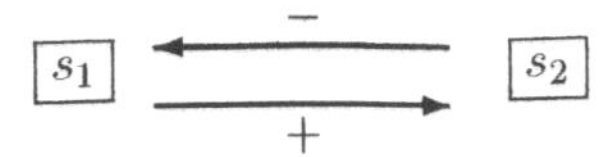

Fig. 4.5. Cycle of length 4: All thresholds are 0 in this case, and the neurons take the values (-1,-1), (-1,1), (1,-1), (1,1) cyclically. This network is necessary for the control of counting and contributes $nec = 2$.

Idea of the Proof ([32]). In order to control the data to and from the head (of the Turing machine), four time steps are necessary: (1 & 2) read & map, (3) write, (4) move. These are provided by the network in Fig. 4.5. A network performing additions according to the algorithm discussed in Sec. 4.4 is of minimal complexity. Thereby, for two digits a and b the new digit is XOR(a, b) and the carry is AND(a, b). Thus, six couplings are necessary for XOR, none for AND because AND(s_1, s_2) is already realized by s_3 (Fig. 4.4), and 2 for control, i.e., 8 for adding a 1 and another one for repeating this process for counting.

4.5.2 On Wittgenstein's Paradox

As was shown in Sec. 4.3, with the aid of the postulate of minimization of complexity, counting can be learned from a finite set of elementary tasks. The above proposition shows that the required complexity is only 9. We conclude that the identification of the uniquely determined correct way of counting ad infinitum *practically* can be performed by a network with the assumed master mechanism, i.e., with the postulate of minimization of complexity. This postulate *emerges* from the Hebb rule under appropriate conditions. The result is relevant for Wittgenstein's paradox [8,9]. The essence of this paradox is that pupils practically learn to count from elementary tasks, although the extension from the given elementary tasks ad infinitum is not uniquely determined. Our result illustrates how a finite series, which by itself is not uniquely extendable, *is extended* uniquely and adequately ad infinitum. Thus, if one assumes that children act according to a master mechanism like that of the principle of minimization of complexity, which can be provided via the Hebb rule, they learn to count.

Furthermore, such a master mechanism cannot be learned without already using a similar master mechanism. The application of a master mechanism of the proposed kind appears to be a part of the nature of children. In this manner, the paradox is solved by naturalization, as is modeled through our cognitive system. More precisely, we have explained how an assumed property of natural nervous systems solves the paradox. The study shows, in agreement with Wittgenstein, that the ability to count cannot be transferred to a cognitive system. But it is *constructed* by the cognitive system according to elementary tasks and to the master mechanism.

Furthermore, Wittgenstein's paradox can be interpreted as an example for the limitations of definability. It is well known that for any mathematical theory *undefined terms* must be included. For instance, in the case of Euclid's geometry, the undefined terms are [44] *point, line, extremities of a line* (i.e., points), *straight line, surface, extremities of a surface* (i.e., lines), *and plane surface.* We already gave a well-defined procedure for enabling the cognitive system to learn undefined terms in a unique manner in Sec. 4.3. Uniqueness requires an instructive trainer who exists according to the postulate, particularly due to minimization. Finally, this solution of Wittgenstein's paradox supports the central idea of intuitionistic logics [11] that humans can count ad infinitum.

4.6 Deductive Inference

How does deductive inference emerge in a cognitive system? In order to study this question, we first formulate our main framework. A question q_i that can be answered in principle, but not in the required time by the application of a rule, is called a *problem about the rule.* Such a problem requires a prediction about a certain future activity of a network. If the network performs straightforwardly, then this activity takes place only after the moment at which the answer is required.

4.6.1 Biological Example

The monkey Sultan is in a cage. At the ceiling is the obligatory banana, too high to reach without a tool. There is a box in the cage; Sultan puts it under the banana, climbs on the box, and gets the desired fruit. Rana, another monkey, watches the scene and is exposed to the same situation afterwards. Rana pulls the box to *some* place in the cage, climbs on it and fails to get the banana. Sultan tries again and gets the banana; Rana watches, tries again, and fails [45]. We interpret this finding as follows: Sultan makes the "ansatz" to increase his height by putting something under his feet. Rana observes that Sultan pulls a box, climbs on the box, and gets the banana; Rana reproduces what she observed. The different actions exhibited by Sultan and Rana are possibly due to different histories of internal changes.

4.6.2 Mathematical Examples

The examples given below are assumed as tasks for a cognitive system. Later, we will discuss in detail how the cognitive system treats them. Our *first* example can be handled by a cognitive system with either inductive or deductive inference. Consider the question of whether, after a year of

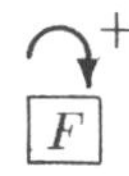

Fig. 4.6. In the network for hypotheses N_H (see text), the relation $F(l-1) = F(l)$ is incorporated by the above network. This is the essence of the proof by complete induction. After irrelevant signals are eliminated from the merged network N_m, the future events are predicted by one neuron only, and they are all the same.

counting, a counting network would still generate a sequence of numbers such that the respective sequence of last digits reads ...0 1 0 1 0 1 .... This and the following question are to be answered within a time shorter than a year, say a day. The *second* example is the question of whether there is a largest prime number.

The above cognitive system that performs according to the postulate will try to answer randomly, and it eventually will correct itself if the answer is wrong. Apparently, this is not the best strategy because the question is to be answered within a day, say and there is a chance to generate a more appropriate answer during that time. An additional master mechanism for problems (MMP) will be introduced that is able to "make the most out of its time."

In the following, problems of a certain format (covering a relatively large set of problems) will be considered. We are given a mapping $M : q_l \to a_l$ for $l \in L$ that is incorporated in N_1 by inductive inference. Let F for any $l \in L$ be another mapping from pairs (q_l, a_l) to ± 1 that can be performed by a finite Turing machine in a finite amount of time. The (question of the) problem is whether $F(q_l, a_l) = 1$ for all $l \in L$. If this is the case, we will say that the elementary tasks of (q_l, a_l) have the *property* P_F. In the first period (learning), F is induced to a network by inductive inference; i.e., the problem is induced to the cognitive system. In the following, we denote this network by N_F. In the second period (internal change), an MMP elaborates an answer. Two MMPs are considered: The *inductive MMP* picks out a finite (limited by the available time) subset of L and checks whether $F = 1$ (for any l in this subset). The *deductive MMP* merges N_1 with N_F such that the answer $\tilde{a}_l$ generated by N_1 is inserted for a_l into N_F. (For simplicity, we do not distinguish whether q_l is taken from the trainer or generated by N_1 as well.) Then, this MMP will (try to) minimize the complexity of the merged network called N_m. This minimization yields one of the following cases. If $c(N_m) = 1$, then either F does not depend on t or F alternates with t; see Fig. 4.6. These two cases are discriminated by the MMP through explicit consideration of the possible values of the neuron(s) at the ends of the coupling. In the first case and $F = 1$ for $l = 0$, the pairs (q_l, a_l) have the property P_F. If one l with $F = -1$ is found, if F alternates, or if $c(N_m) \neq 1$, then the property P_F is not detected.

Fig. 4.7. Relevant signal flow: In our *first* mathematical example, the equality of the last digits of numbers that are generated subsequently by N_1 is checked (two lines merge, i.e., two signals flow to the same vertex v); one such last digit d_1 determines the subsequent d_2.

4.6.3 Relevant Signal Flow

The act of proving a prediction in this framework is performed by the act of minimizing complexity. The act of minimization of complexity plays a key role in the above framework of the deductive MMP and should be investigated systematically. In its full generality, this is beyond the scope of the present investigation, but a straightforward "ansatz" will be outlined; we denote the corresponding MMP as the *standard MMP* in the following.

The value of a mapping F depends on the values of neurons. These, in turn, depend on the signals coming from other neurons, and so forth. This suggests that we consider a *signal diagram* (Fig. 4.7) that contains all flows of signals relevant for the prediction. The events that can occur in the merged network N_m can be described by signals that propagate through the diagram. This signal-flow ansatz allows us to consider the propagation of signals in the signal diagram *as a mapping* M_H. The signals at time t are the questions, and the relevant signals (i.e., all signals that are not yet identified as irrelevant) at time $t + 1$ are the answers.

The mapping M_H can be incorporated into a third network N_H by inductive inference.[24] It is sufficient for N_H to evaluate those signals that are relevant for the evaluation of F, i.e., those that ultimately enter the evaluation of F. This implies that, *if there are identifiable rules in the flow of relevant signals, these will be identified by* N_H. The rules identified by N_H are called *hypotheses.* So far, the procedure is inductive and cannot exclude nonconservative internal changes. However, if we now have rules for the flow of signals incorporated by N_H that can be verified by considering a *finite set of configurations* of the finite set of involved neurons, such rules can be verified explicitly by considering this finite set of configurations.

---

[24]This type of inductive inference is a straightforward generalization of that elaborated in Sec. 4.3. Here, several answers are acceptable, viz. all rules about relevant and also possibly irrelevant signals. The trainer does not provide the correct answer but just yes or no, i.e., whether or not the hypothesized rule is empirically correct (for a few tests). Note that it is straightforward for the deductive MMP to check whether N_H failed to predict a relevant signal.

Let us specify that the standard MMP does consider such finite sets of configurations explicitly.

The standard MMP yields a network N_H of relatively low complexity c that provides exactly the same signals (which are relevant for the question under consideration) as N_m. This reduction of complexity is achieved by inductive inference. The question then is: How do relevant signals flow? This inductive inference is verified or falsified deductively. Thus, deductive inference emerges here. The standard MMP provides an *analysis of incorporated causal relations*, because the signals flow deterministically and already have been incorporated.

4.6.4 Mathematical Examples Revisited

In the *first* example with inductive inference, N_H will find a simple rule for the signal flow in N_m. One such rule is that the value of F (as defined above) at time t is a function of the values of the last digits d_1 and d_2 of two numbers, subsequently generated by N_1 at times $t - t_1 - t_2$ and $t - t_1$, respectively (Fig. 4.7). This mapping F checks whether these digits are unequal. Furthermore, by inductive inference, N_H realizes that the value of such a last digit d_2 at time $t - t_1$ depends on the value of the digit d_1 at time $t - t_1 - t_2$. Hence, F only depends on that digit d_1 at a certain time. These hypotheses formed by N_H can be explicitly checked by the deductive MMP by considering a finite set of signal configiurations. This is the case because N_m is finite. Finally, two cases remain to be considered explicitely. The above last digit d_1 (at time $t - t_1 - t_2$) is 1 or $-$ 1. In both cases, F takes the value 1. Thus, the statement is verified by the cognitive system.

4.6.5 Further Ansatz

The above characterized signal-flow ansatz is not appropriate for *all* problems. Consider our second mathematical example, i.e., the question of whether there is a largest prime number. There are finite networks that check whether numbers are prime or not, or which number of a pair of numbers is larger, or whether for a given prime there is a larger one. However, the answer to the question apparently cannot be found by using the signal-flow ansatz. But it can be found if a new ansatz is provided. For example, one may make a slight *increase in complexity* of the cognitive system[25] as follows.

One assumes a largest prime and, thus, a finite set of primes, and con-

[25] So far, we have explained how the cognitive system can perform inferences by using mechanisms like the Hebb rule, the minimization of complexity, or the signal flow analysis. These mechanisms use data or synapses that are already present. In contrast, an ansatz is relatively new, see Rana and Sultan and the conclusion.

siders the number x, which is the product of these primes plus 1. This construction can be made in terms of a network, and with the standard MMP the cognitive system can conclude that x is prime. The point is that the construction, and in particular x, has not been found from an analysis of the signal flow of N_m but must be regarded as a new ansatz in this framework.

The above discussion suggests that the deductive MMP also can be applied if further ansatz somehow are provided [15]. In neurobiological terms, these ansatz have to be provided by an associative memory, i.e., they are distributed among many synapses. Consequently, it is not expected that one can adequately measure the difference between Sultan and Rana (see above) in terms of single neurons.

4.6.6 Proofs by Complete Induction

In order to illustrate that the studied MMP is widely applicable, we note (without proof) the following. Any proof about a sequence of cases $x_l, l \in \mathbb{N}$, that can be performed by complete induction also can be generated by the standard MMP. In particular, corresponding propositions $A(1)$ and $A(l) \rightarrow A(l+1)$ can be derived within the proof by complete induction. Hence, the corresponding N_m can be reduced to the network of Fig. 4.6.

4.6.7 On Sieves

By a *case* we denote a configuration of signals that represents a pair (q_l, a_l), is generated by N_1, and is transferred to N_F. The network N_F that tests a property P_F of a case is metaphorically called a *sieve for a case.* Those cases that are in accordance with the property P_F fall through the sieve, while the others do not. Conversely, a property P_F that can be checked by a finite Turing machine in finite time for one single case generated by N_1 can be checked by N_F. Alternatively, N_F can be interpreted as a *sieve that discriminates networks* N_1, such that those N_1 that generate cases in accordance with the property P_F fall through the sieve, while the others do not. In this context, we call N_F a sieve.

A series of properties $P_{F_k}, k \in K$, is a unique characterization of certain mathematical objects, and the series is identified as the format of any axiom system for these objects.[26] The corresponding series of sieves N_{F_k} play a key role[27] in the incorporation of axiom systems into networks.

[26]More generally, these objects need not be mathematical objects [3]. For instance, an object that is yellow, lengthy, curved, and tastes like a banana *is* a banana.

[27]A practical advantage of a series of sieves is that in cognitive systems *transformations and compositions of sieves* are possible and essential for deductive reasoning. The detailed investigation thereof is beyond the scope of this chapter

How does a cognitive system generate sieves? Sieves are generated by a variant (the signal flow of N_1 instead of N_m is analyzed) of the standard MMP as follows. Rules about the signal flow in N_1 are hypothetically formed by N_H and verified afterwards. We generalize the property P_F as follows. F is a function from a case to ± 1. If N_H works under the constraint that the identified rules have the format of properties [i.e., these rules tell us whether a case x of the signal flow is generated by N_1 ($F(x) = 1$) or not ($F(x) = -1$)], then we identify N_H as a sieve.

Axiom Systems in Terms of Sieves

Although the ultimate origin of the contents of the cognitive system is seen in the elementary tasks, it is possible to begin with the axioms, as is the case for the axiomatic method [48].

What is an axiom system? A sufficient characterization in terms of properties from which no part can be eliminated, such that the remaining is still sufficient, is called a *minimal sufficient characterization*, or an *axiom system.* Consequently, a series of sieves is a candidate for an axiom system because it is a characterization in terms of properties.

What is a useful axiom system? We call a network N_1 and a series of sieves $N_{F_k}, k \in K$, *self-consistent* if the cognitive system can obtain the series of sieves $N_{F_k}, k \in K$, from N_1, and vice versa. Altogether, we are prepared to show that our cognitive system is able to establish to any consistent task an axiom system in terms of a series of sieves that is self-consistent with the corresponding N_1.

Theorem 2: *From any consistent task, an axiom system can be generated by a cognitive system in a self-consistent manner using the postulate, locations, and the standard MMP.*

Proof: Consider any consistent task. There is a network N_1 that incorporates a corresponding rule; see Theorem 1. We construct a series of sieves as follows.

As a first series, we take the trivial set of sieves consisting of one sieve N_{F_1}, which generates the answer to a given question by (a copy of) N_1. If (and only if) the entering answer equals the answer generated by N_{F_1}, then the entering answer falls through N_{F_1}.

As a second series, we add to the first sieve N_{F_1} additional sieves $N_{F_k}, k > 1$, which we introduced in Sec. 4.6.7.

In order to obtain a third series of sieves $\bar{N}_{F_j}, j \in J$, we eliminate from the second series irrelevant signals via the method introduced in Sec. 4.6.3.

---

but is given in [15, 46]. The description of such transformations in [10] is regarded as inconclusive [47] for feedback networks. Furthermore, sieves can be applied to more than countably many objects which then are represented in a symbolical manner (like π). This is the basis for novel results about logic [18].

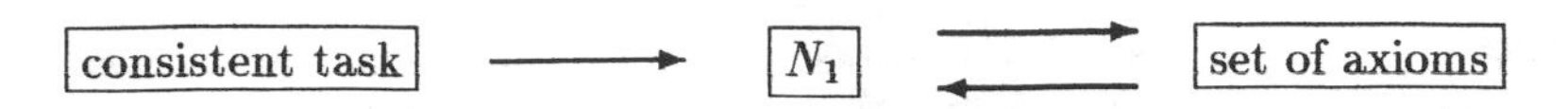

Fig. 4.8. Axioms obtained by the cognitive system. *Left:* N_1 incorporates a rule from a consistent task via inductive inference. *Upper arrow:* The standard MMP generates an axiom system in terms of a series of sieves. Conversely, from a series of sieves the same N_1 is reproduced. Thus, N_1 and the series of sieves are self-consistent.

Finally, we show that N_1 and the third series of sieves $\bar{N}_{F_j}, j \in J$ are self-consistent, i.e., we show that we can reproduce N_1 from $\bar{N}_{F_j}, j \in J$ (Fig. 4.6). For that purpose, we take each question q_l of the dialogue and generate the answer a_l as follows: We generate answers at random, use $\bar{N}_{F_j}, j \in J$, as a sieve, and take the answer a_l that falls through. This answer a_l is the same as that of the dialogue because N_1 makes this answer; hence, N_{F_1} generates this a_l, this a_l falls through the second series of sieves, and then a_l also falls through the third series of sieves.

4.7 Conclusion

Learning. The master mechanism that minimizes the complexity *establishes* an optimal a priori principle, *guarantees* that a trainer can be instructive to the network, *provides* unique inductive inference, and *emerges* from the Hebb rule. The master mechanism can be understood as follows. It *emerges* from the Hebb rule. Its *adaptation* part guarantees that inductive inference takes place. Its *minimization* part guarantees uniqueness. These three properties together solve Wittgenstein's paradox by naturalization.

Internal change. The goal of conservative internal change is to make *predictions* about future activities of (other) networks. Conservative internal changes are *mapped* to formal logic as follows. Networks are interpreted as a model for a logical calculus. Internal changes model transformations of logical formulas. The consistency problem of logic is mapped to the problem to provide conservative internal changes.

Links between learning and internal change. A formal logical calculus does not address learning. In contrast, for Brouwer [11] the (learned) counting ability of humans was the central idea of intuitionistic logic. We study this link with a *learning* network that establishes a *formal* axiom system to any consistent task.

Second link. As a tool for conservative internal change, a master mechanism is investigated that uses inductive *and* deductive inference. The master mechanism extracts possibly relevant signal flows in a hypothetical manner by the former and verifies or rejects these hypotheses by the latter. This combination is advantageous: First, inductive inference is blind for nonconservative internal changes but not for rules inherent to signal flows, while verification is blind for such rules but detects nonconservative internal changes. Second, the conservative internal changes cover infinite objects (Fig. 4.6); this "infinite predictability" is ultimately based on the word "all" in the sentence: "The neuronic dynamics is valid for all neurons at any time."

Third link. New ansatz are efficient tools for internal change in the modeled cognitive systems. Typically, new ansatz are provided by the cultural heritage. The modeled cognitive system is able to learn *and* to change these, i.e., to reject, reconstruct, analyze, reorganize, use, modify, combine, or improve these ansatz.

Acknowledgments. I thank Leo van Hemmen, Gerhard Roth, and Helmut Schwegler for improvements of my manuscript and stimulating discussions.

REFERENCES

[1] D.O. Hebb (1949) *The Organization of Behaviour* (Wiley, New York), p. 62

[2] A.N. Whitehead, B. Russell (1925) *Principia Mathematica* (Cambridge University Press, Cambridge)

[3] D. Hilbert, W. Ackermann (1938) *Grundzüge der theoretischen Logik* (Springer-Verlag, Berlin)

[4] K. Gödel (1931) *Monatshefte für Mathematik und Physik* **38**:173–198

[5] H.D. Ebbinghaus, J. Flum, W. Thomas (1984) *Mathematical Logic* (Springer-Verlag, Berlin)

[6] A. Turing (1937) *Proc. London Math. Soc.* **42**:230–265

[7] J. Piaget (1975) *Gesammelte Werke* (Klett, Stuttgart)

[8] L. Wittgenstein (1964) In: *Remarks on the Foundations of Mathematics*, G.H. von Wright, R. Rhees, G.E.M. Anscombe (Eds.) (Basil Blackwell, Oxford)

[9] W. Stegmüller (1986) *Kripkes Deutung der Spätphilosophie Wittgensteins* (Kröner, Stuttgart)

[10] W.S. Mc Culloch, W.H. Pitts (1943) *Bull. Math. Biophys.* **5**:115–133

[11] L.E.J. Brouwer (1919) *Jahresb. d. Dt. Mathematiker — Vereinigung* **28**:203–208

[12] S.C. Kleene (1952) *Introduction to Metamathematics* (North-Holland, Amsterdam)

[13] S. C. Kleene, R. E. Vesley (1965) *The Foundation of Intuitionistic Mathematics* (North-Holland, Amsterdam) pp. 1–18

[14] J.L. van Hemmen, R. Kühn (1991) In: *Models of Neural Networks*, E. Domany, J.L. van Hemmen, K. Schulten (Eds.) (Springer-Verlag, Berlin) pp. 1–105

[15] H.-O. Carmesin (1993) unpublished

[16] K. Gödel (1930) *Monatshefte für Mathematik und Physik* **37**:349–360

[17] G. Gentzen (1936) *Math. Ann.* **112**:493–565

[18] H.-O. Carmesin (1993) *Consistent Calculus* (Köster, Berlin)

[19] D. Hilbert (1900) *Nachr. v. d. Königl. Ges. d. Wiss. zu Göttingen*, 253–297

[20] H.-O. Carmesin (1992) *Science & Education* **1**:205–215

[21] E.R. Kandel, J.H. Schwarz, T.M. Jessell (1991) *Principles of Neural Science* (Elsevier, New York)

[22] C.R. Legendy (1975) *Intern. J. Neurosci.* **6**:237–254

[23] E.L. Thorndike (1913) *Educational Psychology, Vol. II* (Columbia University, New York)

[24] H.-O. Carmesin, H. Schwegler (1992) In: *Rhythmogenesis in Neurons and Networks*, N. Elsner, D.W. Richter (Eds.) (Thieme, Stuttgart), p. 702

[25] E.R. Caianiello (1961) *J. Theor. Biol.* **2**:204–235

[26] M. Opper (1992) *Phys. Bl.* **48**:569–574

[27] H.-O. Carmesin (1993) *Physics Essays* **8**:38–51

[28] H.-O. Carmesin (1993) In: *Gen — Gehirn — Verhalten*, N. Elsner, D.W. Richter (Eds.) (Thieme, Stuttgart), p. 104

[29] M. Kreyscher, H.-O. Carmesin, H. Schwegler (1993) In: *Gen — Gehirn - Verhalten*, N. Elsner, D.W. Richter (Eds.) (Thieme, Stuttgart), p. 105

[30] H.-O. Carmesin (1993) *Fachberichte Physik* **51**

[31] J. Stoer, R. Bulirsch 1980) *Introduction to Numerical Analysis* (Springer-Verlag, New York)

[32] A. Thoms (1994) *Modellierung Zählender neuronaler Netzwerke* (Diplomarbeit, Universität Bremen)

[33] E.R. Caianiello (1986) In: *Brain Theory*, G. Palm, A. Aertsen (Eds.) (Springer-Verlag, Berlin), pp.147–160

[34] M. Gardner (1977) *Scientific American* **237**:18–25

[35] Solomonoff (1964) *Inform. and Control* **7**:1–22, 224–254; M. Thomas, P. Gacs, R.M. Gray (1989) *Ann. of Prob.* **17**:840–865

[36] J.H. Lin, J.S. Vitter (1991) *Machine Learning* **6**:211–230

[37] H.-O. Carmesin, E. Brinkmann, H. Brügelmann (1994) In: *Am Rande der Schrift, DGLS—Jahrbuch "Lesen und Schreiben 6"*, H. Brügelmann, H. Balhorn, I. Füssenich (Eds.) (Libelle, Bottighofen), pp. 132–141

[38] H.-O. Carmesin, H. Fischer (1992) In: *Didaktik der Physik*, DPG (Ed.) (Physik-Verlag, Weinheim), pp. 197–208

[39] H.-O. Carmesin (1993) In: *Bedeutungen erfinden, im Kopf, mit Schrift und miteinander*, H. Brügelmann, H. Balhorn (Eds.) (Ekkehard Faude, Konstanz), pp. 66–70

[40] L. von Fersen, J.D. Delius (1992) *Spektrum der Wissenschaft* **7/92**:18–22

[41] M. Siemann, J.D. Delius (1994) *Biol. Cyb.* **71**:531–536

[42] R. Sorkin (1983) *Int. J. Theor. Phys.* **22**:1091–1104

[43] M. Koppel, H. Atlan (1991) *Inform. Sci.* **56**:23–37

[44] J.N. Cederberg (1989) *A Course in Modern Geometry* (Springer-Verlag, New York), pp. 26, 201

[45] W. Köhler (1971) *Die Aufgabe der Gestaltpsychologie* (de Gruyter, Berlin)

[46] H.-O. Carmesin (1994) *Theorie neuronaler Adaption* (Köster, Berlin)

[47] G. Palm (1986) In: *Brain Theory*, G. Palm, A. Aertsen (Eds.) (Springer-Verlag, Berlin), pp. 229–230

[48] R. Carnap (1954) *Einführung in die symbolische Logik* (Springer-Verlag, Wien)

5

Statistical Mechanics of Generalization

Manfred Opper and Wolfgang Kinzel[1]

with 18 figures

Synopsis. We estimate a neural network's ability to generalize from examples using ideas from statistical mechanics. We discuss the connection between this approach and other powerful concepts from mathematical statistics, computer science, and information theory that are useful in explaining the performance of such machines. For the simplest network, the perceptron, we introduce a variety of learning problems that can be treated exactly by the replica method of statistical physics.

5.1 Introduction

Neural networks learn from examples. This statement is obviously true for the brain, but artificial networks also adapt their "synaptic" weights to a set of examples. After the learning phase, the system has adopted some ability to generalize; it can make predictions on inputs which it has not seen before; it has learned a rule.

To what extent is it possible to understand learning from examples by mathematical models and their solutions? It is this question that we emphasize in this chapter. We introduce simple models and discuss their properties combining methods from statistical mechanics, computer science, and mathematics.

The simplest model for a neural network is the *perceptron*. It maps an N–dimensional input vector $\boldsymbol{\xi}$ to a binary variable $\sigma \in \{+1, -1\}$, and the function is given by an N–dimensional weight vector $\mathbf{w}$:

$$\sigma = \text{ sign } (\mathbf{w} \cdot \boldsymbol{\xi}) \,. \tag{5.1}$$

Motivated by real neurons, the components of $\mathbf{w}$ may be called *synaptic* weights; i.e., $w(i)$ is a measure of the strength of the influence of the neuron signal $\xi(i)$ to the output neuron σ.

[1]Physikalisches Institut, Universität Würzburg, D-97074 Würzburg, Germany.

For a given $\mathbf{w}$ this function separates the input space by a hyperplane into two parts, $\mathbf{w} \cdot \boldsymbol{\xi} > 0$ and $\mathbf{w} \cdot \boldsymbol{\xi} < 0$, and the hyperplane is normal to $\mathbf{w}$. But also for a given input $\boldsymbol{\xi}$, the space of weights $\mathbf{w}$ is divided into two parts with different outputs σ. Equation (5.1) gives a very limited class of all possible functions from $\boldsymbol{R}^N$ to ± 1. But this limitation is necessary for a good generalization, as we shall show later.

In the simplest case, the perceptron operates in two ways: in a learning and in a generalization phase. In the learning process, the network receives a set of $P = \alpha N$ many examples, i.e., input/output pairs $(\sigma_k, \boldsymbol{\xi}_k)$, $k = 1, ..., \alpha N$, which were generated by some unknown function $\sigma_k = F(\boldsymbol{\xi}_k)$. The weight vector $\mathbf{w}$ is adapted to these examples by some learning algorithm, i.e., the strengths of the synapses are changed when one or more examples are shown to the perceptron. Of course, the aim of learning is to map each pair correctly by Eq. (5.1), and the number of examples for which the network disagrees with the shown output, $\sigma_k \neq \text{sign } (\mathbf{w} \cdot \boldsymbol{\xi}_k)$, is the training error E:

$$E = \sum_{k=1}^{\alpha N} \theta\left(-\sigma_k \, \mathbf{w} \cdot \boldsymbol{\xi}_k\right). \tag{5.2}$$

θ is the step function, $\theta(x) = (\text{ sign } x + 1)/2$. If the examples are generated by another perceptron with weights $\mathbf{w}_t$, then it is possible to obtain zero training error, $\varepsilon_t = 0$, for instance, by using the perceptron learning rule (see [1]).

After the learning phase, the perceptron has achieved some knowledge about the rule by which the examples were produced. Therefore, the network can make predictions on a new input vector $\boldsymbol{\xi}$ that it has not learned before. Let $(\sigma, \boldsymbol{\xi})$ be a new example that the network has *not* seen before. Then the probability that the perceptron gives the wrong answer, $\sigma \neq \text{sign } (\mathbf{w} \cdot \boldsymbol{\xi})$, is given by

$$\varepsilon = \overline{\theta\left(-\sigma \, \mathbf{w} \cdot \boldsymbol{\xi}\right)}, \tag{5.3}$$

where the bar means an average over all possible examples $(\sigma, \boldsymbol{\xi})$.

The calculation of the generalization error ε as a function of the fraction α of the learned examples is the main subject of this chapter. We call the learning network *student* and the example producing rule the *teacher*. Hence, ε is the probability of disagreement between student and teacher on a new input $\boldsymbol{\xi}$. $\varepsilon(\alpha)$ depends on the structure of student and teacher, on the structure of the examples, and on the learning algorithm.

From very general concepts one obtains bounds and relations between different generalization errors. Using methods of statistical mechanics developed from the theory of disordered solids (spin glasses), one obtains exact results on $\varepsilon(\alpha)$ for infinitely large networks ($N \to \infty$). Section 5.2 introduces general results, while the statistical mechanics approach is presented in Sec. 5.3. Section 5.4 discusses scaling ideas, from which the asymptotic behavior of the generalization error can be understood in some cases. A variety of applications for perceptrons are reviewed in Sec. 5.5.

This chapter is not supposed to review the new field of generalization using neural networks. (For a review we recommend the article by Watkin, Rau, and Biehl [2].) But we want to give an introduction to the field with an emphasis on general results and on applications of our own group at Würzburg. We apologize for not referring to a large number of interesting and important results of our colleagues and friends.

5.2 General Results

The theory of learning in neural networks has benefitted from an interplay of ideas that come from various scientific fields; these include *computer science, mathematical statistics, information theory*, and *statistical physics*. In the following, we try to present some of these ideas. We review a variety of general results that can be obtained *without specifying a network architecture*.

5.2.1 PHASE SPACE OF NEURAL NETWORKS

In this section we adress the problem of *noise-free learning* in networks with binary outputs. We assume that an ideal teacher network, with a vector of parameters $\mathbf{w}_t$, exists, who will give answers ($\oplus$ or $\ominus$) on[2] input vectors $\boldsymbol{\xi}$ without making mistakes.

Let us now look at the *phase space of all teachers* $\mathbf{w}_t$, described by a parameter vector $\mathbf{w}_t$, for fixed inputs $\boldsymbol{\xi}_1, \ldots \boldsymbol{\xi}_P$. *Before* knowing the teacher's correct answers to all of these inputs, a learner could partition the phase space into maximally 2^P cells or subvolumes, each cell σ corresponding to one of the 2^P possible labelings (= answers) $\sigma_k = \pm 1$, $k = 1, \ldots, P$. In general, a given type of neural network will not be able to produce all 2^P outputs on the given inputs. If the teacher network has a very complex architecture, we can assume that, by suitable choices of its parameters, more combinations of outputs, in other words, more cells in phase space, can be realized than for a less complex teacher. As we shall see in the next section, this number of cells plays an important role for the learner's ability to understand the teacher's problem.

After the teacher has given the answers, we know to which cell $\mathbf{w}_t$ belongs. In the so-called *consistent* learning algorithms, one trains a student network to respond *perfectly* to the P training inputs. In the following,

[2]In general, we do not assume that the dimensions of parameter space and input space are equal.

we assume that the student belongs to the *same class of networks as the teacher*. Thus, after learning, the student has parameters $\mathbf{w}_s$, which belong to the teacher's cell.

Will the probability of making a mistake on unknown inputs always become small when P grows large, whatever consistent algorithm we choose?

Surprisingly, the answer is yes, if the teacher has a *bounded complexity*. As a measure for this complexity, the so-called *Vapnik–Chervonenkis (VC)* dimension, which comes from mathematical statistics, has been introduced into computer science. We will try to review some of its basic ideas in the next section.

5.2.2 VC Dimension and Worst-Case Results

The *maximal* number of cells in the teacher's space is 2^P for P input vectors. But, due to the teacher's internal structure, the actual number of cells for a set of inputs may not grow exponentially fast in P. A combinatorial theorem, independently proved by Sauer [3] and Vapnik and Chervonenkis [4], gives an upper bound on this number: If d is the size of the *largest set* of inputs realizing *all* 2^d cells, then, for *any set* of $P > d$ inputs, the number $\mathcal{N}(P,d)$ of cells will only grow like a *polynomial* in P. d is called the *VC dimension.*

Formally, *Sauer's lemma* states:
$P \geq d \geq 1$:

$$\mathcal{N}(P,d) \leq \sum_{i=0}^{d} \binom{P}{i} \leq \left(\frac{eP}{d}\right)^d . \tag{5.4}$$

A sketch of the proof of Eq. (5.4) is given in Appendix 5.1. Equation (5.4) shows, that the VC–dimension plays a similar role as the *capacity* of the class of teacher networks. For $P >> d$, only an exponentially small fraction of input–output pairs can be stored in the net. For the perceptron, one has exactly $d = N$, the number of couplings. For general feedforward networks with N couplings and M threshold nodes, the bound $d \leq 2N \cdot \log_2(eM)$ was found in [5].

Using Sauer's lemma, Blumer, Ehrenfeucht, Haussler, and Warmuth [6] showed a worst-case result for the performance of *consistent* algorithms. To understand their result, consider the following learning scenario: After a student has learned a number of P independent random examples perfectly, he or she makes a prediction on an unknown input vector $\boldsymbol{\xi}$, which was drawn from the same distribution as the training examples. The student's probability of making a mistake on the random input $\boldsymbol{\xi}$ defines the generalization error ε. Different students (algorithms) will have different ε. In general, their performance will depend on the random training set, which makes ε a random variable. So we can define the probability $p(\varepsilon)$, that there exists a student, who learns the examples perfectly but makes an error *larger than* ε. In [6] it was shown that, for $P > 8/\varepsilon$,

$$p(\varepsilon) \leq \mathcal{N}(2P, d) \cdot 2^{-\varepsilon P/2} \leq 2 \left(\frac{2eP}{d} \right)^d 2^{-\varepsilon P/2}. \tag{5.5}$$

Statistical physicists often discuss the *thermodynamic limit* $d, P \to \infty$, $\alpha = P/d$ fixed. In this limit, Eq. (5.5) means that *no errors* larger than

$$\varepsilon_{max} = \frac{2 \ln(2e\alpha)}{\alpha}$$

will occur, whatever consistent student we choose.

Due to lack of space, we cannot sketch the proof of Eq. (5.5) here. A simpler theorem, which relates errors and the number of cells within the Bayesian framework of learning, will be proved in Sec. 5.2.4.

The power of the VC method lies in the fact that no specific assumption on the distribution of inputs, other than their independence, must be made. Further, the architecture of the teacher problem to be learned is characterized by *only a single number*, the VC dimension.

As a drawback of the worst-case results, one often finds that "typically," the error bounds are too pessimistic. In the next two sections, we will discuss a more optimistic learning scenario. We show what is gained if, besides the teacher's complexity, more prior knowledge, expressed by a probability distribution on the teacher's parameters, is available.

5.2.3 Bayesian Approach and Statistical Mechanics

The statistical mechanics approach to learning is closely related to concepts established in mathematical statistics and information theory [7, 8, 9, 10, 11]. To explain these connections, let us first briefly remind the reader of some ideas from *density estimation* in mathematical statistics.

A common problem in statistics is to infer a probability density, $\mathcal{P}_\theta(y)$, from a sample of P data values, $y^P \equiv y_1, \ldots, y_P$, independently drawn from this distribution. Here we assume that the class of distributions is known up to an unknown parameter θ. For example, assume $\mathcal{P}_\theta(y) = (2\pi)^{-1/2} \cdot e^{-(1/2)(y-\theta)^2}$, i.e., a Gaussian density, where θ, its mean, is unknown.

One approach to this problem is to estimate the value of θ first and then to approximate the unknown density by

$$\mathcal{P}_{\hat{\theta}}(y),$$

where $\hat{\theta}$ is the estimate. A well-known method is the *maximum likelihood estimation*, which uses a θ that makes the observed data most likely, i.e., which maximizes the *likelihood*

$$\prod_{k=1}^{P} \mathcal{P}_\theta(y_k). \tag{5.6}$$

For the Gaussian density, this leads to the simple arithmetic mean

$$\hat{\theta} = P^{-1} \sum_{k=1}^{P} y_k.$$

In the so-called *Bayesian approach* to density estimation, all prior knowledge (or lack of the same) of the unknown parameter is expressed by a prior distribution $p(\theta)$. For example, if the (Bayesian) statistician knows that the unknown mean of the Gaussian will not be too large or too small, say θ must be between -1 and $+1$, he or she could assume that θ is uniformly distributed in this interval. Rather than giving a single estimate of θ, as in the maximum likelihood case, the Bayesian calculates the *posterior distribution* $p(\theta|y^P)$, which represents his or her knowledge or uncertainty of the parameter after having observed the data values. This is derived by the *Bayes Formula*, which expresses the joint density $\mathcal{P}(y^P, \theta)$ of the data and the parameter in two ways using conditional densities:

$$\mathcal{P}(y^P, \theta) = p(\theta) \cdot \prod_{k=1}^{P} \mathcal{P}_\theta(y_k)$$
$$\mathcal{P}(y^P, \theta) = p(\theta|y^P) \cdot \mathcal{P}(y^P). \tag{5.7}$$

The posterior density is then

$$p(\theta|y^P) = \frac{p(\theta) \cdot \prod_{k=1}^{P} \mathcal{P}_\theta(y_k)}{\mathcal{P}(y^P)}, \tag{5.8}$$

with the normalization

$$\mathcal{P}(y^P) = \int d\theta \prod_{k=1}^{P} \mathcal{P}_\theta(y_k) \cdot p(\theta). \tag{5.9}$$

Note that, for $p(\theta) = const$, the maximum of Eq. (5.8) is just the maximum likelihood estimate.

Then, if the Bayesian is asked to present an estimate of the unknown density, he or she will return the posterior averaged density

$$\mathcal{P}_{\theta,Bayes}(y) = \int d\theta \; p(\theta|y^P) \; \mathcal{P}_\theta(y), \tag{5.10}$$

which in general will not belong to the class of densities originally considered. Besides the most likely value of θ, this estimate includes neighboring values as well.

It can be shown that, if the parameter is actually distributed according to $p(\theta)$, then Eq. (5.10) gives the best approximation to the true density on average [4].

The justification of a prior probability for θ often has been questioned. Even if it is not satisfied, the posterior density Eq. (5.8) will, under some mild conditions, be highly peaked around the true value of θ for $P \to \infty$. The dependence on the actual shape of $p(\theta)$ will disappear asymptotically.

Let us now translate these ideas into the language of supervised learning. The data observed in a learning experiment are the examples consisting of P input–output pairs $\sigma^P \equiv \{(\sigma_1, \boldsymbol{\xi}_1), \ldots, (\sigma_P, \boldsymbol{\xi}_P)\}$. In general, we assume that there is a possibly stochastic relation between the inputs and the outputs, which can be expressed by a relation of the type

$$\sigma = F(\mathbf{w}_t, \boldsymbol{\xi}, \text{"noise"}). \tag{5.11}$$

$\mathbf{w}_t$ is a parameter vector representing an ideal classifier or teacher. In contrast to the previous section, we include the possibility that the observations may contain errors ("noise").

Using a neural network, which can implement functions of the type F (with "noise"=0!), the task of the learner is to find a student vector $\mathbf{w}_s$ that best explains the observed data. This can be understood as an estimation of the parameter $\mathbf{w}$ for the distribution

$$\mathcal{P}_{\mathbf{w}}(\sigma, \boldsymbol{\xi}) = \mathcal{P}_{\mathbf{w}}(\sigma|\boldsymbol{\xi}) \cdot f(\boldsymbol{\xi}), \tag{5.12}$$

where f is the density of the inputs and $\mathcal{P}_{\mathbf{w}}(\sigma|\boldsymbol{\xi})$ is the probability that, given an input $\boldsymbol{\xi}$, one observes an output σ.

The *statistical physics* approach to learning is closely related to the *Bayesian idea.* Based on the pioneering work of Gardner [12], one may study *ensembles* of neural networks to capture a "typical" behavior of their learning abilities. Such ensembles are defined by a Gibbs distribution,

$$p(\mathbf{w}|\sigma^P) = Z^{-1} \cdot p(\mathbf{w}) \cdot \exp\left(-\beta \sum_{k=1}^{P} E(\mathbf{w}; \sigma_k, \boldsymbol{\xi}_k)\right), \tag{5.13}$$

with partition function

$$Z = \int d\mathbf{w} \cdot p(\mathbf{w}) \cdot \exp\left(-\beta \sum_{k=1}^{P} E(\mathbf{w}; \sigma_k, \boldsymbol{\xi}_k)\right). \tag{5.14}$$

E is the *training energy* of the kth example and β^{-1} is the learning temperature in a stochastic learning algorithm. In $p(\mathbf{w})$, all constraints on the possible couplings are summarized.

Equation (5.13) has an interpretation as the *posterior distribution* [Eq. (5.8)] of coupling parameters if we identify

$$\begin{array}{rc} p(\theta) \to & p(\mathbf{w}) \\ \mathcal{P}_\theta(y^P) \to & \mathcal{P}_{\mathbf{w}}(\sigma^P) \propto \prod_{k=1}^{P} \exp\left(-\beta E(\mathbf{w}; \sigma_k, \boldsymbol{\xi}_k)\right) \\ \mathcal{P}(y^P) \to & \mathcal{P}(\sigma^P) \propto Z. \end{array} \tag{5.15}$$

As an example, let us consider a perceptron. We assume that the ideal classification $\sigma = \text{sign}(N^{-1/2}\mathbf{w}_t \cdot \xi)$ is inverted by *output noise*, i.e., $\sigma = \eta \cdot \text{sign}(N^{-1/2}\mathbf{w}_t \cdot \xi)$, where $\eta = -1$ with a probability $e^{-\beta}/1+e^{-\beta}$, and β^{-1} is the noise temperature. Fixing the inputs, the probability of observing P output labels is

$$\mathcal{P}_{\mathbf{w}}(\sigma^P) = \prod_{k=1}^{P} \left\{ \frac{\Theta(\sigma_k N^{-1/2}\mathbf{w} \cdot \boldsymbol{\xi}_k)}{1+e^{-\beta}} + \frac{e^{-\beta}\Theta(-\sigma_k N^{-1/2}\mathbf{w} \cdot \boldsymbol{\xi}_k)}{1+e^{-\beta}} \right\}$$
$$= (1+e^{-\beta})^{-P} \cdot \exp\left[-\beta \sum_{k=1}^{P} \Theta(-\sigma_k N^{-1/2}\mathbf{w} \cdot \boldsymbol{\xi}_k) \right]. \quad (5.16)$$

A second possibility of misclassification arises when the coupling parameters, or network *weights*, of the teacher are uncertain to some degree, i.e., $\mathbf{w}_t$ is replaced by $\mathbf{w}_t + \mathbf{v}$, where $\mathbf{v}$ is Gaussian with 0 mean and $\mathbf{v} \cdot \mathbf{v} = \beta^{-1}$. Now,

$$\mathcal{P}_{\mathbf{w}}(\sigma^P) = \prod_{k=1}^{P} H(-\beta^{1/2}\sigma_k N^{-1/2}\mathbf{w} \cdot \boldsymbol{\xi}_k), \quad (5.17)$$

where

$$H(x) = \int_x^\infty Dt$$

and

$$Dt = dt \cdot (2\pi)^{-\frac{1}{2}} \cdot \exp(-\tfrac{1}{2}t^2)$$

is the Gaussian measure.

To summarize, we obtain for the training energies

$$E(\mathbf{w}; \sigma, \boldsymbol{\xi}) = \begin{cases} \Theta(-\sigma N^{1/2}\mathbf{w} \cdot \boldsymbol{\xi}) & \text{for output noise} \\ -\beta^{-1} \ln\left(H(-\beta^{1/2}\sigma N^{-1/2}\mathbf{w} \cdot \boldsymbol{\xi})\right) & \text{for weight noise.} \end{cases} \quad (5.18)$$

The case of output noise also can be formulated for general neural networks and leads to a total training energy that is just the number of inputs, for which the noisy outputs σ and the student's answer disagree.

From the Bayesian viewpoint, the posterior distribution could be used to make predictions on new inputs $\boldsymbol{\xi}$ by calculating the output with the largest posterior probability. This is the *Bayes algorithm*, which, for binary outputs [cf. Eq. (5.10)], answers

$$\sigma = \text{sign}\left[\int d\mathbf{w}\, p(\mathbf{w}|\sigma^P) F(\mathbf{w}, \boldsymbol{\xi}, \text{"noise"} = 0) \right]. \quad (5.19)$$

Unfortunately, this represents a superposition of many neural networks, each given by a coupling vector $\mathbf{w}$. In general, this output cannot be realized by a single network of the same architecture, but requires a more complicated machine.

An algorithm that also uses the entire posterior density is the *Gibbs algorithm* [9, 10, 13], which draws a *a single* vector $\mathbf{w}$ *at random* according to the posterior Eq. (5.13). This is precisely what we would call the "typical" neural network in statistical physics.

This should be contrasted with a *maximum likelihood strategy*, which simply chooses a student vector [for $p(\mathbf{w}) = const$] that minimizes the training energy E. In the case of noisy outputs, the student would try to learn perfectly as many examples as it can, even if a fraction of them contains wrong classifications.

In general, perfect knowledge of the prior distribution of teachers will not be available. Nevertheless, the Gibbs distribution [Eq. (5.13)] is a natural device for defining learning algorithms, even if they are not optimally matched to the learning problem. We will discuss some examples in the section on perceptrons.

5.2.4 Information-Theoretic Results

In this section we explore in more detail what would happen if the Bayesian assumption was perfectly realized. That is, we assume that "nature" actually selects teacher problems at random, and that their prior distribution is completely known by the student.[3]

Learning more and more examples, the student's knowledge of the unknown teacher parameters grows. This knowledge gained by learning a new example is expressed in the so-called *information gain*. As was shown by Haussler, Kearns, and Shapire [13], this quantity can be related to the average error made by a student using the Gibbs and Bayes algorithms. Finally, using information theory and the VC approach, inequalities for errors can be derived. We restrict ourselves to the case of noise-free learning. A more general treatment can be found in [14].

We assume for this section that the inputs are fixed, so that the only randomness is in the choice of the teacher, and, for the Gibbs algorithm, in the choice of the student.

Having observed P classified inputs, we know that the teacher is constrained to one of the $\mathcal{N}(P, d)$ nonempty cells. Thus, the posterior density for the teacher's parameters is 0 outside the cell (see Fig. 5.1 for a perceptron) and equals

$$p(\mathbf{w})/V(\sigma^P) \tag{5.20}$$

inside the cell, where

[3]This a natural assumption for physics students, who, in preparing their exams, often use a catalog of the professor's questions from previous exams.

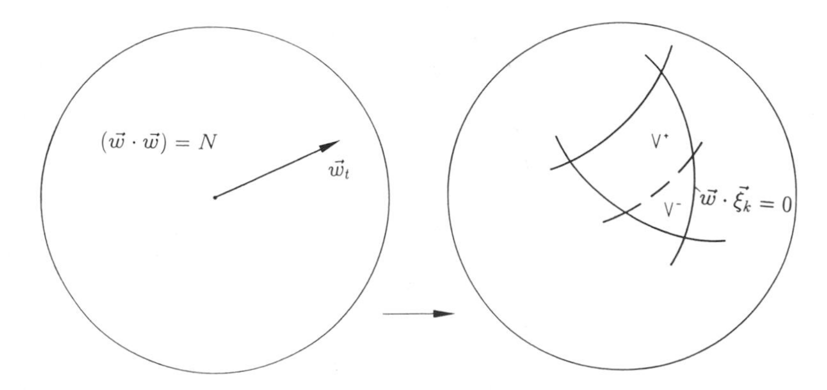

Fig. 5.1. Sketch of the phase space of weights for a perceptron. *Left:* Before learning, the vector $\mathbf{w}_t$ is completely unknown and assumed to be randomly distributed on the surface of an N-dimensional sphere. *Right:* After learning of P input–output examples $\boldsymbol{\xi}_k$, σ_k, the teacher $\mathbf{w}_t$ must be in a smaller cell of the phase space with boundaries given by the planes $\sigma_k \mathbf{w} \cdot \boldsymbol{\xi}_k = 0,\ k = 1, \ldots, P$. A new input (dashed line) divides the cell into two new subcells, V^+ and V^-, corresponding to the two possible answers.

$$V(\sigma^P) \equiv V_P = \int_{cell} p(\mathbf{w})\, d\mathbf{w} \tag{5.21}$$

is its (weighted) volume, satisfying $\sum_{\sigma_1 \ldots \sigma_P = \pm 1} V(\sigma^P) = 1$.

Let us begin with the Gibbs algorithm and fix the teacher for a moment. The learner chooses a vector $\mathbf{w}_s$ at random, with density [Eq. (5.20)]. If a new input is added, the cell is divided into two subcells (Fig. 5.1). If an output cannot be realized, we will formally assume a new cell with zero volume.

Let us compare the student's prediction on the new input with the teacher's answer. Both *agree only* if the student vector $\mathbf{w}_s$ is in the same cell as the teacher's. Averaging over $\mathbf{w}_s$, this will happen with probability

$$Y = \frac{V_{P+1}}{V_P}, \tag{5.22}$$

where V_{P+1} is the volume of the teacher's new cell. The probability of making a mistake thus is given by $1 - Y$.

The Bayesian prediction would weight the answers of the two subcells with their corresponding posterior probabilities and vote for the output

$$\sigma = \mathrm{sign}[V^+ - V^-].$$

Thus, the answer of the largest cell wins. Since the Bayesian gives the answer with largest posterior probability, he or she will, on average, have the lowest number of mistakes over all of the algorithms.[4] The Bayesian will only make a mistake if the teacher is in the smaller volume, i.e., if $Y < \frac{1}{2}$. To this algorithm we can assign the number

$$\Theta(1 - 2Y) \in \{0, 1\}, \tag{5.23}$$

which counts as a "1" when the algorithm makes a mistake.

Finally, by observing a new classified input, our uncertainty on the teacher's couplings will be reduced if the volume of the teacher's cell shrinks.[5] Formally, this corresponds to an *information gain*,

$$\Delta I = -[\ln(V_{P+1}) - \ln(V_P)] = -\ln(Y). \tag{5.24}$$

Obviously, Y, the volume ratio, is a random variable with respect to the random teacher and the inputs. Performing the average over the teacher only, simple and useful relations between the information gain and the probabilities of mistakes may be derived next.

Clearly, Y does not change if the teacher is moved inside a cell. Thus, we can average any function $F(Y)$ over the space of teachers, first by integrating over all teachers *inside* a cell, and then summing over all cells, labeled by their configuration σ^{P+1} of outputs:

$$\langle F(Y) \rangle = \sum_{\sigma_1 \ldots \sigma_{P+1} = \pm 1} V(\sigma^{P+1}) \cdot F(V(\sigma^{P+1})/V(\sigma^P)). \tag{5.25}$$

The factor $V(\sigma^{P+1})$ is the integral over the new cell [Eq. (5.21)]. Thus, outputs, which cannot be realized, are counted with zero weight.

We first will show the useful relation

$$\langle F(Y) \rangle = \langle\, YF(Y) + (1 - Y)F(1 - Y)\, \rangle. \tag{5.26}$$

Beginning with the right-hand side, and using the definition in Eq. (5.25), we fix the first P labels and sum over the σ_{P+1}. Let V^+ and V^- be the two possible subvolumes and $Y^+ = V^+/V(\sigma^P)$, $Y^- = 1 - Y^+$. The summation over σ_{p+1} gives a contribution

$$\begin{aligned} V^+[Y^+F(Y^+) + Y^-F(Y^-)] + V^-[Y^-F(Y^-) + Y^+F(Y^+)] = \\ V(\sigma^P)[Y^-F(Y^-) + Y^+F(Y^+)] = V^-F(Y^-) + V^+F(Y^+) \end{aligned} \tag{5.27}$$

---

[4]We always assume that the teacher actually was drawn from the assumed prior distribution.

[5]An interpretation of $-\ln(V_P)$ in terms of the *stochastic complexity* of Rissanen [15] has been discussed in [16]. Viewing the learning of the examples as an encoding of the outputs in the network's parameters $\mathbf{w}$, this quantity measures how many bits we need to describe the parameters if we use only a finite set of discrete values for the components of $\mathbf{w}$.

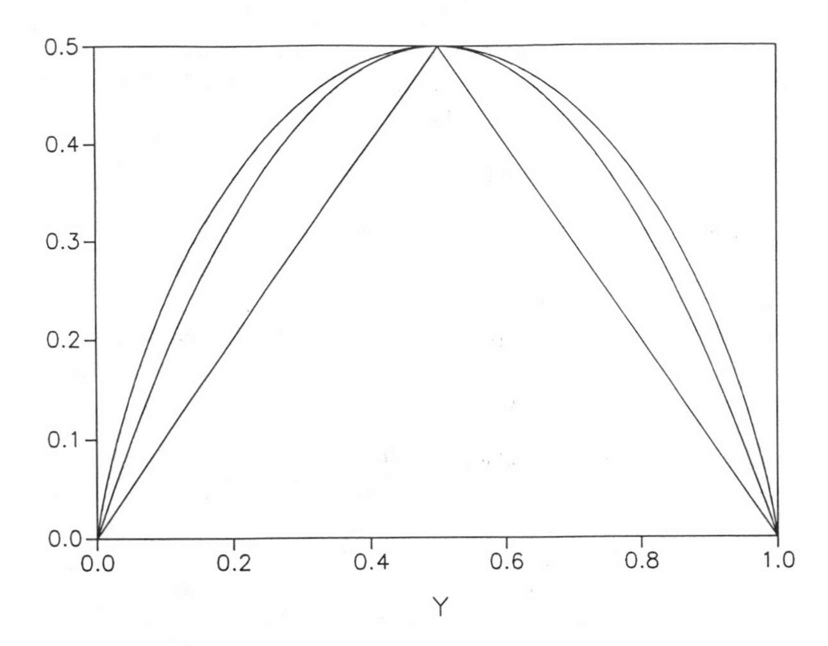

Fig. 5.2. Graphic demonstration of the inequalities $\min(Y, 1-Y)$ (*lower curve*) $\leq 2(Y-Y^2)$ (*middle curve*) $\leq -1/2\ln 2(Y\ln Y+(1-Y)\ln(1-Y))$ (*upper curve*).

to Eq. (5.26). Here we have used $V^+ + V^- = V(\sigma^P)$. Summing over the remaining labels, we obtain Eq. (5.26).

Using relation (5.26), the total probabilities of mistakes, in other words, the errors averaged over all teachers (but for fixed inputs), are given by

$$\begin{aligned} \varepsilon_{Gibbs} &= \langle 1-Y\rangle = 2\langle\, Y - Y^2\,\rangle \\ \varepsilon_{Bayes} &= \langle\Theta(1-2Y)\rangle = \langle\min(Y, 1-Y)\rangle, \end{aligned} \tag{5.28}$$

where the first equality is from Eq. (5.22) and the second from Eq. (5.23). The average information gain is rewritten as

$$\langle\Delta I\rangle = -\langle\ln(Y)\rangle = -\langle\, Y\ln Y + (1-Y)\ln(1-Y)\,\rangle. \tag{5.29}$$

By comparing the three curves in Fig. 5.2, we conclude that

$$\begin{aligned} \varepsilon_{Gibbs} &\leq 2\varepsilon_{Bayes} \\ \varepsilon_{Gibbs} &\leq \frac{1}{2\ln(2)}\langle\Delta I\rangle. \end{aligned} \tag{5.30}$$

Although the random Gibbs algorithm is not optimal, its error is of the same order of magnitude as that of the optimal Bayes algorithm.

The second inequality (5.30) indicates that, in order to gain a lot of information on the teacher, a student should select inputs on which his or her performance is bad. This can be utilized in the so-called query algorithms (see Sec. 5.5.4).

Using the VC method and Eq. (5.30), an estimate of the decrease of the generalization error for the Gibbs algorithm can be obtained [13].

Summing the second inequality (5.30) from $P = 0$ to $P = M - 1$, we can bound the average *cumulative* number of mistakes,

$$\sum_{P=0}^{M-1} \epsilon_{Gibbs}(P) \leq -\frac{1}{2\ln(2)} \sum_{\sigma_1 \ldots \sigma_M = \pm 1} V(\sigma^M) \ln(V(\sigma^M)), \tag{5.31}$$

where we have used the fact that the individual terms

$$\Delta I = -[\ln(V_{P+1}) - \ln(V_P)]$$

sum up to $-\ln(V_M) \equiv -\ln(V(\sigma^M))$. Since the volume of each cell equals its probability, the sum in the last expression equals the *entropy* of the distribution of outputs.

As is well known from information theory, the entropy is maximal if all probabilities are equal. In other words, this happens if the total unit volume of the phase space is equally divided under the $\mathcal{N}(M, d)$ cells. Thus we obtain the inequalities

$$\sum_{P=0}^{M-1} \epsilon_{Gibbs}(P) \leq \frac{1}{2\ln(2)} \ln\left(\mathcal{N}(M, d)\right) \leq \frac{1}{2\ln(2)} \cdot d(\ln(M/d) - 1). \tag{5.32}$$

The logarithmic growth in M indicates a faster decay of errors than the worst-case result $\epsilon(P) \simeq \ln(\alpha)/\alpha$, with $\alpha = P/d$. Rather, the estimate is consistent with a faster decay $\epsilon_{Gibbs}(P) \propto \alpha^{-1}$, asymptotically. In fact, using more refined techniques, it is shown in [13] that

$$\epsilon_{Gibbs}(P) \leq 2/\alpha. \tag{5.33}$$

Since this bound holds for arbitrary distributions of inputs, even very artificial ones, one might expect that, for "typical" distributions, learning might be even faster. Using the statistical mechanics approach, we will see, however, in the section on perceptrons, that the α^{-1} decay also holds for a natural distribution of inputs.

A greater speed of generalization only can be achieved if the asymptotic information gain from new new inputs can be enlarged. We will come back to this idea in Sec. 5.5.4.

5.2.5 Smooth Networks

Most parts of this chapter deal with networks that have binary outputs and the sign transfer function. Often in technical applications of neural nets, the transfer functions between in– and outputs are highly nonlinear, but they nevertheless are *smooth* functions. This property is utilized in the so-called backpropagation algorithm [28], where a training energy is minimized via gradient descent. This requires the calculations of derivatives of the energy with respect to the coupling parameters.

It turns out that the asymptotic behavior of the generalization errors can be calculated easier than for binary outputs.

We assume a learning algorithm that is defined by the Gibbs ensemble

$$p(\mathbf{w}|\sigma^P) = Z^{-1} \cdot \exp\left(-\beta \sum_{k=1}^{P} E(\mathbf{w}; \sigma_k, \xi_k)\right). \tag{5.34}$$

We assume that σ_k is a function of the inputs and a teacher parameter vector. In the following we will not assume that the problem is completely learnable. Then the aim of a learner will be to find a network that minimizes the training energy *averaged over the space of all examples.* If P, the number of examples, grows large, we expect that the final state of the network converges to the optimal value $\mathbf{w}_0$, for which

$$\partial_i \, \overline{E(\mathbf{w}; \sigma, \xi)}_{\mathbf{w}=\mathbf{w}_0} = 0, \tag{5.35}$$

for all i. The bar denotes the average over the examples, and the derivative is with respect to the components $w(i)$.

The generalization error after learning P examples is

$$\varepsilon = \int d\mathbf{w} \; \overline{p(\mathbf{w}|\sigma^P) \, E(\mathbf{w}; \sigma, \xi)}. \tag{5.36}$$

We further expect that the posterior density is strongly peaked at its maximum $\hat{\mathbf{w}}$, the *maximum likelihood estimate.* The fluctuations around this value are, to the lowest order, Gaussian with zero mean and covariance:

$$\langle (w(i) - \hat{w}(i)) \, (w(j) - \hat{w}(j)) \rangle \simeq (\beta \; P)^{-1} \; (U^{-1})_{ij}, \tag{5.37}$$

where

$$U_{ij} = P^{-1} \; \partial_i \partial_j \sum_{k=1}^{P} E(\mathbf{w}; \sigma_k, \xi_k)_{\mathbf{w}=\hat{\mathbf{w}}} \simeq \partial_i \partial_j \; \overline{E(\mathbf{w}; \sigma, \xi)}. \tag{5.38}$$

Expanding Eq. (5.36) around ($\mathbf{w} = \hat{\mathbf{w}}$), and averaging over the Gaussian fluctuations in Eq. (5.37), we get

$$\varepsilon \simeq \overline{E(\hat{\mathbf{w}}; \sigma, \xi)} + \frac{1}{2} (\beta P)^{-1} \sum_{ij} U_{ij} (U^{-1})_{ij}. \tag{5.39}$$

The sum on the right-hand side simply equals N, the number of weights.

For P large, $\hat{\mathbf{w}}$ will be close to the optimum $\mathbf{w}^0$. To estimate the difference between $\hat{\mathbf{w}}$ and $\mathbf{w}^0$, we use the fact that $\mathbf{w} = \hat{\mathbf{w}}$ extremizes the learning energy, i.e., it fulfills

$$0 = P^{-1/2}\partial_i \sum_{k=1}^{P} E(\mathbf{w}; \sigma_k, \xi_k)_{\mathbf{w}=\hat{\mathbf{w}}} \simeq$$

$$\underbrace{P^{-1/2}\partial_i \sum_{k=1}^{P} E(\mathbf{w}; \sigma_k, \xi_k)_{\mathbf{w}=\mathbf{w}^0}}_{\gamma_i} + \sum_j U_{ij}\sqrt{P}(\hat{w}(j) - w(j)^0), \tag{5.40}$$

where we have expanded to the first order at $\mathbf{w} = \mathbf{w}^0$. We also neglected the dependence of U_{ij} on $\mathbf{w}$. The first term, γ_i, is a sum of independent random variables and is, in the limit, Gaussian distributed. We find from Eq. (5.35) that $\overline{\gamma_i} = 0$ and

$$\overline{\gamma_i\, \gamma_j} \simeq \overline{\partial_i E \cdot \partial_j E} \equiv I_{ij}. \tag{5.41}$$

Using this information, we can solve Eq. (5.40) to get

$$P\ \overline{(w(i) - w^0(i))\ (w(j) - w^0(j))} \simeq (U^{-1}\ I\ U^{-1})_{ij}. \tag{5.42}$$

Finally, we expand the first term of Eq. (5.39) at $\mathbf{w}^0$ up to the second order in $(w(i) - w(i)^0)$; the first order clearly vanishes. Using Eq. (5.42), we get

$$\varepsilon \simeq \varepsilon_{min} + \frac{1}{2P}Tr(U^{-1}\ I) + \frac{N}{2\beta P}. \tag{5.43}$$

ε_{min} is the minimal error achieved by the parameter $\mathbf{w}^0$.

This result has been shown in [17] using the replica method. In [18], a similar result has been proved using the analogy to density estimation in mathematical statistics. In this framework, the matrix I is proportional to the so-called *Fisher Information*, defined as

$$I_{ij} = \int dy\ \partial_i \ln(\mathcal{P}_\theta(y)) \cdot \partial_j \ln(\mathcal{P}_\theta(y)). \tag{5.44}$$

Here we have used the terminology of Sec. 5.2.3 and we assumed that the parameter θ is a vector. I plays an important role in the asymptotics of statistical estimation procedures [4].

The result in Eq. (5.43) has the same $\propto P^{-1}$ behavior as the decay of the Gibbs errors in Eq. (5.33). It should be noted, however, that the definition (5.36) of the generalization error does not correspond to a binary classification problem like the ones treated in the previous sections. If we would force a smooth network to give "straight" answers $\oplus$ or $\ominus$, by clipping its outputs after training, the generalization error may be different. As we will see in Sec. 5.5.2, for the ADALINE algorithm, a slower performance $\varepsilon \propto 1/\sqrt{P}$ can be observed in such a case.

5.3 The Perceptron

5.3.1 Some General Properties

The perceptron shows many interesting features that distinguish it from other neural networks.

One of the oldest rigorous results for perceptrons is the number of possible output combinations or *cells*. Besides the estimate of Sauer's lemma, we know a precise result for the perceptron, given by Cover [19] in 1965: For any set of P inputs in general position,[6] one has *exactly*

$$\mathcal{N}(P,N) = 2\sum_{i=0}^{N-1}\binom{P-1}{i}, \tag{5.45}$$

where N is the number of weights. Equation (5.45) also yields $P = N$ for the largest number of input vectors with $\mathcal{N}(P,N) = 2^P$. Thus, the VC dimension equals N.

The independence of $\mathcal{N}(P,N)$ from the location of input vectors is no longer valid when we look at other networks. Perceptrons with binary weights already show large fluctuations for this quantity. Based on exact enumerations on small systems [20], but for many samples of random inputs, we have obtained lower bounds on the VC dimension d for this model. Finite-size scaling (see Fig. 5.3) indicates that for $N \to \infty$ we will have $d \simeq N/2$.

Another striking feature is the simple geometric picture (Fig. 5.4) of the perceptron's classification ability. In the space of the inputs, the vector of couplings defines a separating plane perpendicular to $\mathbf{w}$. Inputs on the side of this normal vector are classified as $\oplus$, while those on the other side are classified as $\ominus$. Perceptrons realize *linear-separable* functions.

As a consequence of this geometric picture, we can easily find the generalization error ε (= probability of making a mistake) when the inputs have a spherical distribution. Such a distribution can be realized from independent, normally distributed cartesian components $\xi(j)$, $j = 1, \ldots, N$, with density

$$f(\boldsymbol{\xi}) = (2\pi)^{-N/2}\exp(-\tfrac{1}{2}\boldsymbol{\xi}\cdot\boldsymbol{\xi}). \tag{5.46}$$

For fixed teacher and student, one finds

$$\varepsilon = \frac{1}{\pi}\arccos\left(\frac{\mathbf{w}_s\cdot\mathbf{w}_t}{|\mathbf{w}_s||\mathbf{w}_t|}\right). \tag{5.47}$$

Equation (5.47) will be used extensively in the following sections. Although this theorem can be derived by averaging over the Gaussian random variables, it is immediately clear from the geometric construction of Fig. 5.4.

[6] Any subset of inputs containing no more than N input vectors is linearly independent.

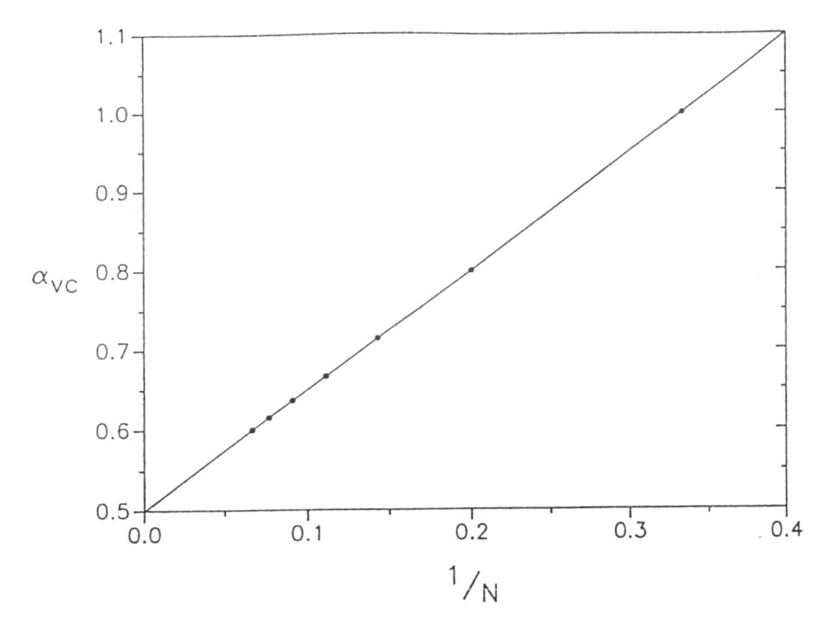

Fig. 5.3. VC dimension for the perceptron with binary weights. The curve gives a lower bound as a function of N, the number of weights. The data were obtained from large samples of input sets upon calculating the number of possible labelings by scanning all 2^N weight vectors.

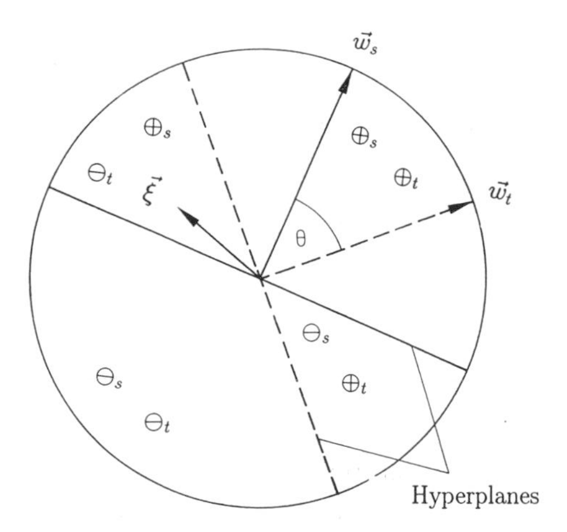

Fig. 5.4. For a perceptron, teacher $\mathbf{w}_t$ and student $\mathbf{w}_s$ determine separating planes in the input space. Inputs are mapped onto $\oplus$ if they are in the same half-space as $\mathbf{w}_{t,s}$. Thus, the generalization error equals the ratio $\varepsilon = \theta/\pi$ of area with different outputs and total areas.

Note that, for $N \to \infty$, any distribution with the same first two moments will give the above result. A popular choice is $\xi_k(j) = \pm 1$ with probability $\frac{1}{2}$.

5.3.2 Replica Theory

In this section we develop a general framework that will allow us to treat a variety of perceptron learning problems using the replica method.

Following Gardner's approach, we will consider a Gibbs ensemble of perceptrons defined by the distribution

$$p(\mathbf{w}_s|\sigma^P) = Z^{-1} \cdot p(\mathbf{w}_s) \cdot \exp\left(-\beta \sum_{k=1}^{P} E(\mathbf{w}_s; \sigma_k, \boldsymbol{\xi}_k)\right) \tag{5.48}$$

with partition function

$$Z = \int d\mathbf{w}\; p(\mathbf{w}) \cdot \exp\left(-\beta \sum_{k=1}^{P} E(\mathbf{w}; \sigma_k, \boldsymbol{\xi}_k)\right).$$

In the following we will keep the form of Eq. (5.48) rather general: We will only assume that E depends on the internal fields $N^{-1/2}\sigma_k \mathbf{w} \cdot \boldsymbol{\xi}_k$. Thus, we consider partition functions of the type

$$Z(\sigma^P) = \int d\mathbf{w}\; p(\mathbf{w}) \cdot \prod_{k=1}^{P} \Phi(N^{-1/2}\sigma_k \mathbf{w} \cdot \boldsymbol{\xi}_k), \tag{5.49}$$

with an arbitrary Φ.

We constrain the coupling vectors to the surface of a sphere, i.e.,

$$p(\mathbf{w}) = V_0^{-1}\delta(\mathbf{w} \cdot \mathbf{w} - q_0 N),$$

with $V_0 = e^{N/2(\ln 2\pi + 1)} \simeq \int \delta(\mathbf{w} \cdot \mathbf{w} - N) d\mathbf{w}$. Finally, $d\mathbf{w} = \prod_{j=1}^{N} dw(j)$ is the volume element in cartesian coordinates.

One of the basic assumptions of the statistical mechanics approach can be stated as follows: The free energy per coupling $\mathcal{F}$, defined by

$$\mathcal{F} = N^{-1} \ln Z(\sigma^P), \tag{5.50}$$

is a self-averaging quantity for $N \to \infty$ and most "natural" distributions of the random examples. This means that it equals its average

$$\mathcal{F} = N^{-1} \cdot \sum_{\sigma_1 \ldots \sigma_P = \pm 1} \overline{\mathcal{P}(\sigma^P) \ln Z(\sigma^P)} \tag{5.51}$$

for almost all realizations of the random examples. Here,

$$\mathcal{P}(\sigma^P) = \mathcal{P}(\sigma_1, \ldots, \sigma_P | \boldsymbol{\xi}_1, \ldots, \boldsymbol{\xi}_P)$$

is the total probability over all teachers (and noise) that, *given the inputs*, the binary classifications σ^P will be observed. The bar denotes the average over the distributions of inputs. If Eq. (5.48) was actually the posterior distribution corresponding to a prior distribution of random teachers (see Sec. 5.2.3), we would have

$$\mathcal{P}(\sigma^P) = Z(\sigma^P)/\mathcal{C}, \tag{5.52}$$

where the normalization

$$\mathcal{C} = \sum_{\sigma_1 \ldots \sigma_P = \pm 1} Z(\sigma^P)$$

is assumed to be independent of the inputs. In general, we will not restrict ourselves to Eq. (5.52), but rather we will use the more general ansatz,

$$\mathcal{P}(\sigma^P) = Z_t(\sigma^P)/\mathcal{C}_t$$

$$Z_t(\sigma^P) = \int d\mathbf{w}_t \, p_t(\mathbf{w}) \cdot \prod_{k=1}^{P} \Phi_t(N^{-1/2}\sigma_k \mathbf{w}_t \cdot \boldsymbol{\xi}_k), \tag{5.53}$$

where Φ_t can be different from Φ and

$$p_t(\mathbf{w}) = V_0^{-1}\delta(\mathbf{w}_t \cdot \mathbf{w}_t - N).$$

To perform the average over the inputs, the *replica trick* is utilized:

$$\mathcal{F} = N^{-1} \cdot \sum_{\sigma_1 \ldots \sigma_P = \pm 1} \overline{\mathcal{P}(\sigma^P) \ln Z(\sigma^P)} = \lim_{n \to 0} \frac{\partial}{\partial n} N^{-1} \ln \Xi_n, \tag{5.54}$$

where

$$\Xi_n = \sum_{\sigma_1 \ldots \sigma_P = \pm 1} \overline{Z_t(\sigma^P) Z^n(\sigma^P)} \tag{5.55}$$

is the weighted and averaged n-times replicated partition function. Equation (5.55) will be calculated for integer n, and the result then will be continued to reals.

Since all inputs are assumed to be statistically independent and drawn from the same distribution, we get

$$\Xi_n = \int d\mathbf{w}_t \, p(\mathbf{w}_t) \prod_{a=1}^{n} d\mathbf{w}_a \, p(\mathbf{w}_a)$$

$$\times \left(\sum_{\sigma = \pm 1} \overline{\Phi_t(N^{-1/2}\sigma \mathbf{w}_t \cdot \boldsymbol{\xi}) \prod_{a=1}^{n} \Phi(N^{-1/2}\sigma \mathbf{w}_a \cdot \boldsymbol{\xi})} \right)^P . \tag{5.56}$$

For the inputs, we use the Gaussian distribution (5.46). Φ and Φ_t depend on $\boldsymbol{\xi}$ only via $u_a = N^{-1/2}\sigma \mathbf{w}_a \cdot \boldsymbol{\xi}$, $a = 1, \ldots, n$, and $u_{n+1} = u_t = N^{-1/2}\sigma \mathbf{w}_t \cdot \boldsymbol{\xi}$.

For fixed couplings, these are joint Gaussian random variables with 0 means and second moments $Q_{ab} = \overline{u_a u_b} = N^{-1}\mathbf{w}_a \cdot \mathbf{w}_b$. Thus, we have, for $P = \alpha N$,

$$N^{-1}\ln(\Xi_n) = N^{-1}\ln \int \prod_{a=1}^{n+1} d\mathbf{w}_a \, p_a(\mathbf{w}_a) \exp[\alpha N \mathcal{G}_1(n)] \tag{5.57}$$

with

$$e^{\mathcal{G}_1(n)} = \overline{2\Phi_t(u_{n+1}\{Q\}) \prod_{a=1}^{n} \Phi(u_a\{Q\})}. \tag{5.58}$$

The average over the u_a can be performed with the help of the following basic assumption of mean–field theory.

For $N \to \infty$, the integrals over $\mathbf{w}_a$ will be dominated by regions in the phase space where the matrix elements Q_{ab} assume *nonfluctuating values.* These are the *order parameters*, which determine the macroscopic physics of the network. Assuming that *replica symmetry* is valid, the order parameters will obey $Q_{ab} = q$, for $a \neq b$ and $a, b \leq n$. Further, $Q_{a,n+1} = R$.

$q = N^{-1}\mathbf{w}_a \cdot \mathbf{w}_b$ is the typical overlap between any two student vectors $\mathbf{w}_a$ and $\mathbf{w}_b$, which are drawn randomly from the Gibbs distribution (5.48). Accordingly, $R = N^{-1}\mathbf{w}_t \cdot \mathbf{w}_s$ is the overlap between a teacher and a student. By Eq. (5.47), the knowledge of the order parameters enables us to obtain the generalization error via

$$\varepsilon = \frac{1}{\pi}\arccos(R/\sqrt{q_0}). \tag{5.59}$$

Using the replica-symmetric ansatz, the Gaussian fields can be constructed explicitly,

$$u_a = z_a(q_0 - q)^{1/2} - tq^{1/2} \tag{5.60}$$

for $a \leq n$ and

$$u_{n+1} = y(1 - R^2/q)^{1/2} - tR/q^{1/2}, \tag{5.61}$$

where z_a, y, t are independent Gaussian variables with variance 1. Obviously, these variables yield the correct second moments. Now the Gaussian average is easily performed, yielding

$$e^{\mathcal{G}_1(n)} = 2\int_{-\infty}^{\infty} Dt \int_{-\infty}^{\infty} Dy \, \Phi_t\left(y(1 - R^2/q)^{1/2} - tR/q^{1/2}\right) \times \left[\int_{-\infty}^{\infty} Dz \, \Phi\left(z\sqrt{q_0 - q} - t\sqrt{q}\right)\right]^n. \tag{5.62}$$

Again, $Dt = (2\pi)^{-1/2}e^{-1/2t^2}\,dt$ is the Gaussian measure. Using the saddle-point method, for $N \to \infty$, we finally get

$$\lim_{N\to\infty} N^{-1}\ln \Xi_n = Extr_{q,R}\,[\alpha\mathcal{G}_1 + \mathcal{G}_2].$$

The second term is an entropic term coming from the phase-space integral, where the order parameters q and R are fixed:

$$e^{N\mathcal{G}_2(n)} = V_0^{-(n+1)} \int \prod_{a=1}^{n+1} d\mathbf{w}_a \prod_{a\leq b} \delta(\mathbf{w}_a \cdot \mathbf{w}_b - NQ_{ab}). \tag{5.63}$$

In replica symmetry, it is not hard to evaluate this expression, giving the result

$$\mathcal{G}_2 = \frac{n-1}{2} \ln(q_0 - q) + \frac{1}{2} \ln\left[(q_0 - q) + n(q - R^2)\right]. \tag{5.64}$$

Finally, performing the derivative with respect to n yields

$$\mathcal{F} = Extr_{q,R}\left[\alpha \mathcal{F}_1 + \mathcal{F}_2\right], \tag{5.65}$$

where

$$\begin{aligned}\mathcal{F}_1 &= \int_{-\infty}^{\infty} Dt \quad \frac{\int_{-\infty}^{\infty} Dy\ \Phi_t\left(y(1-R^2/q)^{1/2} - tR/q^{1/2}\right)}{\int_{-\infty}^{\infty} Dy\, \Phi_t(y)} \\ &\quad \times \ln\left[\int_{-\infty}^{\infty} Dz\, \Phi\left(z\sqrt{q_0 - q} - t\sqrt{q}\right)\right]\end{aligned} \tag{5.66}$$

and

$$\mathcal{F}_2 = \frac{1}{2} \ln(q_0 - q) + \frac{1}{2} \frac{q - R^2}{(q_0 - q)}. \tag{5.67}$$

Extremizing the free energy in Eq. (5.65), we will get the physical values of the order parameters q and R, which in turn determine the generalization error ε.

5.3.3 Results for Bayes and Gibbs Algorithms

Before we come to specific deterministic learning algorithms, we will study the performance of the Gibbs algorithm for a perceptron. As in Secs. 5.2.3 and 5.2.4, we will assume that the prior distribution of the teacher is known to the student.

However, it should be noted that, for the spherical density of inputs in Eq. (5.46), by symmetry, the order parameters will not depend on the actual teacher. Thus, for this special density, the following results will hold not only on average, but also for any specific teacher. If noise is present in the teacher's classifications, we also will assume that the student will know the type of noise and its strength.

The interpretation of the Gibbs ensemble as the posterior distribution in the Bayesian sense simplifies the algebra. In this case we always have $\Phi = \Phi_t$.

Then, teacher and student will enter the replica theory in a completely symmetric way. The teacher is just another replica, so that, from the beginning, we can set $q = R$ and $q_0 = 1$.

Now Eq. (5.65) is replaced by

$$\mathcal{F} = Extr_q \left\{ \frac{\alpha}{A_0} \int_{-\infty}^{\infty} A(t;q) \ln\left[A(t;q)\right] + \frac{1}{2}\ln(1-q) + \frac{q}{2} \right\}, \tag{5.68}$$

where

$$A(t;q) = \int_{-\infty}^{\infty} Dz\, \Phi\left(z\sqrt{1-q} - t\sqrt{q}\right) \tag{5.69}$$

and

$$A_0 = \int_{-\infty}^{\infty} Dz\, \Phi(z).$$

It is interesting to note that this expression could have been derived by a slight modification of the standard replica trick, where we replace the limit $n \to 0$ by $n \to 1$. Setting $Z_t = Z$, we get

$$\mathcal{F} = \lim_{n\to 1} \frac{\partial}{\partial n} N^{-1} \ln \sum_{\sigma_1 \ldots \sigma_P = \pm 1} \overline{Z^n(\sigma^P)}. \tag{5.70}$$

We now give explicit expressions for noise-free and noisy teachers [see Eq. (5.18)]:

$$\Phi(u) = \begin{cases} \Theta(u) & \text{no noise} \\ \exp[-\beta\Theta(-u)] & \text{output noise} \\ H(-\beta^{-\frac{1}{2}}u) & \text{weight noise,} \end{cases} \tag{5.71}$$

leading to

$$A(t;q) = \begin{cases} H(\gamma t) & \text{no noise} \\ e^{-\beta} + (1-e^{-\beta})H(\gamma t) & \text{output noise} \\ H(\hat{\gamma} t) & \text{weight noise} \end{cases} \tag{5.72}$$

with $\gamma = \sqrt{q/1-q}$ and $\hat{\gamma} = \sqrt{q/1-q+1/\beta}$. For output noise, $A_0 = \frac{1}{2}(1+e^{-\beta})$, and $A_0 = \frac{1}{2}$ in the other cases. Calculating the order parameter q from Eq. (5.68), yields the Gibbs error as:

$$\varepsilon_{Gibbs} = \frac{1}{\pi} \arccos(q). \tag{5.73}$$

For noisy outputs, this is the probability that the student will find the *ideal* output of the teacher.

Solving the order parameter equation asymptotically for $\alpha \to \infty$, i.e., $q \to 1$, one obtains

$$\epsilon_{Gibbs} \simeq \begin{cases} 0.62 \cdot \alpha^{-1} & \text{without noise} \\ C_1(\beta) \cdot \alpha^{-1} & \text{output noise} \\ C_2(\beta) \cdot \alpha^{-1/2} & \text{weight noise} \end{cases} \tag{5.74}$$

C_1, C_2 are functions of the temperature. C_1 converges to the value 0.62 for $\beta \to \infty$, whereas C_2 goes to 0, indicating the crossover to the faster decay in the noise-free limit.

The decay $\propto \alpha^{-1}$ in the noise-free case is of the same order as the bound (5.33) discussed in Sec. 5.2.4. Remarkably, this asymptotic decay still persists if output noise is included. When the noise temperature grows large (i.e., $\beta \to 0$), the coefficient C_1 diverges like $4/\beta^2$.

The case of weight noise also has been studied in [8, 21]. However, the authors calculated the Gibbs error for a different algorithm, which uses the sum of mistakes [the *first* line in Eq. (5.18)] as the learning energy. With an optimized learning temperature, $\varepsilon_{Gibbs} \simeq \alpha^{-1/4}$ was found. With a 0 temperature learning, which corresponds to minimizing the training energy (maximum likelihood), the behavior is even worse. This shows that the generalization ability can be remarkably enhanced if more information on the teacher is included in the learning algorithm.

The error of the Bayes algorithm has been calculated in [9, 10]. We will give a different derivation by looking at the volume ratio,

$$Y = \frac{V(\sigma^{P+1})}{V(\sigma^P)}, \tag{5.75}$$

previously defined in Sec. 5.2.4. Equation (5.75) describes the reduction of the volume of the teacher's cell when a new input is learned. As was shown in Sec. 5.2.4, Y can be used to find Gibbs and Bayes errors, as well as the information gain.

Obviously, Y is an average of the function $\Theta(N^{-1/2}\sigma_{P+1}\mathbf{w}_t \cdot \boldsymbol{\xi}_{P+1})$ over all couplings of the teacher's old cell $V(\sigma^P)$. We can write

$$Y = \langle \Theta(N^{-1/2}\sigma_{P+1}\mathbf{w}_t \cdot \boldsymbol{\xi}_{P+1})\rangle. \tag{5.76}$$

One of the basic assumptions of the replica-symmetric mean-field theory is the *clustering hypothesis* [22], which states that the thermodynamic fluctuations of different components $w_t(j)$ of $\mathbf{w}_t$ are uncorrelated in the thermodynamic limit. From the central limit theorem, we conclude that, for *fixed* input $\boldsymbol{\xi}_{P+1}$, the field $N^{-1/2}\sigma_{P+1}\mathbf{w}\cdot\boldsymbol{\xi}_{P+1}$ can be written as $N^{-1/2}\sigma_{P+1}\langle\mathbf{w}\rangle\cdot\boldsymbol{\xi}_{P+1}+v$, where the fluctuating part v is Gaussian distributed and has variance

$$\langle v^2\rangle = N^{-1}(\langle \mathbf{w}\cdot\mathbf{w}\rangle - \langle\mathbf{w}\rangle^2) = 1-q. \tag{5.77}$$

Here, we have again used the clustering hypothesis, yielding

$$q = N^{-1}\mathbf{w}_a \cdot \mathbf{w}_b = N^{-1}\langle\mathbf{w}\rangle^2. \tag{5.78}$$

Performing the average over v gives the expression

$$Y = H\left(\frac{N^{-1/2}\sigma_{P+1}\langle\mathbf{w}\rangle\cdot\boldsymbol{\xi}_{P+1}}{\sqrt{1-q}}\right), \tag{5.79}$$

which holds for a fixed input pattern *and* classification label in the thermodynamic limit! The Bayesian algorithm votes for that value of σ_{P+1} which gives the largest volume, in other words, the largest value for Y. By its definition, $H(x) = \int_x^\infty Dt > \frac{1}{2}$, if $x > 0$. This has the consequence that a student with coupling vector $\mathbf{w}_s = \langle \mathbf{w} \rangle$ will always make the optimal Bayes decision.

This was first shown in [23] by means of a slightly different argument. It is nontrivial because, in general, the "Bayesian student" does not belong to the phase space of the teachers. Finally, to get the Bayes error, we simply have to find the average overlap between the student and a random teacher [Eq. (5.47)]:

$$\frac{\langle \mathbf{w}_t \rangle \cdot \langle \mathbf{w}_t \rangle}{|\langle \mathbf{w}_t \rangle| \cdot \sqrt{N}} = \sqrt{q}. \tag{5.80}$$

Hence,

$$\varepsilon_{Bayes} = \frac{1}{\pi} \arccos(\sqrt{q}). \tag{5.81}$$

Solving the order parameter equation (5.68), we get an asymptotic decay,

$$\varepsilon_{Bayes} \simeq 0.44 \cdot \alpha^{-1},$$

for large α. A comparison of Gibbs and Bayes errors is given in Fig. 5.5. Different algorithms to achieve the performance of the Bayes prediction can be found in [2, 9, 24].

We will end this section by calculating the density of Y. We first need the probabilities of the classification labels $\sigma_{P+1} = \pm 1$. These probabilities are proportional to the volumes of the two new cells. Thus, they simply equal $Y(\sigma_{P+1})$! Using that $N^{-1/2}\langle \mathbf{w} \rangle \cdot \boldsymbol{\xi}_{P+1}$ is Gaussian distributed with respect to the random input $\boldsymbol{\xi}_{P+1}$, with variance equal to $\langle \mathbf{w} \rangle^2 = q$, we find

$$f(Y) = 2 \int_{-\infty}^{\infty} Dt\, H(\gamma t)\, \delta(Y - H(\gamma t)). \tag{5.82}$$

Here, $\gamma = \sqrt{q/1-q}$ and $\delta(\cdot)$ is Dirac's δ–function. Equation (5.82) is depicted in Fig. 5.6 for different values of q. This density is also valid for discrete couplings as long as replica symmetry is exact. Figure 5.7 gives a result for $f(Y)$ obtained from simulations of perceptrons with binary weights. Here, the volumes of the cells were found by *counting* the number of discrete coupling vectors belonging to each cell.

The smooth behavior of $f(Y)$ somehow seems to contradict the VC results. Since the number of cells grows only like a polynomial in P, most of the cells will not be split into two pieces by adding a new input. Thus, $f(Y)$ should contain δ–functions at $Y = 0$ and $Y = 1$, corresponding to one new cell with the old volume, and one with 0 volume. This would in fact be true if all of the cells had the same volume. We conclude that those cells which are not cut into two pieces have neglectable volume (probability) in the thermodynamic limit.

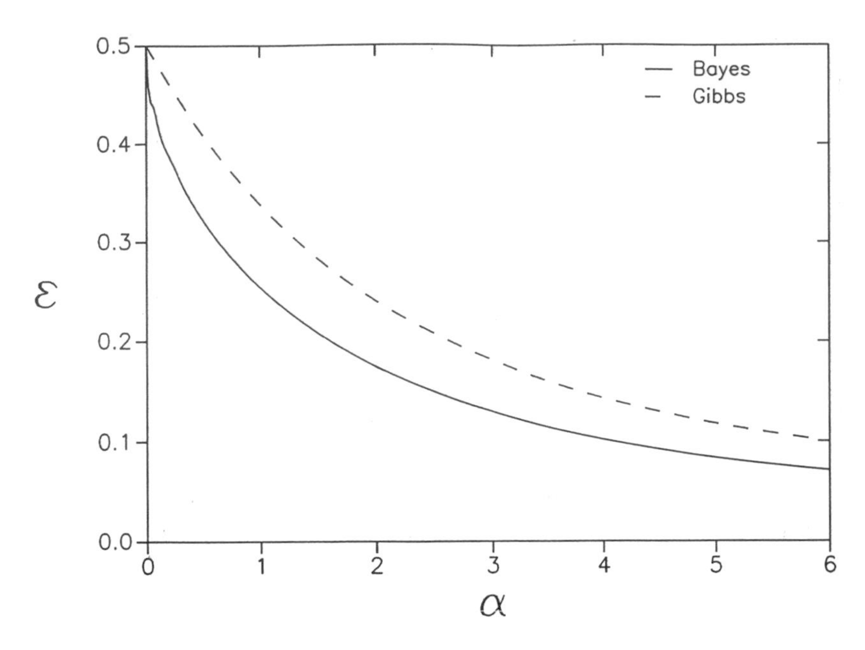

Fig. 5.5. Comparison of Gibbs and Bayes generalization errors.

5.4 Geometry in Phase Space and Asymptotic Scaling

The result of the replica theory for the Gibbs algorithm shows an asymptotic decay of the error $\varepsilon_{Gibbs} \simeq \alpha^{-1}$. The same power law was obtained as an upper bound from the VC theory in Sec. 5.2.4. While for the replica calculation a specific distribution of inputs was assumed, in the VC approach only the VC dimension of the network entered the theory. Thus, arises the question of whether the asymptotic scaling of the generalization error can be explained using only a few parameters of a network.

As we will see, simple geometric scaling arguments will bring us a step closer to this idea of universality. We begin with the perceptron. The phase space of all perceptrons is a simple manifold — the surface of a sphere. The generalization error,

$$\varepsilon = \frac{1}{\pi} \arccos(\mathbf{w}_s \cdot \mathbf{w}_t), \tag{5.83}$$

which is valid for normalized teacher and student vectors, is just the arclength of the shortest line (the geodesic) between $\mathbf{w}_s$ and $\mathbf{w}_t$, and ε is a natural distance between perceptron networks.

A second contribution to a geometry in phase space comes from the information theoretic results of Sec. 5.2.4. We remember that the average information gain for any new input is an upper bound for the Gibbs error

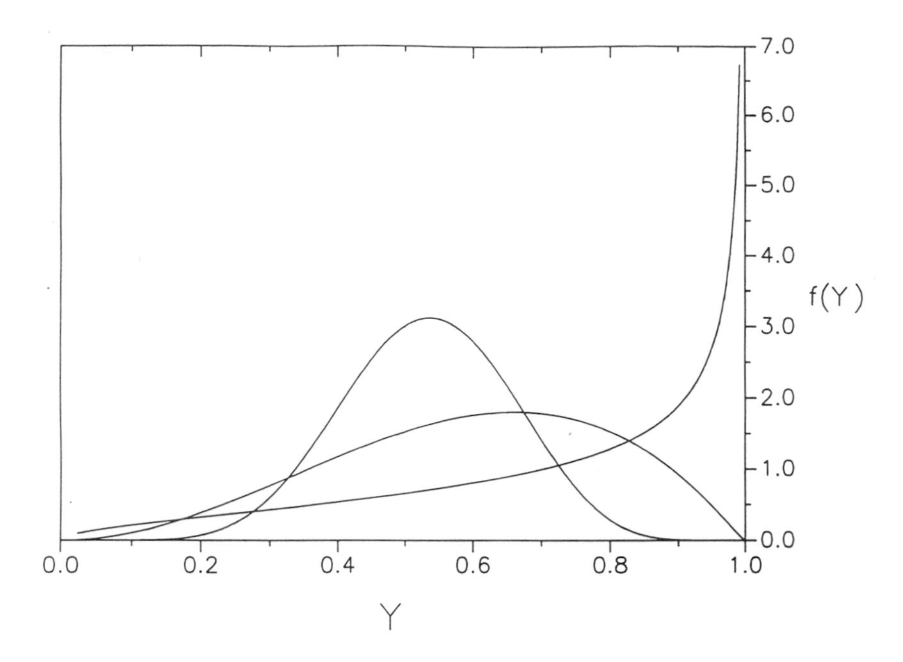

Fig. 5.6. Density of the volume ratio Y for $q = 0.1$ (bell–shaped curve), $q = 0.3$ (flatter curve), and $q = 0.7$ (curve peaked at 1).

on that input. Assuming that both quantities will be of the same order asymptotically, we set

$$\langle \Delta I \rangle = -\langle \ln(V_{P+1}/V_P) \rangle \simeq \varepsilon. \tag{5.84}$$

V_P is *the volume* of the teacher's cell and ε is, as we have shown, a *typical distance* in the cell. Since the number of couplings N is the dimension of the manifold, we expect that

$$V_P \simeq \varepsilon^N. \tag{5.85}$$

Then, with $P = \alpha N$, Eq. (5.84) can be written as

$$\varepsilon(\alpha) = -\frac{\partial}{\partial \alpha} \ln(\varepsilon(\alpha)), \tag{5.86}$$

from which the asymptotic relation

$$\varepsilon(\alpha) \simeq \alpha^{-1}$$

follows.

As a further consequence we see that, if the learner can select examples such that the asymptotic information gain becomes a constant for each new input, then a faster decay of the generalization error like

$$\varepsilon \simeq \exp(-\alpha N \langle \Delta I \rangle) \tag{5.87}$$

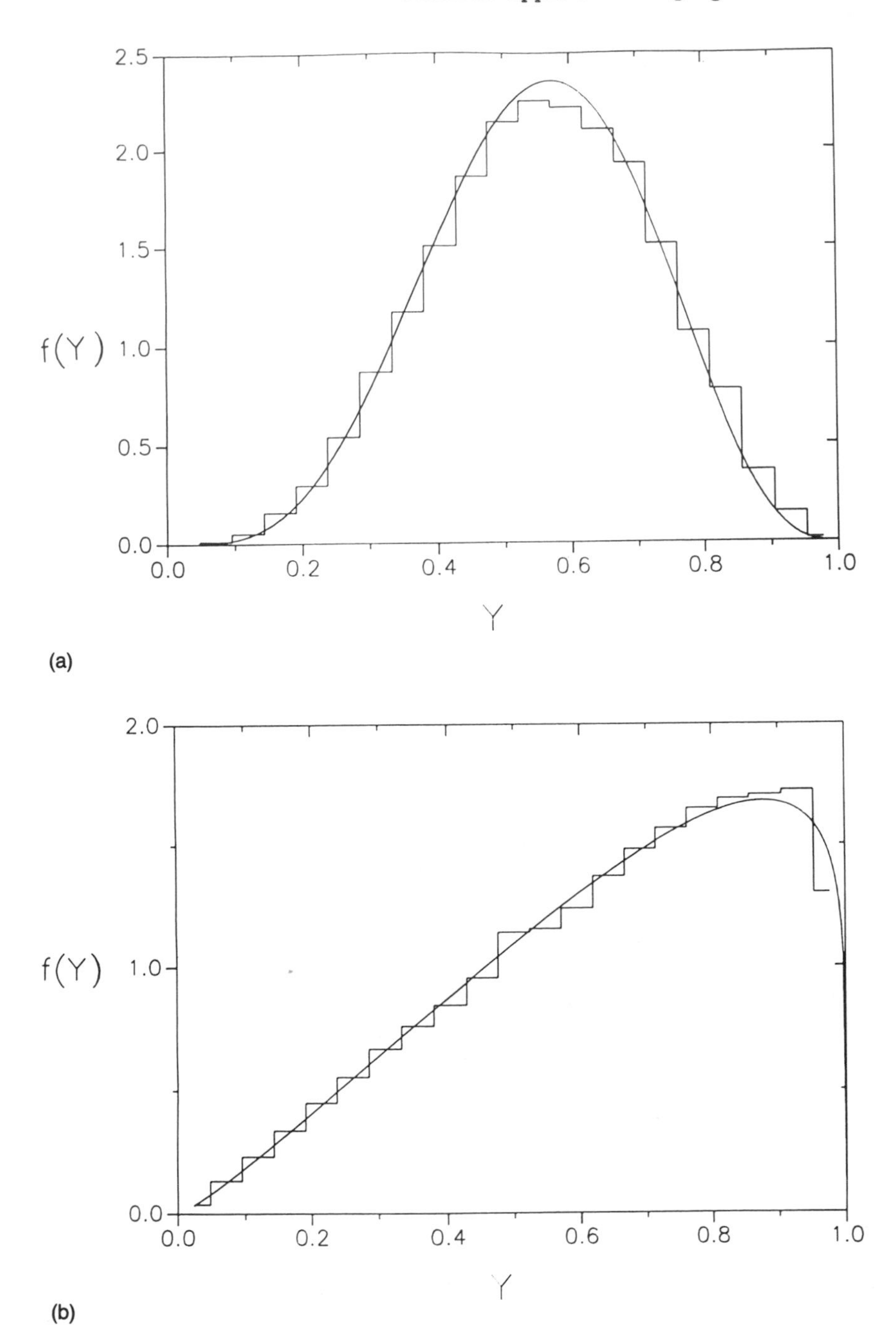

Fig. 5.7. Density of volume ratio Y from simulations of a perceptron with binary weights. (a) $P = 4, N = 14$. (b) $P = 16, N = 20$. The smooth curves are the theoretical predictions.

Table 5.1. Volumes in input space

	Ω_1	Ω_2	Ω_3	Ω_4	Ω_5	Ω_6	Ω_7	Ω_8
σ_a	-1	-1	-1	-1	1	1	1	1
σ_b	-1	-1	1	1	-1	-1	1	1
σ_c	-1	1	-1	1	-1	1	-1	1

is expected. In fact, such behavior is observed for query algorithms (see Sec. 5.5.4).

The interpretation of the generalization error as a distance between networks is no artefact of the perceptron. Generalizing Eq. (5.83) to arbitrary networks, we will show that the probability $\Delta(t, s)$ (over all inputs) that two networks with parameters $\mathbf{w}_t$ and $\mathbf{w}_s$ *do not give the same answer* defines a metric in the space of networks (of a given type). The only nontrivial part[7] is the triangular inequality. Consider three parameter vectors $\mathbf{w}_a$,$\mathbf{w}_b$,$\mathbf{w}_c$ and divide the *input space* into 8 sets with volumes $\Omega_1, \ldots, \Omega_8$, $\sum_i \Omega_i = 1$, according to the outputs $\sigma_{a,b,c}$ (see Table 5.1). Then, $\Delta(a, b)$ Probability of all $\boldsymbol{\xi}$, for which $\mathbf{w}_a$ and $\mathbf{w}_b$ have different outputs $= \Omega_3 + \Omega_4 + \Omega_5 + \Omega_6$. Similarly, $\Delta(b, c) = \Omega_2 + \Omega_3 + \Omega_6 + \Omega_7$ and $\Delta(a, c) = \Omega_2 + \Omega_4 + \Omega_5 + \Omega_7$. Thus,

$$\Delta(a, b) + \Delta(b, c) = \Omega_2 + 2\Omega_3 + \Omega_4 + \Omega_5 + 2\Omega_6 + \Omega_7 \geq \Delta(a, c).$$

This completes the proof of the triangular inequality.

So we can expect that the asymptotic scaling of the learning error based on the simple geometric picture is valid for more general types of networks or learning machines.

Based on similar ideas, an asymptotic scaling of the information gain $\langle \Delta I \rangle \simeq \alpha^{-1}$ for noise-free learning was predicted in [25]. Using this simple geometric picture, we now derive an asymptotic result for the Gibbs error in the case of learning with strong output noise [26]. This will be shown for *a general network*, where only the number N of free adjustable parameters enters the calculation.

We consider a teacher network with a noisy output,

$$\sigma = \eta \cdot F(\mathbf{w}_t, \boldsymbol{\xi}). \tag{5.88}$$

The teacher's ideal answer is inverted, i.e., $\eta = -1$, with probability $e^{-\beta}/1 + e^{-\beta}$ independent of the inputs. The task of the learner is to construct a deterministic, i.e., noise-free, student network $\mathbf{w}_s$,

$$\sigma = F(\mathbf{w}_s, \boldsymbol{\xi}), \tag{5.89}$$

[7] We neglect the possibility that two different parameters $\mathbf{w}_a$ and $\mathbf{w}_b$ will give the same outputs on *all* inputs.

who will be able to give the teacher's ideal answers [$\eta = +1$ in Eq. (5.88)]. We will use the Gibbs algorithm to construct such a network. This algorithm will draw a $\mathbf{w}_s$ randomly from the posterior distribution of the unknown teacher, after having seen P noisy examples. Using the ideas of Sec. 5.2.3, the students will have probability

$$p(\mathbf{w}_s|\sigma^P) = Z^{-1} \exp\left(-\beta \sum_{k=1}^{P} E(\mathbf{w}_s; \sigma_k, \boldsymbol{\xi}_k)\right), \tag{5.90}$$

where

$$Z(\sigma^P) = \int d\mathbf{w} \, \exp\left(-\beta \sum_{k=1}^{P} E(\mathbf{w}; \sigma_k, \boldsymbol{\xi}_k)\right). \tag{5.91}$$

$E(\mathbf{w}_s; \sigma_k, \boldsymbol{\xi}_k)$ equals 1 if $\sigma_k \neq F(\mathbf{w}_s, \boldsymbol{\xi}_k)$, i.e., if the student does not learn the outputs correctly.

By using the Gibbs algorithm, the student will not simply try to minimize his or her learning error, but instead will make mistakes on the observed labels with probability $e^{-\beta}/1 + e^{-\beta}$. This is precisely the rate at which the teacher produces wrong outputs. Using the temperature β^{-1}, the student assumes a priori that a fraction of the teacher's answers are not correct.

Fixing teacher and student for a moment, the probability that the student's and the teacher's ideal answer disagree on a new input $\boldsymbol{\xi}$, i.e., that

$$F(\mathbf{w}_t, \boldsymbol{\xi}) \neq F(\mathbf{w}_s, \boldsymbol{\xi}), \tag{5.92}$$

is given by

$$\Delta(t, s) = 1 - \sum_{\sigma=\pm 1} \overline{E(\mathbf{w}_t; \sigma, \boldsymbol{\xi}) E(\mathbf{w}_s; \sigma, \boldsymbol{\xi})}. \tag{5.93}$$

Given the P outputs, the teacher and the student have the same distribution [Eq. (5.90)] by the definition of the algorithm. Using this fact, and weighting all possible output configurations with their probability [Eq. (5.52)], $\mathcal{P}(\sigma^P) = \mathcal{C}^{-1} \cdot Z(\sigma^P)$, we can average Eq. (5.93) over teachers and students:

$$\langle \Delta(t, s) \rangle = \sum_{\sigma_1 \ldots \sigma_P = \pm 1} \mathcal{C}^{-1} \cdot Z(\sigma^P) \int d\mathbf{w}_t \, d\mathbf{w}_s \, \Delta(t, s) \cdot p(\mathbf{w}_t|\sigma^P) \cdot p(\mathbf{w}_s|\sigma^P). \tag{5.94}$$

The total Gibbs error is obtained by averaging this expression over the training inputs. This can be done with the replica method, in a form similar to Eq. (5.70). One finds, using Eqs. (5.90) and (5.91),

$$\varepsilon_{Gibbs} = \lim_{\substack{n \to 1 \\ \gamma \to 0}} \frac{\partial^2}{\partial \gamma \partial n} \ln \int \prod_{a=1}^{n} d\mathbf{w}_a \exp[-P G_n(\{\mathbf{w}_a\}) + \gamma \sum_{a \neq b} \Delta(a, b)] \tag{5.95}$$

with the replica Hamiltonian

$$G_n = -\ln[\overline{\sum_{\sigma=\pm 1} \exp(-\beta \sum_{a=1}^{n} E(\mathbf{w}_a, \sigma, \boldsymbol{\xi})}). \tag{5.96}$$

This result has an interesting limit for strong noise, i.e., small β:

$$G_n(\{\mathbf{w}_a\}) = -\ln 2 + \frac{n\beta}{2} - \frac{\beta^2 n^2}{8} + \frac{\beta^2}{4} \sum_{a \neq b} \Delta(a,b) + \mathcal{O}(\beta^3). \tag{5.97}$$

Here we have made use of the fact that $(E(\mathbf{w}_a; \sigma, \boldsymbol{\xi}))^2 = E(\mathbf{w}_a; \sigma, \boldsymbol{\xi})$ and $\sum_{S \pm 1} E(\mathbf{w}_a; \sigma, \boldsymbol{\xi}) = 1$. Inserting this into Eq. (5.95), we get

$$\epsilon_{Gibbs} \simeq -\lim_{n \to 1} \frac{\partial^2}{\partial n \partial B} \ln \int \prod_{a=1}^{n} d\mathbf{w}_a \, \exp[-B \sum_{a \neq b} \Delta(a,b)], \tag{5.98}$$

where the derivative with respect to B has to be taken at $B = P\beta^2/4$. The phase-space integral in Eqs. (5.98) is the partition function for n classical "particles" at temperature B^{-1} interacting with the pair potential $\Delta(a,b)$.

If the number of examples P grows large, the effective temperature B^{-1} goes to 0 and the particles are close together at the minimum of the potential. In other words, $\Delta(t,s)$ vanishes, and we have perfect generalization!

To estimate the speed of generalization, we fix one of the couplings, e.g., $\mathbf{w}_n$. If all distances $\Delta(n,b)$ are small for large B, then the triangular inequality will enforce all other distances $\Delta(a,b)$ to be small as well.

Our basic assumption is that for small distances the manifold of parameters $\mathbf{w}$ is locally flat. In suitably chosen coordinates, with $\mathbf{w}_n$ at the origin, $\mathbf{w}_a \equiv w_a(i), \quad i = 1, \ldots, N$, the volume element ($\mathbf{w}_n$ is fixed)

$$d\mathbf{w}_a \simeq \prod_{i=1}^{N} dw_a(i) \tag{5.99}$$

is locally cartesian. Also, the distances $\Delta(a,b)$ are expected to be of the form $\Delta[\{w_a(i) - w_b(i)\}]$ for $w_a(i) - w_b(i) << 1$, and Δ should obey the "regular scaling" of a length,

$$B \cdot \Delta[\{w_a(i) - w_b(i)\}] = \Delta[\{B \cdot (w_a(i) - w_b(i)\}]. \tag{5.100}$$

Then we can simply scale the inverse temperature B out of Eq. (5.98) by using $Bw_a(i)$ as new coordinates. We get

$$\epsilon_{Gibbs} \simeq -\lim_{n \to 1} \frac{\partial^2}{\partial n \partial B} \ln[B^{-(n-1)N}] = \frac{N}{B}. \tag{5.101}$$

Setting $B = P\beta^2/4$, we get

$$\varepsilon_{Gibbs} \simeq \frac{4}{\beta^2 \alpha}. \tag{5.102}$$

This coincides with the known result in the case of the perceptron. Note, however, that in the present approach we have made no assumptions on the distribution of inputs and the special architecture of the network.

Since exact replica calculations for multilayer networks become technically very involved, we expect that the geometric approach will provide a useful alternative, at least in asymptotic regions. It would be interesting to establish a connection with the VC results.

5.5 Applications to Perceptrons

In this section we discuss several applications of the statistical mechanics of generalization. In particular, we concentrate on the simplest case: the teacher as well as the student are simple one–layer perceptrons, with one input layer $\boldsymbol{\xi}$, one weight layer $\mathbf{w}_t$ or $\mathbf{w}_s$, respectively, and one output bit σ. As before, we normalize the teacher weight vector to $\mathbf{w}_t \cdot \mathbf{w}_t = N$:

$$\sigma = \text{ sign } \left(\frac{1}{\sqrt{N}} \, \mathbf{w}_{t/s} \cdot \boldsymbol{\xi} \right) . \tag{5.103}$$

The student tries to learn a set of $\alpha N = P$ input–output examples $\sigma_k, \boldsymbol{\xi}_k$, $k = 1, \ ..., \alpha N$, given by the teacher network. In the following, several learning rules are considered; in addition, the structure of the teacher may be different from that of the student, or it may even change with time. It turns out that the simplest case — perceptron learns from perceptron — already shows many interesting phenomena.

The advantage of simplicity is the fact that one obtains exact mathematical relations, for example, the generalization error ε as a fuction of the number αN of learned examples. Furthermore, the simple structure is always a part of more complex networks, and from understanding the perceptron it may be possible to derive results for multilayer networks.

5.5.1 Simple Learning: Hebb Rule

The learning rule that easily can be analyzed [27] is the Hebb rule: At each presentation of a new example $(\sigma_k, \boldsymbol{\xi}_k)$ the product of input and output bits is added to the corresponding weight,

$$\vec{w_s} \ (t+1) = \vec{w_s}(t) + \frac{1}{\sqrt{N}} \, \sigma_k \, \boldsymbol{\xi}_k \ . \tag{5.104}$$

If each example is presented once, and if the initial weight vector is 0, then the final weights are given by

$$\vec{w_s} = \frac{1}{\sqrt{N}} \sum_{k=1}^{P} \sigma_k \, \boldsymbol{\xi}_k \ . \tag{5.105}$$

Note that σ_k is given by the teacher,

$$\sigma_k = \text{sign}\left(\frac{1}{\sqrt{N}}\,\mathbf{w}_t\cdot\boldsymbol{\xi}_k\right). \tag{5.106}$$

Now we study the case of random inputs $\boldsymbol{\xi}_k$. We are interested in the generalization error ε, which, following Eq. (5.47), is given by the overlaps $R = \mathbf{w}_t \cdot \mathbf{w}_s/N$ and $q_0 = \mathbf{w}_s \cdot \mathbf{w}_s/N$:

$$\varepsilon = \frac{1}{\pi}\arccos\left(\frac{R}{\sqrt{q_0}}\right). \tag{5.107}$$

At each step of presenting a new example $(\sigma_k, \boldsymbol{\xi}_k)$ the teacher–student overlap $R = \mathbf{w}_t \cdot \mathbf{w}_s/N$ changes by an amount ΔR given by

$$\Delta R = \frac{1}{N}\,\frac{1}{\sqrt{N}}\ \text{sign}\ (\mathbf{w}_t\cdot\boldsymbol{\xi}_k) \qquad \mathbf{w}_t\cdot\boldsymbol{\xi}_t = \frac{1}{N}\frac{|\mathbf{w}_t\cdot\boldsymbol{\xi}_k|}{\sqrt{N}}\ . \tag{5.108}$$

However, for different input patterns $\boldsymbol{\xi}_k$, the variable $u = \mathbf{w}_t \cdot \boldsymbol{\xi}_k/\sqrt{N}$ is Gaussian distributed (u = sum of independent random numbers) with

$$\overline{u} = 0 \qquad \text{and} \qquad \overline{u^2} = \frac{1}{N}\,\mathbf{w}_t\cdot\mathbf{w}_t = 1\ . \tag{5.109}$$

Hence, with $\overline{|u|} = \sqrt{2/\pi}$, on average, the teacher–student overlap changes by the amount $\Delta R = \sqrt{2/\pi}/N$, which gives

$$R = \sqrt{\frac{2}{\pi}}\ \alpha\ . \tag{5.110}$$

The square of Eq. (5.104) gives the change of the student–student overlap, and one has

$$\begin{aligned}\Delta q_0 &= \frac{1}{N}\left(\frac{2}{\sqrt{N}}\ \sigma_k\ \boldsymbol{\xi}_k\cdot\mathbf{w}_s(t)\ +\ 1\right)\\ &= \frac{1}{N}\,(2\ \text{sign}\ (u)\cdot z + 1)\,.\end{aligned} \tag{5.111}$$

The variable $z = \mathbf{w}_s(t)\cdot\boldsymbol{\xi}_k/\sqrt{N}$ is again Gaussian distributed with

$$\overline{z^2} = q_0 \qquad \text{and} \qquad \overline{zu} = R\ .$$

The correlations between z and u are taken into account by the substitution $z = Ru + \sqrt{q_0 - R^2}t$ with $\overline{t^2} = 1$ and $\overline{tu} = 0$. One obtains for the average of Δq_0

$$\Delta q_0 = \frac{1}{N}\left(2R\ \overline{|u|}\ + 1\right) = \Delta R^2 + \frac{1}{N}\ . \tag{5.112}$$

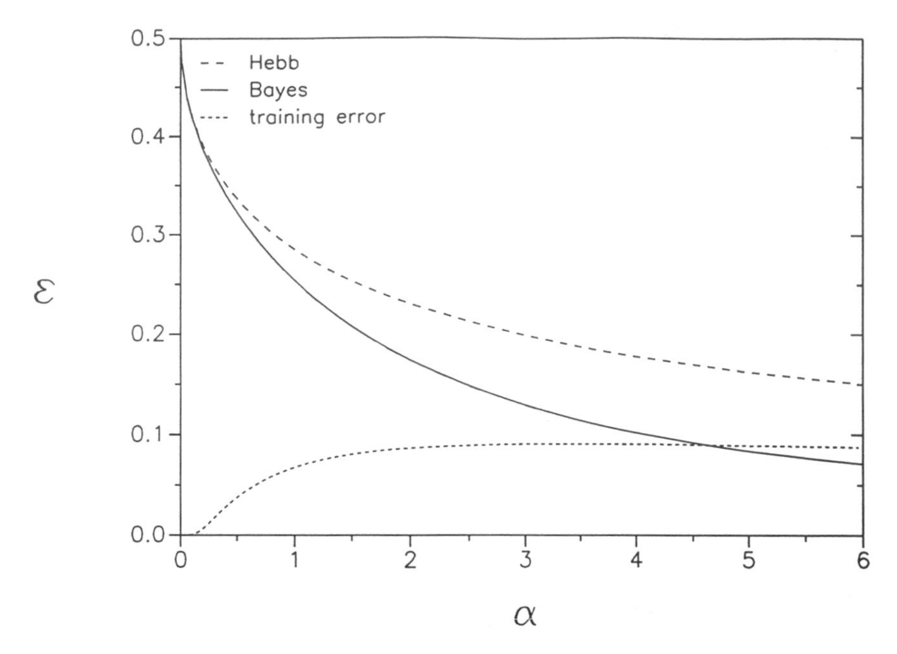

Fig. 5.8. Generalization error for Hebbian learning. The other two curves are the Bayesian error and the Hebbian training error.

This gives

$$q_0 = \alpha + R^2, \tag{5.113}$$

and, as the final result,

$$\varepsilon = \frac{1}{\pi} \arccos\left(\sqrt{\frac{2}{\pi}} \frac{\alpha}{\sqrt{\alpha + \frac{2}{\pi}\alpha^2}}\right) = \frac{1}{\pi} \arctan\left(\sqrt{\frac{\pi}{2\alpha}}\right) . \tag{5.114}$$

Hebbian learning also may be considered as a drifting random walk of $\mathbf{w}_s$ in an N-dimensional vector space [2]. The component of $\mathbf{w}_s$ in the direction of the teacher increases like $\sqrt{2/\pi\alpha}$ while, perpendicular to the teacher, the student performs a random walk with mean-square displacement α. The ratio of the two lengths determines $\tan(\pi\varepsilon)$, according to Fig. 5.4.

Figure 5.8 shows the generalization error ε [Eq. (5.114)] as a function of the number of learned examples α. If only a finite number of examples has been stored ($\alpha = 0$), the network cannot generalize, and one has $\varepsilon = 0.5$. But if the number of examples is of the order of the number of weights, ε decreases. For large values of α, Eq. (5.114) gives

$$\varepsilon \propto 1/\sqrt{\alpha} \, . \tag{5.115}$$

Hence, asymptotically, the Hebbian rule is worse than the Bayesian optimum $\varepsilon \simeq 0.44/\alpha$. Nevertheless, it is surprising that the rule gives a reasonably low error ε. That is, the Hebbian network cannot learn perfectly; its training error

$$\varepsilon_t = \overline{\theta\left[-\left(\mathbf{w}_t \cdot \boldsymbol{\xi}_k\right)\left(\mathbf{w}_s \cdot \boldsymbol{\xi}_k\right)\right]} \tag{5.116}$$

is nonzero for any $\alpha > 0$. With the Gaussian variables u and t as before, one has

$$\varepsilon_t = \overline{\theta\left[-u\left(Ru + \sqrt{q_0 - R^2}\; t + (u)\right)\right]}, \tag{5.117}$$

which gives [27]

$$\varepsilon_t = \frac{1}{2} - \int_0^\infty Du\; erf\left(u\sqrt{\frac{\alpha}{\pi}} + \frac{1}{\sqrt{2\alpha}}\right) . \tag{5.118}$$

Hence, for $\alpha \simeq 5$, one finds a maximal training error of about 10%, which appears to be rather large. Nevertheless, the Hebbian network is able to generalize reasonably well.

5.5.2 Overfitting

If one has a cost function E that depends continuously on the weight vector $\mathbf{w}_s$, then a learning rule may be defined as a gradient descent in the N-dimensional weight space:

$$\mathbf{w}_s\;(t+1) = \mathbf{w}_s(t) - \gamma\; \nabla\; E\left(\mathbf{w}_s(t)\right). \tag{5.119}$$

In many applications, the cost function is defined as the quadratic deviation between student and teacher output. In a multilayer feedforward network with continuous activation functions, the gradient rule is called *error backpropagation* [28].

Unfortunately, a gradient cannot be defined for binary student output. But one may try to learn the binary teacher output by a linear student network, minimizing the cost function

$$E = \sum_k \left(\frac{1}{\sqrt{N}}\; \sigma_k \mathbf{w}_s \cdot \boldsymbol{\xi}_k - 1\right)^2 \tag{5.120}$$

with σ_k given by the teacher network, $\sigma_k = \operatorname{sign}\left(\mathbf{w}_t \cdot \boldsymbol{\xi}_k/\sqrt{N}\right)$. This gives the learning algorithm

$$\mathbf{w}_s(t+1) = \mathbf{w}_s(t) + \frac{\gamma}{\sqrt{N}}\left(1 - \frac{1}{\sqrt{N}}\; \sigma_k\; \mathbf{w}_s(t)\; \boldsymbol{\xi}_k\right) \sigma_k\; \boldsymbol{\xi}_k\;. \tag{5.121}$$

This algorithm has been studied for more than 30 years [29]; it is called ADALINE. For attractor networks it improves the storage capacity for random patterns from $\alpha_c = 0.14$ (Hebbian weights) to $\alpha_c = 1$ [30].

For $E = 0$, Eq. (5.120) gives αN many linear equations for the N unknown coefficients of $\mathbf{w}_s$:

$$\frac{1}{\sqrt{N}} \mathbf{w}_s \boldsymbol{\xi}_k = \text{sign}\left(\frac{1}{\sqrt{N}} \mathbf{w}_t \cdot \boldsymbol{\xi}_k\right); \qquad k = 1, ..., \alpha N . \tag{5.122}$$

If the input patterns $\boldsymbol{\xi}_k$ are linearly independent, one can solve this equation for $\alpha < 1$. But, for $\alpha > 1$, it is obvious that Eq. (5.122) cannot be fulfilled; although the rule is realizable, the ADALINE algorithm cannot learn it perfectly. The training error E_t increases for $\alpha > 1$ to a nonzero value.

Although the learning algorithm is defined by the linear network, its training and generalization errors still are defined by the nonlinear network $\sigma = \text{sign}(\mathbf{w} \cdot \boldsymbol{\xi})$. Both of the errors can be calculated analytically using the replica method of Sec. 5.3.2 [31]. Using the Gibbs weight $\exp[-\beta E]$, one finds the properties of the stationary state of the weight vector $\mathbf{w}_s$ (i.e. after having learned for infinitely many timesteps t) from the limit $\beta \to \infty$.

In Eq. (5.66) we replace $\Phi(u)$ by

$$\Phi(u) = \sqrt{\beta} \exp\left[-\tfrac{1}{2}\, \beta\, (u-1)^2\right] \tag{5.123}$$

and Φ_t by

$$\Phi_t(u) = \Theta(u). \tag{5.124}$$

Then we solve the saddle-point equations for the order parameters q and R. For the limit $\beta \to \infty$, we have to consider two cases:

$\alpha < 1$: In this case, one has $E = 0$, and the length $\sqrt{q_0}$ of the student $\mathbf{w}_s$ is a free parameter that is maximized by $q_0 \to q$. One finds

$$R = \alpha\sqrt{\frac{2}{\pi}}; \qquad q_0 = \frac{\alpha - R^2}{1 - \alpha} . \tag{5.125}$$

$\alpha > 1$: One has only one minimum of E, and q converges to q_0 automatically. However, the quantity $\beta(q_0 - q)$ remains finite and nonzero. One finds

$$R = \sqrt{2/\pi}; \qquad q_0 = \frac{1 + \frac{2}{\pi}(\alpha - 2)}{\alpha - 1} . \tag{5.126}$$

These equations show that the length of the student vector diverges when α approaches the value 1. This means — since the overlap R between the teacher and the student remains finite — that the generalization error ε increases to 1/2. *At $\alpha = 1$, the network cannot generalize, although it has learned perfectly!*

The linear network tries to learn a nonlinear problem; for $\alpha \leq 1$, it does so by increasing the length of the weight vector. This gives a low performance of generalization; this effect has been named *overfitting* [27]. For $\alpha > 1$, the network cannot learn perfectly, and its generalization error decreases. Figure 5.9 compares the ADALINE rule with other learning rules.

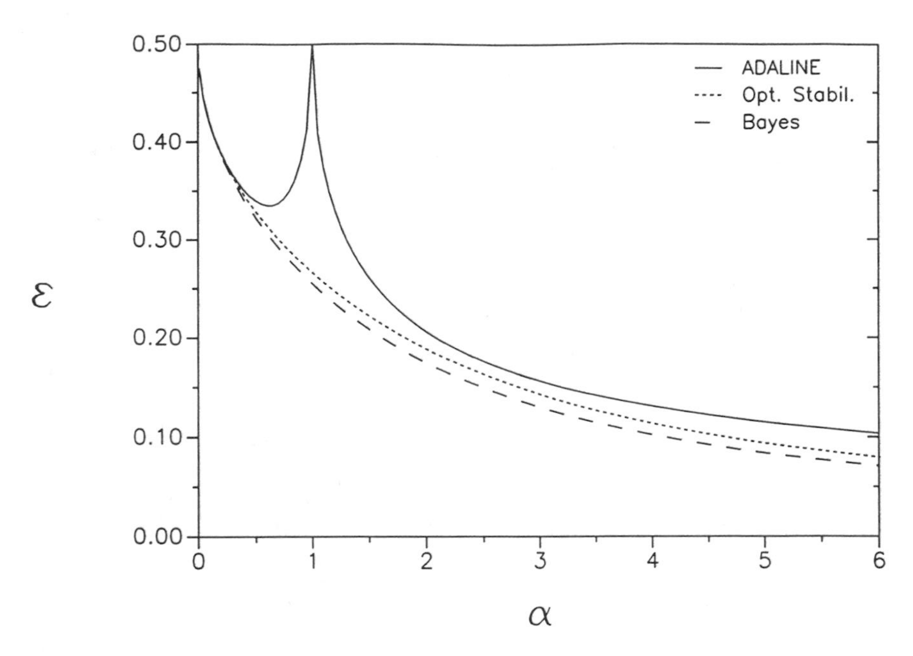

Fig. 5.9. Comparison of generalization errors for ADALINE learning, perceptron learning with optimal stability, and the optimal Bayes prediction.

For small α values, Eq. (5.125) agrees with the Hebbian rule equations (5.110) and (5.113). In fact, the ADALINE algorithm only adds an additional weight to the Hebbian term $\sigma_k\, \boldsymbol{\xi}_k/\sqrt{N}$ that is small for small α. For large α, one finds

$$\varepsilon \propto 1/\sqrt{\alpha}, \tag{5.127}$$

which is again the result of the Hebbian network. Note that in both cases the training error is nonzero.

It is interesting to note that the results for the order parameters [Eqs. (5.125) and (5.126)] can be obtained without the replica method, by an explicit calculation of the coupling vectors [60]. This will be shown in Appendix 5.2.

Finally, we want to mention that the *dynamics* of ADALINE learning can be solved exactly, in contrast to nonlinear learning rules [32, 33, 34, 35]. It can easily be shown that the dynamics is governed by the spectrum of eigenvalues of the matrix

$$B_{ij} = \frac{1}{N} \sum_{k=1}^{P} \xi_k(i)\, \xi_k(j). \tag{5.128}$$

B measures correlations between different input bits; note that at each input unit i the P different training vectors define a P–dimensional vector, and Eq. (5.128) gives the product of those vectors. B is a kind of random matrix; its spectrum is a distorted semicircle between the values $(1 \pm \sqrt{\alpha})^2$ with an additional degenerate eigenvalue 0 for $\alpha < 1$, [32, 1]. One finds that for $\alpha \to 1$ the longest relaxation time diverges like $|\sqrt{\alpha} - 1|^{-2}$. Hence, one obtains a critical slowing down at the transition to perfect learning.

5.5.3 Maximal Stability

The simple perceptron $\mathbf{w}_s$ has learned an example $\boldsymbol{\xi}_k$ if

$$\operatorname{sign}\left(\frac{1}{\sqrt{N}}\ \mathbf{w}_t \cdot \boldsymbol{\xi}_k\right) = \operatorname{sign}\left(\frac{1}{\sqrt{N}}\ \mathbf{w}_s \cdot \boldsymbol{\xi}_k\right) . \tag{5.129}$$

Its ability to generalize is related to the fact that the sign function maps *similar* input vectors $\boldsymbol{\xi}$ to the *same* output bit σ_k. But from the above equation it is obvious that this property is optimal if the quantity

$$\Delta_k = \frac{\sigma_k}{\sqrt{N}}\ \mathbf{w}_s \cdot \boldsymbol{\xi}_k \tag{5.130}$$

is as large as possible (for fixed norm $\mathbf{w}_s \cdot \mathbf{w}_s / N$). The quantity

$$\Delta = \min_k \frac{\sigma_k \sqrt{N}\ \mathbf{w}_s \cdot \boldsymbol{\xi}_k}{|\mathbf{w}_s|} \tag{5.131}$$

is called the *stability* of the perceptron, and a good learning algorithm should maximize the stability Δ. For attractor networks, a similar relation is assumed between the stability and the size of the basin of attraction [12].

Equation (5.131) can be related to quadratic optimization with boundary conditions:

$$\begin{gathered}\text{Minimize } \mathbf{w} \cdot \mathbf{w} \\ \text{with the conditions } \tfrac{\sigma_k}{\sqrt{N}}\ \mathbf{w} \cdot \boldsymbol{\xi}_k \geq 1 \\ \text{for all patterns } \boldsymbol{\xi}_k.\end{gathered}$$

It turns out that the optimal perceptron $\mathbf{w}_s$ classifies the training examples into two classes [36]: One set of patterns is right at the boundary $\Delta_k = 1$, and the second set is in the interior of the allowed region. But only the first set has to be learned by the perceptron, namely, one has

$$\mathbf{w}_s = \frac{1}{\sqrt{N}} \sum_k x_k\ \boldsymbol{\xi}_k \tag{5.132}$$

with coefficients x_k that are 0 for the second set. The number $\alpha_{eff} N$ of the examples belonging to the first set can be calculated by the replica method; it is shown in Fig. 5.10 as a function of αN many random examples $\boldsymbol{\xi}_k$. α_{eff} remains smaller than 1, even for $\alpha \to \infty$. Only $\alpha_{eff} N$ many examples

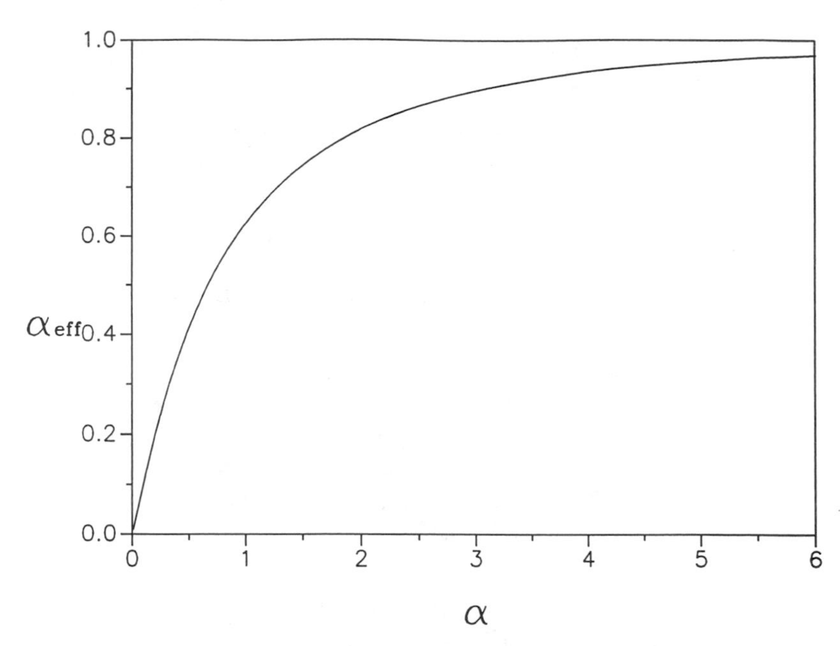

Fig. 5.10. Effective number of inputs per weight to be learned by the perceptron with optimal stability.

have to be learned by the network; and for these the ADALINE learning rule is sufficient, leading to $E = 0$ in Eq. (5.120). For large α, most of the added training examples are useless since they give $\Delta_k \geq 1$ and do not change the student $\mathbf{w}_s$ (they are not learned). Of course, this is related to the fact that the generalization error is small.

Optimal stability has a surprising geometric implication: Consider a set of αN many random points $\boldsymbol{\xi}$ on the unit hypersphere in αN-dimensional space. Label each point black or white randomly. Then a two-dimensional projection of the points looks like Fig. 5.11(a). For $N \to \infty$ and $\alpha < 2$, a perceptron with optimal stability $\Delta > 0$ exists (Sec. 5.3.1). This means that there is a weight vector $\mathbf{w}_s$, and a two–dimensional projection on a plane containing $\mathbf{w}_s$ looks like Fig. 5.11(b). Now black points are separated from white ones and there is a gap Δ between the two clouds. Precisely at the boundary planes of the gap there are $\alpha_{eff} N$ many points. Hence, just by rotating the cloud of random points one can find a view where the black and white points are clearly separated.

There is an interesting general relation between α_{eff} and ε, which holds independently of the distribution of the inputs. Consider the case where a $P + 1$st example $(\boldsymbol{\xi}, \sigma)$ is added to the training set of $P = \alpha N$ inputs. If we run our algorithm on this new, enlarged set, the coupling vector of the P input problem is only changed if

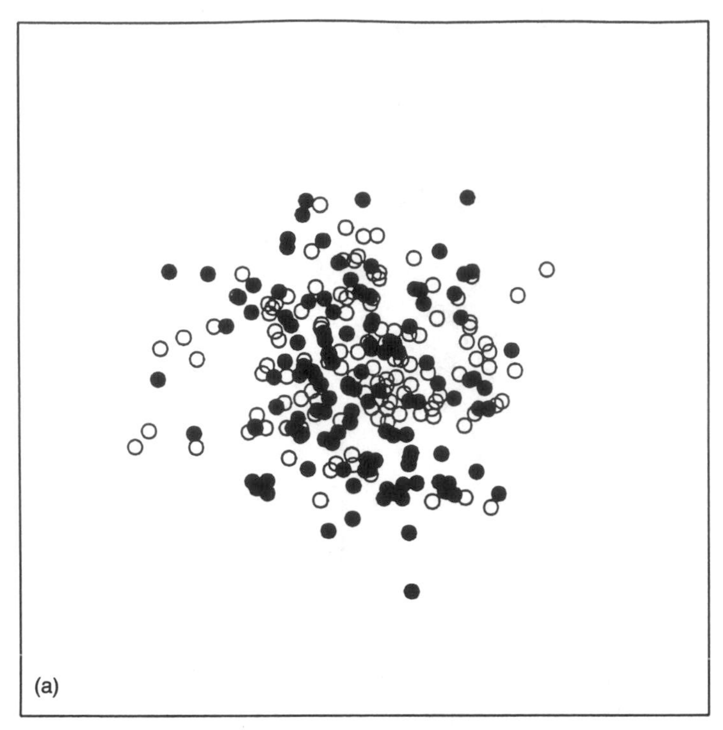

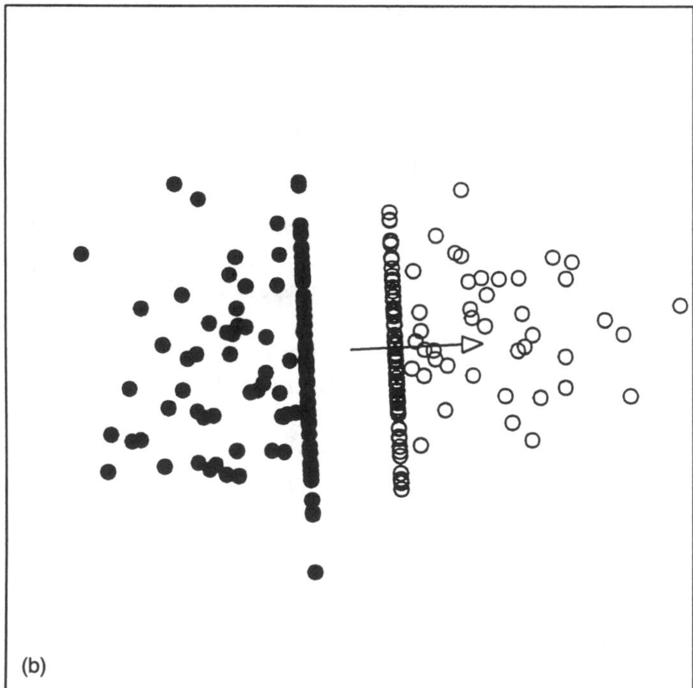

Fig. 5.11. Separating random inputs with a perceptron of maximal stability. The classifications are random (no teacher). (a) Projection onto a random plane. (b) Projection onto a plane containing $\mathbf{w}_s$. The arrow shows the student vector $\mathbf{w}_s$.

$$N^{-1/2}\sigma \mathbf{w}\cdot\boldsymbol{\xi} < 1,$$

where $\mathbf{w}$ is the vector of the old couplings. If, on the other hand,

$$N^{-1/2}\sigma \mathbf{w}\cdot\boldsymbol{\xi} \geq 1,$$

the old couplings also provide optimal stability for the $P+1$ pattern system. If this happens, then the new pattern is *uncorrelated* to $\mathbf{w}$.

Let $\mathcal{P}_0$ be the probability over the distribution of the new input for this event. It turns out that $\mathcal{P}_0$ and the probability G for a *correct* generalization on the new input are rather similar:

$$\begin{aligned} \mathcal{P}_0 &= \Pr(N^{-1/2}\sigma \mathbf{w}\cdot\boldsymbol{\xi} \geq 1) \\ G &= \Pr(N^{-1/2}\sigma \mathbf{w}\cdot\boldsymbol{\xi} \geq 0). \end{aligned} \tag{5.133}$$

Since $1 > 0$, it is clear that $P_0 \leq G$, so that we have for the generalization error

$$\varepsilon = 1 - G \leq 1 - \mathcal{P}_0. \tag{5.134}$$

Now, $\alpha\mathcal{P}_0$ is the average, relative number of patterns that need not be learned explicitly. Conversely, $\alpha_{eff} = \alpha(1-\mathcal{P}_0)$ is the average fraction of patterns that must be learned. Since we always have $\alpha_{eff} \leq 1$, we get from Eq. (5.134)

$$\alpha\varepsilon = \alpha(1-G) \leq \alpha_{eff} \leq 1. \tag{5.135}$$

Thus, we will always have $\varepsilon \leq 1/\alpha$.

There are several algorithms that are guaranteed to find the optimal perceptron for αN many random examples $\boldsymbol{\xi}_k$. Unfortunately, one cannot classify the examples according to Eq. (5.132) *in advance*; hence, one has to learn all of them instead of a fraction α_{eff}/α of them, which becomes very small for $\alpha \to \infty$.

One algorithm (Minover [38]) is an extension of the standard Rosenblatt [37] rule; another faster algorithm (Adatron [39]) is related to the quadratic optimization discussed above. But algorithms derived from standard optimization theories also have been developed [40].

All of these algorithms converge to the perceptron with maximal stability. Its properties have been calculated in [31] using the replica method of Gardner [12] introduced in Sec. 5.3.2. Now the function Φ of Eq. (5.66) is

$$\Phi(u_a) = \Theta(u_a - \kappa), \tag{5.136}$$

and $\Phi_t(u) = \Theta(u)$. Maximizing κ shrinks the volume in student space $\mathbf{w}_s$ to a single point, and the overlap $q = \mathbf{w}_a \cdot \mathbf{w}_b/N$ approaches the square of the norm, i.e., $q \to q_0 = \mathbf{w}_s \cdot \mathbf{w}_s/N$.

Figure 5.9 shows the generalization error as a function of the size α of the training set. ε decreases monotonically and behaves like

$$\varepsilon \simeq \frac{0.50}{\alpha} \tag{5.137}$$

for $\alpha \to \infty$. Hence, asymptotically, the perceptron with optimal stability can generalize much better than the Hebbian or ADALINE rule (for which $\varepsilon \simeq 1/\sqrt{\alpha}$). It performs only slightly worse than the Bayesian lower bound of Sec. 5.3.3, although this difference means that maximal stability does not imply optimal generalization.

5.5.4 Queries

In the previous applications of the statistical mechanics of neural networks only random input patterns were considered. However, it seems obvious that the student network can improve its generalization performance if it selects input patterns according to its present state of knowledge. In particular, if the fraction α of learned examples is large, a new random plane is unlikely to cut the (small) version space into two parts of roughly the same size; hence, the gain of information about the teacher is very small (see Sec. 5.2.4).

Much more information can be obtained if the student selects a question according to its present state [41]. For the simple perceptron, a good choice seems to be a pattern $\boldsymbol{\xi}_k$ that is perpendicular to the weight vector $\mathbf{w}_s$. Such a pattern is at the border of knowledge; tiny chances of $\mathbf{w}_s$ produce different answers.

For the simplest learning rule, e.g., the Hebbian algorithm discussed in Sec. 5.5.1, one easily obtains a differential equation for the overlap R and the length q_0, which determine the generalization error. Equation (5.111) gives $\Delta q_0 = 1/N$ since $\boldsymbol{\xi}_k \cdot \mathbf{w}_s = 0$ by construction. Hence, one has

$$q_0 = \alpha \; . \tag{5.138}$$

But $\overline{|\mathbf{w}_t \cdot \boldsymbol{\xi}_k|}$ of Eq. (5.108) also can be easily calculated. With $|\mathbf{w}_t| = \sqrt{N}$, the component of $\mathbf{w}_t$ perpendicular to $\mathbf{w}_s$ has a length $\sqrt{N}\sin\theta$, where θ is the angle between the teacher and the student vectors. If $\boldsymbol{\xi}_k$ is chosen randomly in the plane perpendicular to $\mathbf{w}_s$, then $\mathbf{w}_t\ \boldsymbol{\xi}_k$ is Gaussian distributed with variance

$$\begin{aligned} \overline{(\mathbf{w}_t \cdot \boldsymbol{\xi}_k)^2} &= N \sin^2 \theta \\ &= N(1-\cos^2 \theta) \; . \end{aligned} \tag{5.139}$$

With $\cos\theta = R/\sqrt{q_0} = R/\sqrt{\alpha}$, one finds

$$\frac{1}{\sqrt{N}}\overline{|\mathbf{w}_t\ \boldsymbol{\xi}_k|} = \sqrt{\frac{2}{\pi}}\sqrt{1-\frac{R^2}{\alpha}}, \tag{5.140}$$

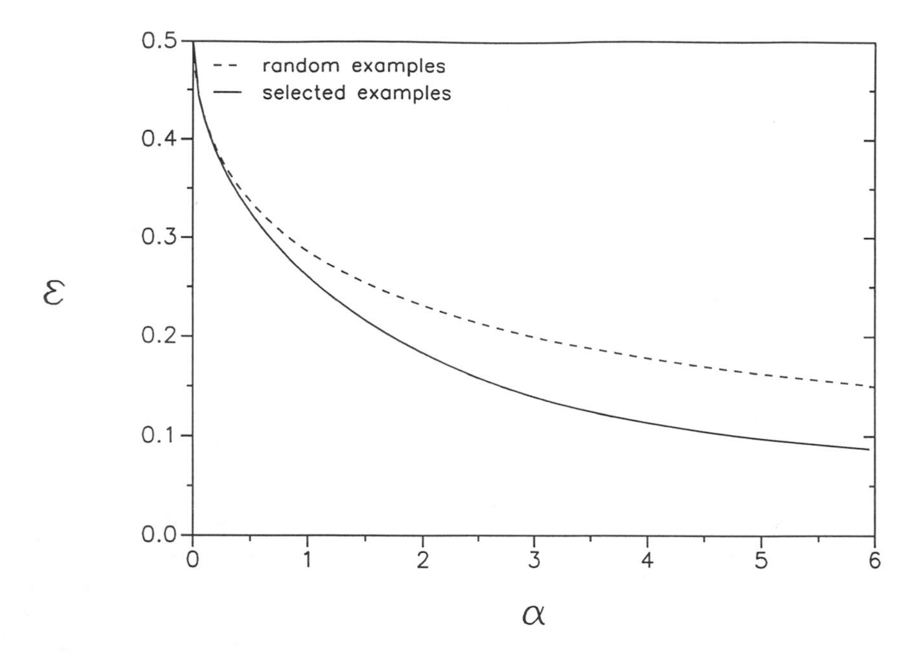

Fig. 5.12. Comparison of learning with selected and random inputs, using Hebb's rule.

which gives, with Eq. (5.108),

$$\frac{dR}{d\alpha} = \sqrt{\frac{2}{\pi}}\,\sqrt{1 - \frac{R^2}{\alpha}}\,. \tag{5.141}$$

The solution $R(\alpha)$ determines the generalization error $\varepsilon(\alpha)$ by $\varepsilon = \arccos\,(R/\sqrt{\alpha})/\pi$.

Figure 5.12 compares the results of random and selected examples. Although the generalization error is lower for "intelligent" questions, the asymptotic decay for large values of α is $\varepsilon \propto 1/\sqrt{\alpha}$ for both cases.

This is different if the whole set of examples is relearned after a new pattern was selected. Then the perceptron with maximal stability gives an exponential decay of the generalization error with an increasing fraction α of the number of learned examples [41, 42]. For *random* patterns, the Bayesian bound of Sec. 3.3.3 as well as the optimal perceptron give $\varepsilon \propto 1/\alpha$. Hence, in this case, selected examples give much better performance.

For more complicated networks it may be difficult to find patterns at the border of knowledge: An algorithm has been investigated that uses the principle of maximal disagreement between several students as a selection process [43]. Several students are trained on the same set of examples by an algorithm that selects students randomly in the version space. Then an algorithm starts which selects a new example for the training set: Many random input vectors are presented to the students, and one is chosen on which the students disagree most. This problem has been solved using the replica theory. For large α, the gain of information becomes constant, yielding an exponential decay of the generalization error; this even holds for only two students.

Selecting examples according to the weight vector $\mathbf{w}_s$ (or several vectors $\mathbf{w}_s$) may not be the best way of selecting examples. If the student learns a new example that is perpendicular to all of the previous ones, the generalization error is much lower than for the examples perpendicular to the actual $\mathbf{w}_s$ (α) [2]. However, this algorithm works only for $\alpha < 1$.

5.5.5 Discontinuous Learning

If one increases the number of examples, one expects that the generalization error of a network continuously decreases to its minimal possible value for $\alpha \to \infty$. If the rule is completely learnable, then the asymptotic error is 0, at least for perfect learning. However, a different behavior is observed for perceptrons with binary weights: For small α, ε decreases; but at a critical value α_c, ε jumps discontinuously to a lower value that is 0 for a realizable rule [44, 45]. This transition occurs even for high-temperature learning. In this case, it can be easily described analytically, since one does not need replicas [47]. We consider the case where both the teacher and the student are simple perceptrons with binary weights $\mathbf{w}_s, \mathbf{w}_t \in \left\{1/\sqrt{N},\ -1/\sqrt{N}\right\}^N$. The student learns a set of αN many examples $(\boldsymbol{\xi}_k,\ \sigma_k)$ from the teacher, and the training algorithm is a Monte Carlo procedure. After learning each weight vector, $\mathbf{w}_s$ occurs with probability

$$p(\mathbf{w}_s) \ \propto \ \exp\left[-\beta \sum_k \ \Theta\left[-(\mathbf{w}_t \cdot \boldsymbol{\xi}_k)(\mathbf{w}_s \cdot \boldsymbol{\xi}_k)\right]\right] \ . \tag{5.142}$$

For high temperatures, $T = 1/\beta$, the free energy f per synapse of the thermal equilibrium after learning is only a function of the overlap R between the teacher and the student:

$$-\beta f = -\frac{\alpha\beta}{\pi} \arccos\ R - \frac{1-R}{2} \ln \frac{1-R}{2} - \frac{1+R}{2} \ln \frac{1+R}{2} \ . \tag{5.143}$$

The first term is the generalization error, and the second term is the entropy of Ising variables with magnetization R. Note that T and α appear only as

$$\alpha_{eff} = \alpha / T \ . \tag{5.144}$$

Hence, in the limit $T \to \infty$, the network has to learn $\alpha \to \infty$ many examples. The minimum of $f(R)$ gives the equation that determines the overlap R:

$$R = \tanh \frac{\alpha\beta}{\pi\sqrt{1-R^2}} \ . \tag{5.145}$$

Solving these equations, one finds three different regimes of α_{eff}:

1. For $\alpha_{eff} < 1.7$, a state with $R < 1$ is the minimum of f. The generalization error decreases from $\varepsilon = \frac{1}{2}$ at $\alpha_{eff} = 0$ to $\varepsilon \simeq 0.2$ at $\alpha_{eff} = 1.7$.

2. Between $1.7 < \alpha_{eff} < 2.1$, the state with $R < 1$ is a local minimum, only; the state $R = 1$ has lower free energy. Hence, the system has a first-order transition to perfect generalization.

3. For $\alpha_{eff} > 2.1$, the metastable state with $R < 1$ disappears.

Note that for large α the network collapses to its ground state at *high temperatures*! To understand this, consider a small deviation $\delta R = 1 - R$ from the state of perfect generalization. The energy increases like $E \propto N\sqrt{\delta R}$. This increase cannot be compensated for by the entropy increase $\delta S \propto N(\delta R)\ln(\delta R)$. Hence, the state $R = 1$ is always a local minimum of $f(R)$; and if the initial state of the student is identical to the teacher, then no Monte Carlo algorithm can move the student out of this state of perfect generalization. Increasing the complexity of the student network by using a multilayer architecture with binary weights leads to even more phases and discontinuous transitions of the generalization error [46].

At zero temperature, i.e., for perfect learning, the first-order transition for the binary weight perceptron occurs at $\alpha_c = 1.245$ [45]. Approaching the transition from below, $\alpha \rightarrow \alpha_c$, the entropy S obtained from the number of weight vectors $\mathbf{w}_s$ that learn αN many examples perfectly goes to 0. S has been calculated by the replica method.

Figure 5.13 shows the phase diagram of the binary perceptron obtained from replica calculation including replica symmetry breaking (RSB) [47]. In addition to the three phases discussed above, there is a spin-glass phase where a solution with one-step RSB exists; this solution is metastable. The spin-glass phase indicates a complex space of students $\{\mathbf{w}_s\}$ who learn perfectly. Its implications for a dynamics of the binary weight perceptron are still unclear. A direct treatment of the dynamics gives new types of freezing transitions [48].

How many questions does one have to ask in order to obtain a complete knowledge about the N unknown weights of the student $\mathbf{w}_s$? For binary weights, one needs at least N questions; hence, the minimal possible number of patterns for which a transition to perfect generalization occurs is N. This gives a lower bound

$$\alpha_c > 1 \ . \tag{5.146}$$

Therefore, learning random patterns with a transition $\alpha_c = 1.245$ is not the optimal way of asking questions. A better strategy seems to be learning patterns at the border of knowledge, as was discussed in Sec. 5.5.4. In fact, a replica calculation gives $\alpha_c \simeq 1.14$ [42].

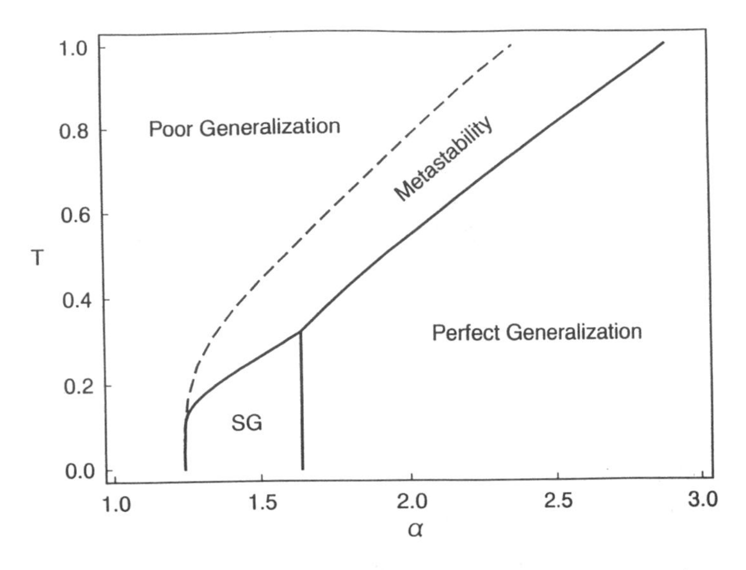

Fig. 5.13. Phase diagram for the perceptron with binary weights (taken from [47]). To the left of the dashed line, the equilibrium state has $R < 1$. To the right, the state of perfect generalization ($R = 1$) is the absolute minimum of the free energy. Between the dashed line and the solid spinodal line, the $R < 1$ state is metastable. In the region marked by "SG," a one-step replica symmetry breaking predicts a metastable spin-glass phase.

For $T = 0$, all results so far have been obtained by phase-space calculations. This means that one calculates the volume of all students who learn perfectly. However, a practicable training algorithm does not exist yet. In fact, finding a $\mathbf{w}_s$ may be an NP–hard problem of combinatorial optimization [48, 49], at least for $\alpha < 1.63$ (the upper limit of the spin-glass phase), for which an algorithm converging in a time that is a polynomial in N does not exist. Then even simulated annealing does not yield perfect learning.

5.5.6 Learning Drifting Concepts

In the previous sections the examples were given by a rule (= teacher) defined by a perceptron with a stable weight vector $\mathbf{w}_t$. All of the examples were learned iteratively, that is, the training algorithm was repeated for all of the examples until it converged.

But neural networks also may be useful for situations where the rule slowly changes with time, and the network tries to follow the changes by learning only the most recent examples. Hence, the teacher continuously changes his opinion and the student tries to adapt to such a dynamic process by learning the examples and, if possible, predicting $\mathbf{w}_t$ for the next time step.

In the simplest case, the teacher vector $\mathbf{w}_t$ is performing a random walk in the N-dimensional space [50, 51] with

$$\mathbf{w}_t(t+1) \cdot \mathbf{w}_t(t) = 1 - \frac{\eta}{N}, \tag{5.147}$$

where η is a measure of the *drift velocity.* The student learns only one example $(\boldsymbol{\xi}, \sigma)$ with $\sigma = \text{sign}\,(\mathbf{w}_t \cdot \boldsymbol{\xi})$ given by the teacher at time t. The learning rule uses only information about the output bit σ and the field $h(t)$ of the student. One defines

$$\mathbf{w}_s(t+1) = \mathbf{w}_s(t)\left(1 - \frac{\lambda}{N}\right) + \frac{1}{\sqrt{N}} f\,(\sigma(t), h(t))\;\sigma(t)\;\mathbf{w}_s(t)\,. \tag{5.148}$$

f is a function that has to be optimized, and $h(t)$ is the field generated by the student, $h(t) = (1/\sqrt{N})\;\mathbf{w}_s(t) \cdot \boldsymbol{\xi}$; λ gives an additional weight decay which reduces the length of the student vector $\mathbf{w}_s$.

Again we need the overlaps $R = \mathbf{w}_t \cdot \mathbf{w}_s/N$ and $q_0 = \mathbf{w}_s \cdot \mathbf{w}_s/N$ to determine the generalization error $\varepsilon = (1/\pi)\;\arccos\;\left(R/\sqrt{q_0}\right)$. But, since only the latest example is learned, one obtains simple differential equations for $R(t)$ and $q_0(t)$, in analogy to Sec. 5.5.1. The changes of R and q are given by

$$\begin{aligned}
\Delta R &= \frac{1}{N}\left[f(\sigma, h(t))\frac{\mathbf{w}_t \cdot \boldsymbol{\xi}}{\sqrt{N}}\;\sigma_t - (\lambda + \eta)\;R\right] \\
\Delta q_0 &= \frac{2}{N}\left[f(\sigma, h(t))\;\sigma\;h(t) + \frac{1}{2}f^2(\sigma, h(t)) - \lambda q_0\right]\,.
\end{aligned} \tag{5.149}$$

These equations have to be averaged over different examples $\boldsymbol{\xi}$ and different random walks of the teacher $\mathbf{w}_t$. For random examples one obtains

$$\begin{aligned}
\frac{dR}{d\alpha} &= \overline{f(\sigma, h)\;\frac{\mathbf{w}_t\;\boldsymbol{\xi}}{\sqrt{N}}\;\sigma - (\lambda + \eta)R} \\
\frac{dq_0}{d\alpha} &= \overline{2f(\sigma, h)\frac{\mathbf{w}_t \cdot \boldsymbol{\xi}}{\sqrt{N}} + \frac{1}{2}f^2(\sigma, h) - 2\lambda q_0}\,.
\end{aligned} \tag{5.150}$$

The fields $\mathbf{w}_s \cdot \boldsymbol{\xi}$ and $\mathbf{w}_t \cdot \boldsymbol{\xi}$ are correlated Gaussian variables that allow an easy calculation of the average values $\overline{(\ldots)}$ similar to Sec. 5.5.1. The "time t" has been replaced by αN, the number of learned examples.

As before, the simplest learning rule is the Hebbian one, with $f = 1$. In this case, one finds without decay ($\lambda = 0$):

$$\begin{aligned}
R(\alpha) &= \sqrt{\frac{2}{\pi}}\,\frac{1}{\eta}\left(1 - e^{-\eta\alpha}\right) \\
q_0(\alpha) &= \left(1 + \frac{4}{\pi\eta}\right)\alpha + \frac{4}{\pi\eta^2}\;e^{-\eta\alpha} + \left(1 - \frac{4}{\pi\eta^2}\right)\,.
\end{aligned} \tag{5.151}$$

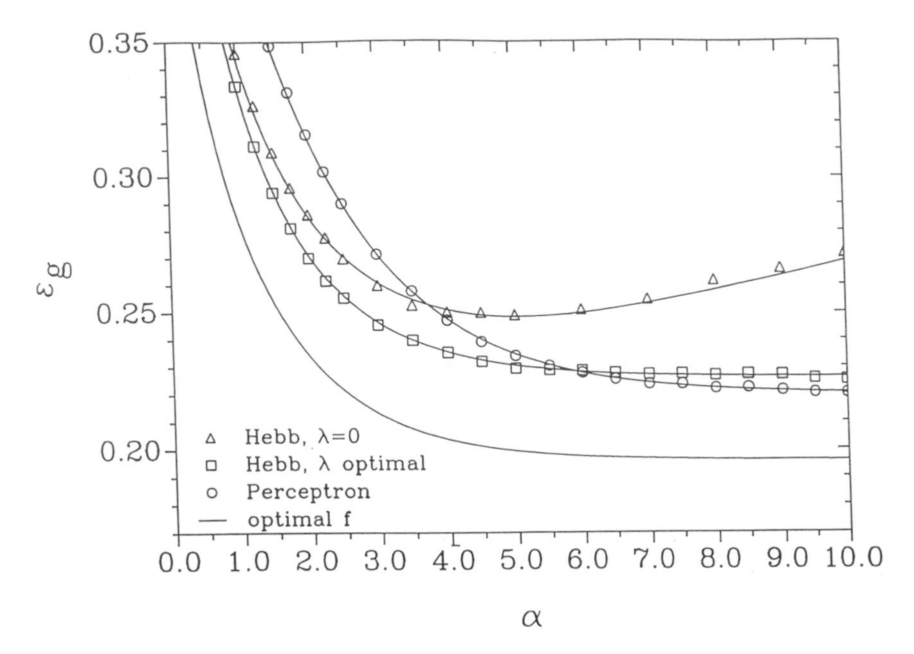

Fig. 5.14. Generalization errors for the learning of drifting concepts, using Hebbian learning (with and without weight decay), the perceptron algorithm, and the on-line algorithm with optimal f given in [53].

Figure 5.14 shows the generalization error $\varepsilon(\alpha)$ given by these equations. It has a minimum at some α value but then increases to $\varepsilon = 1/2$. Hence, the student cannot generalize if he or she has learned too much!

The reason for this surprising feature is the fact that the Hebbian couplings have equal strengths for all of the examples. But, since the teacher changes his or her direction, the examples produced some time ago destroy the most recent information that is important for generalization.

In fact, a weight decay $\lambda > 0$ produces *forgetting* [52]; hence, the error $\varepsilon(\alpha)$ decreases to a stationary value $\varepsilon(\infty)$ that can be minimized with respect to λ; the result is shown in Fig. 5.15. The minimal asymptotic error increases with small drift parameters η as

$$\varepsilon_{opt}(\infty) \simeq \frac{1}{\pi^{3/4}} \, \eta^{1/4} \, . \tag{5.152}$$

A better training algorithm is the perceptron learning rule [1], with $f(\sigma, h) = \theta\,(\kappa - \sigma h/q_0)$. Now, $\varepsilon(\infty)$ can be minimized with respect to the two parameters κ and λ. One finds [50, 51] for small η values

$$\varepsilon(\infty) \simeq 0.51 \, \eta^{1/3} \, . \tag{5.153}$$

The same power of η is found if the learner knows $\varepsilon(\alpha)$ and uses this information to derive an optimal function $f(\sigma, h)$ [53].

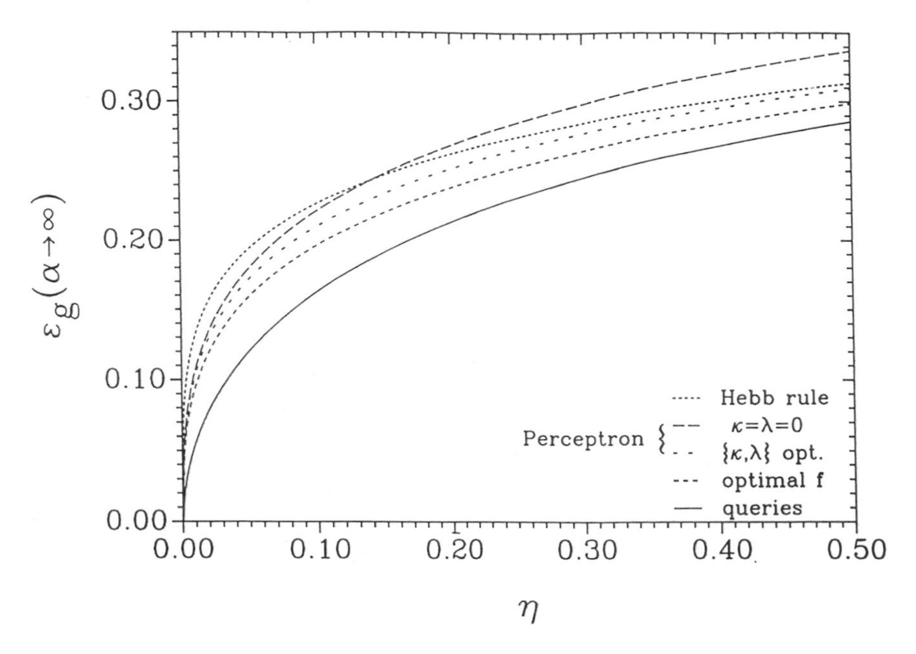

Fig. 5.15. Stationary value $\varepsilon(\infty)$ for learning of drifting concepts using the learning algorithms mentioned in Fig. 5.15. Also, the effect of queries is included.

An additional improvement is obtained by selecting examples as in Sec. 5.5.4. For the Hebb rule with optimal decay, one finds

$$\varepsilon(\infty) = \frac{1}{\pi} \arccos \frac{1}{\sqrt{1+\eta\pi}} \simeq \sqrt{\frac{\eta}{\pi}} \,. \tag{5.154}$$

Using $\varepsilon(\alpha)$ with an optimal f, the generalization error decays exponentially fast to the same asymptotic error [Eq. (5.154)].

Of course, a random walk cannot be predicted by definition. But for deterministic changes of the teacher or for biased random walks it should be possible to predict future actions of the teacher by studying the history of the presented examples. The statistical mechanics of such problems still have to be formulated.

5.5.7 Diluted Networks

In the previous examples the student had the same structure as the teacher. But it may be interesting to study cases where the student has to deduce the *structure* of the teacher from the set of presented examples. A simple case is the *diluted* teacher: Both teacher and student are simple perceptrons, but a certain fraction f of the couplings is erased. This means that the teacher has a fixed set of weights that are equal to 0, and the student also has a fixed fraction of 0 weights, but he or she is allowed to choose which bonds are to be erased.

Hence, the student has additional dynamic variables $\mathbf{c} \in \{0,1\}^N$, which are multiplied with the weights $\mathbf{w}_s \in \boldsymbol{R}^N$. For this problem, the perceptron of optimal stability can be calculated using the phase-space integral of Gardner, but now with the additional discrete variables $\mathbf{c}$ [54]. One obtains the generalization error ε as a function of α, f_s, f_t, where f_s and f_t are the fraction of nonzero bonds of the student and teacher, respectively. One finds that ε has a minimum as a function of f_s, and for large α this minimum approaches $f_s \simeq f_t$.

Again, the replica calculation does not provide us with a learning algorithm. Finding the optimal configuration $\mathbf{c}$ is presumably an NP–hard problem of combinatorial optimization, similar to the binary perceptron. Therefore, a practicable algorithm does not exist, yet. However, one might guess that, by learning the complete network and by erasing the weak bonds, one may obtain a good approximation of the optimal perceptrons. In fact, this is the case for attractor networks ($\hat{=}$ random teacher) [55].

A fast and effective dilution algorithm is given by the Hebbian couplings:

$$c_i = 0 \quad \text{if} \quad \left| \frac{1}{N} \sum_{\nu=1}^{\alpha N} \xi_i^\nu \, \sigma^\nu \right| < s, \tag{5.155}$$

where s is determined by f_s. For this fixed dilution vector $\mathbf{c}$, the remaining weights are determined by the standard algorithms for the perceptron of optimal stability [1].

The generalization error ε has been calculated analytically [54]. The order parameters now are defined by

$$\begin{aligned} q &= \frac{1}{N f_s} \sum_{i=1}^{N} c_i w_a(i) \; w_b(i) \\ R &= \frac{1}{N\sqrt{f_s f_t}} \sum_{i=1}^{N} c_i w_t(i) \; w_s(i). \end{aligned} \tag{5.156}$$

R and q determine ε as usual. Figure 5.16 shows the result. For $f_s < f_t$, the target rule is unrealizable; the student cannot reproduce the teacher perfectly, even for $\alpha \to \infty$. For $f_s > f_t$, the student has too many degrees of freedom, which deteriorates his or her ability to generalize. Hence, ε has a maximum that approaches $f_s \to f_t$ for a large fraction α of learned examples.

Note that f_s is a fixed parameter. What remains is to find an algorithm that determines the optimal dilution fraction f_s of the student. By such a learning rule the student would be able to explore the structure of the teacher.

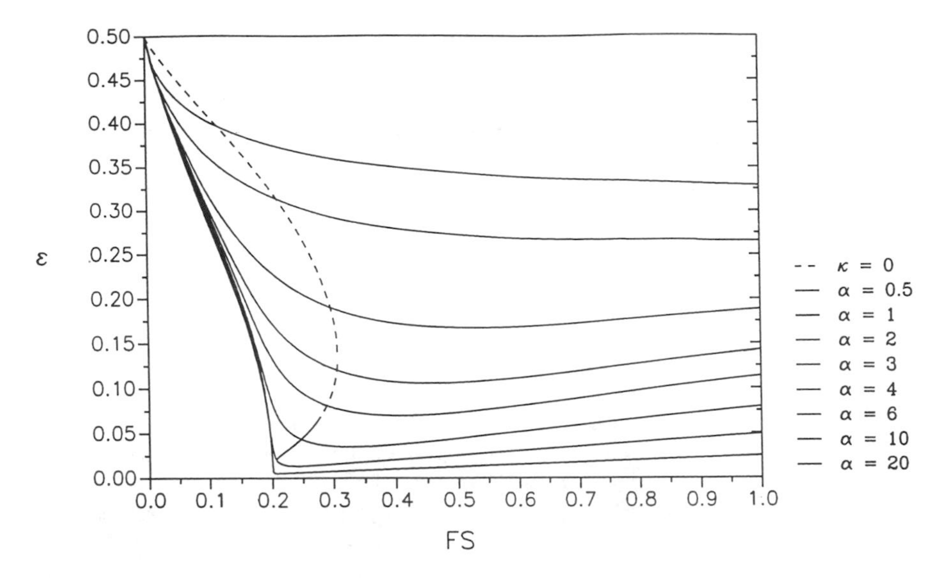

Fig. 5.16. Generalization error as a function of student dilution f_s for a teacher with dilution $f_t = 0.2$. The dashed curve separates training with errors (*left*) and without errors (*right*).

5.5.8 Continuous Neurons

Up to now we have mainly discussed output neurons with the step transfer function sign (x). The teacher as well as the student is a network with binary output $\sigma \in \{+1, -1\}$. But functions with continuous output also are interesting. First, they model a continuous firing rate as a function of excitation potential for real neurons; and, second, tasks for neurocomputers may involve analog signals, and learning rules like gradient descent work only for continuous functions [56].

In the context of statistical mechanics, continuous neurons have been studied for a simple perceptron [57]. The teacher and the student are perceptrons with weight vectors $\mathbf{w}_t$ and $\mathbf{w}_s$, respectively. But now the output signal is given by

$$\sigma = \tanh\left(\frac{\gamma}{\sqrt{N}}\, \mathbf{w} \cdot \boldsymbol{\xi}\right), \tag{5.157}$$

where γ is a parameter that measures the degree of nonlinearity of the transfer function. An increase of the student's length $q_0 = \mathbf{w}_s \cdot \mathbf{w}_s / N$ can be compensated for by a decrease of the slope γ_s of Eq. (5.157). Hence, only the product $\gamma_s^2\, q_0$ has a physical meaning, and the student has the freedom to adjust its slope γ_s during learning. In [57], $q_0 = 1$ was chosen.

Learning again is expressed as minimizing a cost function E, which is chosen as the quadratic deviation

$$E = -\sum_k \left[\tanh\left(\frac{\gamma_t}{\sqrt{N}}\,\mathbf{w}_t \cdot \mathbf{w}_k\right) - \tanh\left(\frac{\gamma_s}{\sqrt{N}}\,\mathbf{w}_s \cdot \boldsymbol{\xi}_k\right)\right]^2 . \quad (5.158)$$

By defining

$$\sigma_k = \left(\frac{\gamma_t}{\sqrt{N}}\mathbf{w}_t \cdot \boldsymbol{\xi}_k\right), \quad (5.159)$$

one observes that $E = 0$ implies $\tilde{E} = 0$ for the function

$$\tilde{E} = -\sum_k \left[\sigma_k - \frac{\gamma_s}{\sqrt{N}}\,\mathbf{w}_s \cdot \boldsymbol{\xi}_k\right]^2 . \quad (5.160)$$

This is just the cost function for a linear network! For $\alpha < 1$, $\tilde{E} = 0$ gives less equations than unknowns, and the solution with minimal γ_s (corresponding to minimal norm) is given by the pseudoinverse as in Sec. 5.5.2. Using the replica method, one finds [57]

$$\gamma_s = \sqrt{\alpha}\gamma_t, \qquad R = \sqrt{\alpha} . \quad (5.161)$$

For $\alpha > 1$, $E = \tilde{E} = 0$ gives perfect generalization with $\gamma_s = \gamma_t$ and $R = 1$.
The generalization error ε can be defined by

$$\varepsilon = \frac{1}{2}\overline{\left[\tanh\left(\frac{\gamma_t}{\sqrt{N}}\,\mathbf{w}_t \cdot \boldsymbol{\xi}\right) - \tanh\left(\frac{\gamma_s}{\sqrt{N}}\,\mathbf{w}_s \cdot \boldsymbol{\xi}\right)\right]^2}, \quad (5.162)$$

i.e., the quadratic deviation between the answers of the teacher and the student for random patterns.

One finds for $\alpha < 1$

$$\varepsilon = \frac{1}{2}\int_{-\infty}^{\infty} Dx \int_{-\infty}^{\infty} Dy \left[\tanh(\gamma_t x) - \tanh\left(\gamma_t\sqrt{\alpha(1-\alpha y)} + \gamma_t \alpha x\right)\right]^2 \quad (5.163)$$

while $\varepsilon = 0$ for $\alpha > 1$. Figure 5.17 shows the results $\varepsilon(\alpha)$ for different teacher slopes γ_t. Surprisingly, for $\gamma_t > 1.33$, the generalization error increases with α when only a small number α of examples has been learned.

5.5.9 Unsupervised Learning

In the previous sections, a teacher function presented answers to random inputs to a student network. Hence, the teacher classified the input patterns.

However, sometimes one would like to find a classification of input patterns *without* knowing the answer of a teacher; hence, the input patterns

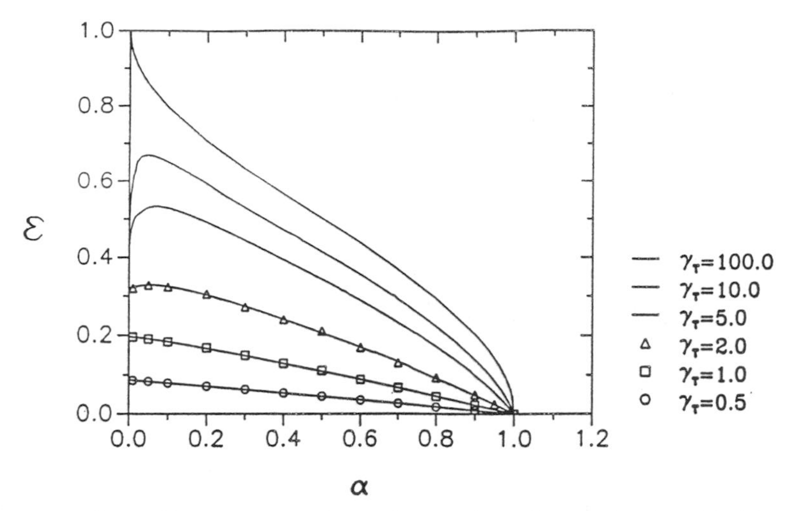

Fig. 5.17. Generalization errors for student and teacher with a "tanh($\gamma\cdot$)" transfer function. γ_T is the gain factor of the teacher.

$\boldsymbol{\xi}_k$ have a structure that the student network has to find out. Recently this problem of unsupervised learning was studied in the framework of the statistical mechanics of simple perceptrons [58].

The inputs are no longer completely random, but they have an internal structure defined by a teacher vector $\mathbf{w}_t$: The patterns $\boldsymbol{\xi}_k$ belong to two "clouds" with respect to the overlap to the teacher. The distribution of $u = \mathbf{w}_t \cdot \boldsymbol{\xi}_k/\sqrt{N}$ is a double Gaussian, that is, each peak has a width 1 and the two peaks are separated by $2\rho/\sqrt{N}$ with a parameter $\rho = 0(1)$. Note that the patterns have only a very weak overlap $\rho/\sqrt{N}$ with the teacher vector $\mathbf{w}_t$. In contrast to the previous problems, the student does not know the sign of u.

Learning again is expressed as minimizing a cost function E, and statistical mechanics of the phase space of all students $\mathbf{w}_s$ is used to calculate the overlap $R = \mathbf{w}_t \cdot \mathbf{w}_s/N$ after having learned αN many examples $\boldsymbol{\xi}_k$ taken from the double peak distribution.

Two cost functions have been considered:

$$E_A = -\frac{1}{N}\sum_k (\mathbf{w}_s \cdot \boldsymbol{\xi}_k)^2 \tag{5.164}$$

$$E_B = -\sum_k \theta(\kappa - |\mathbf{w}_s \cdot \boldsymbol{\xi}_k|/\sqrt{N}) \; . \tag{5.165}$$

The first corresponds to principal component analysis [56] and the second to finding the perceptron of maximal stability κ, i.e., one maximizes κ with $E_B = 0$, if possible.

In both cases, one finds a critical value α_c below which the student cannot generalize ($R = 0$). Only if the number of learned examples is larger

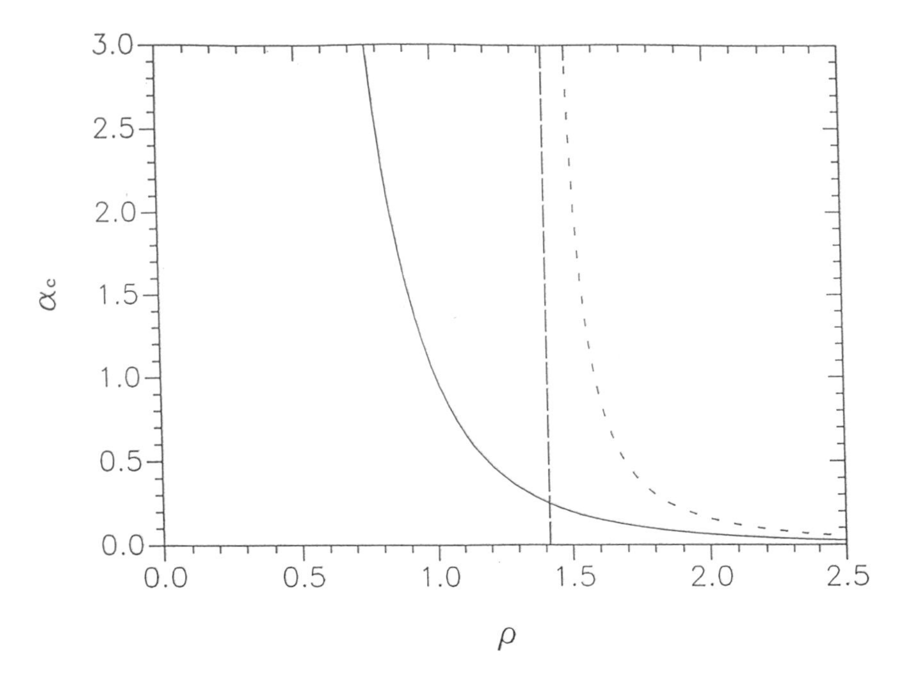

Fig. 5.18. Critical number of inputs/weight, below which unsupervised learning is impossible for: principal component analysis (solid curve) and maximal stability (dashed curve). The vertical line is $\rho = \sqrt{2}$.

than $\alpha_c N$ can the student develop an overlap with the teacher direction. Of course, the sign of the classification cannot be deduced, since it is not shown by the teacher (unsupervised learning). Figure 5.18 shows the critical number α_c as a function of ρ, which measures how strong the double peak structure of the cloud of patterns shows up. For the first case, α_c diverges with $\rho > 0$, and one finds

$$\alpha_c = \rho^{-1/4} . \tag{5.166}$$

But, surprisingly, the perceptron with maximal stability cannot generalize if the distinction ρ of the two classes of input patterns is smaller than $\rho_c = \sqrt{2}$. And, even for $\alpha \to \infty$, the generalization error does not decrease to 0, but one has

$$R(\alpha \to \infty) = \pm \sqrt{1 - 2/\rho^2} \qquad \text{for} \;\; \alpha > \sqrt{2} . \tag{5.167}$$

5.6 Summary and Outlook

We set out to convince the reader that the study of simple mathematical models is a promising way to understand at least part of a neural network's abilities to learn from examples. Thus, in the first part of this chapter we tried to review a few of the basic theoretical ideas and tools which are currently discussed in the computer science and statistical physics literature on neural networks.

The *Vapnik–Chervonenkis* method, well known in theoretical computer science, is able to bound the generalization error using only a single parameter of the class of networks, rather than their complete architecture.

The statistical physicist's tools, which mainly are based on the *replica method*, are designed for very large nets and allow for the exact calculation of learning curves in a variety of circumstances. Here, however, one is practically restricted to simple architectures and some hopefully "natural" probability distributions for the examples to be learned.

In the second part of the chapter we concentrated on the statistical physicist's methods and presented a variety of learning problems that can be treated *exactly* for a special network, the perceptron, which is far from being a toy model. Although there is a great interest to study more complicated, multilayer nets [2], the amount of recent results for perceptrons suggests that there are still more interesting facts to be discovered for this machine.

We found a rich structure of *learning curves* that may not be easily recovered within the VC framework. This stems from the fact that problems such as *overfitting, discontinuous learning*, or *intelligent dilution* are essentially related to either specific learning algorithms or specific features of the network architecture. On the other hand, comparing the VC predictions and the concrete learning curves for the perceptron, we found that the VC bounds match the correct order of magnitude for the typical *asymptotic* behavior in many cases. Thus, it seems that the asymptotic region can be estimated correctly by using only a *few parameters* of a neural network.

It would be a challenge to combine statistical physics methods based on the replica trick and the VC techniques. Such an approach may be helpful and important in treating multilayer nets when the complex structure of the network's phase space makes exact replica calculations a hard task.

Acknowledgments. We thank Andreas Mietzner for assistance and acknowledge support from the Deutsche Forschungsgemeinschaft and the Volkswagenstiftung.

Appendix 5.1: Proof of Sauer's Lemma

As can be easily seen, the first inequality of Eq. (5.4) is proved if we can show the following *theorem:*

Consider a sequence of inputs $\boldsymbol{\xi}^P = \boldsymbol{\xi}_1, \ldots, \boldsymbol{\xi}_P$. If there is an integer d, such that the number $\mathcal{N}(\boldsymbol{\xi}^P)$ of cells or output configurations fulfills

$$\mathcal{N}(\boldsymbol{\xi}^P) > \sum_{i=0}^{d} \binom{P}{i}, \tag{5.168}$$

then we can find a subsequence of these inputs, of length $d+1$, for which

$$\mathcal{N}(\boldsymbol{\xi}^{d+1}) = 2^{d+1}.$$

The proof is by induction on P and d. It is easy to see that the theorem holds $d = 0$. It also holds for any $P \leq d$ because, in this case, the premise (5.168) can never be fulfilled: The sum of binomials is then $\geq 2^P$. But $\mathcal{N}(\boldsymbol{\xi}^P)$ must always be $\leq 2^P$.

Let the assertion be true for all $d \leq d_0$ and all numbers of inputs. Now, assume that the theorem is also true for $d = d_0 + 1$ and for $P < P_0$ inputs. We then will show that it holds for all P.

We add a $P_0 + 1$st input $\boldsymbol{\xi}$ and assume the premise (5.168):

$$\mathcal{N}(\boldsymbol{\xi}^{P_0+1}) > \sum_{i=0}^{d_0+1} \binom{P_0+1}{i}. \tag{5.169}$$

If, on the *first* P_0 *inputs*, we had $\mathcal{N}(\boldsymbol{\xi}^{P_0}) > \sum_{i=0}^{d_0+1} \binom{P_0}{i}$, then, by the induction assumption, the theorem is true.

So let us discuss the other case:

$$\mathcal{N}(\boldsymbol{\xi}^{P_0}) \leq \sum_{i=0}^{d_0+1} \binom{P_0}{i}.$$

We divide the old cells (those for the first P_0 inputs) into two groups: a group M_2, which contains cells that will split into *two subcells* on presenting the new input, i.e., both outputs are possible on $\boldsymbol{\xi}$. The remaining cells, i.e., those which do not split, are contained in M_1. Obviously,

$$\mathcal{N}(\boldsymbol{\xi}^{P_0+1}) = |M_1| + 2|M_2|$$

$$\mathcal{N}(\boldsymbol{\xi}^{P_0}) = |M_1| + |M_2| \leq \sum_{i=0}^{d_0+1} \binom{P_0}{i}. \tag{5.170}$$

The bars denote the number of cells in the groups. Now, we study two possibilities: If $|M_2| \leq \sum_{i=0}^{d_0} \binom{P_0}{i}$, then, by Eq. (5.170), we would have

$$\mathcal{N}(\boldsymbol{\xi}^{P_0+1}) \leq \sum_{i=0}^{d_0+1} \binom{P_0}{i} + \sum_{i=0}^{d_0} \binom{P_0}{i} = \sum_{i=0}^{d_0+1} \binom{P_0+1}{i}, \tag{5.171}$$

by a standard addition theorem for binomials. But this contradicts our condition (5.169). So we are left with the second possibility:

$$|M_2| > \sum_{i=0}^{d_0} \binom{P_0}{i}.$$

By the induction assumption we can find a subsequence of length $d_0 + 1$ out of the first P_0 inputs, such that the teachers of the cells in M_2 produce 2^{d_0+1} cells. Since these are able to give *both* possible answers on the new input $\boldsymbol{\xi}$, we have constructed a subsequence of length $d_0 + 2$ with 2^{d_0+2} output combinations. This completes the proof.

Appendix 5.2: Order Parameters for ADALINE

For $\alpha < 1$ it is well known [59] that the coupling vector can be explicitly written as

$$\mathbf{w}_s = N^{-1/2} \sum_{kl} \sigma_k (C^{-1})_{kl} \boldsymbol{\xi}_l \tag{5.172}$$

with $C_{kl} = N^{-1} \boldsymbol{\xi}_k \cdot \boldsymbol{\xi}_l$. The length of the coupling vector is then

$$q_0 = N^{-1} \mathbf{w} \cdot \mathbf{w} = N^{-1} \sum_{kl} \sigma_k (C^{-1})_{kl} \sigma_l. \tag{5.173}$$

The basic idea is to calculate the order parameters from an average over the teacher. Technically, it is useful to choose Gaussian distributed teacher vectors with density

$$g(\mathbf{w}_t) = (2\pi)^{-N/2} \cdot \exp(-\tfrac{1}{2} \mathbf{w}_t \cdot \mathbf{w}_t).$$

This realizes a homogeneous distribution on the surface of a sphere. The outputs are then $\sigma_k = \mathrm{sign}(u_k)$, where the fields $u_k = N^{-1/2} \mathbf{w}_t \cdot \boldsymbol{\xi}_k$ are Gaussian variables with

$$\langle u_k \, u_l \rangle = C_{kl}. \tag{5.174}$$

For random inputs, C_{kl} is typically of order $N^{-1/2}$ for $k \neq l$, and

$$\langle \sigma_k \sigma_l \rangle = \begin{cases} 1 & \text{for } k = l \\ \frac{2}{\pi} C_{kl} + \mathcal{O}(\frac{1}{N}) & \text{for } k \neq l. \end{cases} \tag{5.175}$$

One can show [1] that for random inputs and $N \to \infty$,

$$N^{-1} \sum_k (C^{-1})_{kk} = \frac{\alpha}{1 - \alpha}. \tag{5.176}$$

Using this equation, and inserting Eq. (5.175) into Eq. (5.173), we get

$$q_0 = \frac{\alpha - \frac{2}{\pi}\alpha^2}{1 - \alpha}. \tag{5.177}$$

Finally, for the second order parameter, we get

$$R = N^{-1}\langle \mathbf{w}_t \cdot \mathbf{w}_s \rangle = N^{-1} \sum_{kl} (C^{-1})_{kl} \langle u_k \ \mathrm{sign}(u_l) \rangle = \alpha \sqrt{\frac{2}{\pi}}. \qquad (5.178)$$

The case $\alpha > 1$ can be treated by the same method. We will not give the details here [60]. We only mention that $\mathbf{w}_s$ is the minimum of the quadratic learning error

$$\sum_k (\sigma_k - N^{-1/2} \mathbf{w} \cdot \boldsymbol{\xi}_k)^2. \qquad (5.179)$$

Taking the gradient, we get explicitly for the ith component

$$w_s(i) = \sum_k (B^{-1})_{ij} f_j, \qquad (5.180)$$

with

$$B_{ij} = N^{-1} \sum_k \xi_k(i) \xi_k(j)$$

and

$$f_j = N^{-1/2} \sum_k \sigma_k \xi_k(j).$$

Again, the order parameters can be claculated by averaging over the teacher vector.

References

[1] W. Kinzel, M. Opper (1991) Dynamics of learning, In: *Physics of Neural Networks*, J. L. van Hemmen, E. Domany, K. Schulten (Eds.) (Springer-Verlag, New York), p. 149

[2] T.L.H. Watkin, A. Rau, M. Biehl (1993) *Rev. Mod. Phys.* **65**:499

[3] N. Sauer (1972) *J. Comb. Theory A* **13**:145

[4] V.N. Vapnik (1982) *Estimation of Dependences Based on Empirical Data* (Springer-Verlag, New York)

[5] E. Baum, D. Haussler (1989) *Neural Comput.*. **1**(1):151–160

[6] A. Blumer, A. Ehrenfeucht, D. Haussler, M.K. Warmuth (1989) *J. Assoc. Comp. Mach.* **36**:929

[7] E. Levin, N. Tishby, S. Solla (1989) A statistical approach to learning and generalization in neural networks, In: *Proc. 2nd Workshop on Computational Learning Theory* (Morgan Kaufmann)

[8] G. Gyorgyi, N. Tishby (1990) Statistical theory of learning a rule, In: *Neural Networks and Spin Glasses*, (World Scientific)

[9] M. Opper, D. Haussler (1991) *Phys. Rev. Lett.* **66**:2677

[10] M. Opper, D. Haussler (1991) In: *IVth Annual Workshop on Computational Learning Theory (COLT91)* (Santa Cruz, 1991) (Morgan Kaufmann, San Mateo, CA), pp. 75–87

[11] D. Haussler, M. Kearns, M. Opper, R.E. Schapire (1991) Estimating average — Case learning curves using Bayesian, statistical physics and VC dimension methods, In: *Neural Information Processing (NIPS 91)*

[12] E. Gardner (1988) *J. Physics A* **21**:257–270

[13] D. Haussler, M. Kearns, R. Schapire (1991) In: *IVth Annual Workshop on Computational Learning Theory (COLT91)* (Santa Cruz, 1991) (Morgan Kaufmann, San Mateo, CA), pp. 61–74

[14] D. Haussler, A. Barron (1992) How well do Bayes methods work for on-line prediction of $\{+1,-1\}$ values? In: *Proc. Third NEC Symposium on Computation and Cognition* (SIAM, Philadelphia, PA)

[15] J. Rissanen (1986) *Ann. Stat.* **14**:1080

[16] R. Meir, J.F. Fontanari (1993) *Proc. IVth International Bar–Ilan Conference on Frontiers in Condensed Matter Physics*, published in *Physica A* **200**:644

[17] H. Sompolinsky, N. Tishby, H.S. Seung (1990) *Phys. Rev. Lett.* **65**:1683

[18] S. Amari, N. Murata (1993) *Neural Computation* **5**:140

[19] T.M. Cover (1965) *IEEE Trans. El. Comp.* **14**:326–334

[20] G. Stambke (19XX) diploma thesis

[21] G. Gyorgyi (1990) *Phys. Rev. Lett.* **64**:2957

[22] M. Mezard, G. Parisi, M.A. Virasoro (1987) *Spin Glass Theory and Beyond*, Lecture Notes in Physics, 9 (World Scientific)

[23] T.L.H. Watkin (1993) *Europhys. Lett.* **21**:871

[24] R. Meir, J.F. Fontanari (1992) *Phys. Rev. A* **45**:8874

[25] S. Amari (1993) *Neural Networks* **6**:161

[26] M. Opper, D. Haussler, in preparation

[27] F. Vallet, J. Cailton, P. Refregier (1989) *Europhys. Lett.* **9**:315–320

[28] D.E. Rumelhart, J.L. McClelland, eds. (1986) *Parallel Distributed Memory* (MIT Press, Cambridge, MA)

[29] B. Widrow, M.E. Hoff (1960) WESCON Convention, Report IV, 96

[30] I. Kanter, H. Sompolinsky (1987) *Phys. Rev. A* **35**:380

[31] M. Opper, W. Kinzel, J. Kleinz, R. Nehl (1990) *J. Phys. A* **23**:L581

[32] M. Opper (1989) *Europhys. Lett.* **8**:389

[33] A.J. Hertz, A. Krogh, G.I. Thorbergsson (1989) *J. Phys. A* **22**:2133

[34] A. Krogh, J. Hertz (1991) In: *Advances in Neural Information Processing Systems III* (Morgan Kaufmann, San Mateo, CA)

[35] Y. LeCun, I. Kanter, S. Solla (1991) *Phys. Rev. Lett.* **66**:2396

[36] M. Opper (1988) *Phys. Rev. A* **38**:3824

[37] F. Rosenblatt (1961) *Principles of Neurodynamics — Perceptrons and the Theory of Brain* (Spartan Books, Washington DC)

[38] W. Krauth, M. Mezard (1987) *J. Phys. A* **20**:L745

[39] J. Anlauf, M. Biehl (1989) *Europhys. Lett.* **10**:687

[40] P. Ruján (1993) *J. de Phys. (Paris) I* **3**:277

[41] W. Kinzel, P. Ruján (1990) *Europhys. Lett.* **13**:473

[42] T.L.H. Watkin, A. Rau (1992) *J. Phys. A* **25**:113

[43] H.S. Seung, M. Opper, H. Sompolinsky (1992) In: *Vth Annual Workshop on Computational Learning Theory (COLT92)* (Pittsburgh 1992) pp. 287–294 (Assoc. for Computing Machinery, New York)

[44] E. Gardner, B. Derrida (1989) *J. Phys. A* **22**:1983

[45] G. Gyorgyi (1990) *Phys. Rev. A.* **41**:7097

[46] H. Schwarze, M. Opper, W. Kinzel (1992) *Phys. Rev. A* **46**:6185

[47] H. Seung, H. Sompolinsky, N. Tishby (1992) *Phys. Rev. A* **45**:6056

[48] H. Horner (1992) *Z. Phys. B* **87**:371

[49] H.K. Patel (1993) *Z. Physik B* **91**:257

[50] M. Biehl, H. Schwarze (1992) *Europhys. Lett.* **20**:733

[51] M. Biehl, H. Schwarze (1993) *J. Phys. A* **26**:2561

[52] M. Biehl (19XX) diploma thesis, University of Giessen

[53] O. Kinouchi, N. Caticha (1992) *J. Phys. A* **25**:6243

[54] P. Kuhlmann, K.R. Müller (1994) *J. Phys. A* **27**:3759

[55] R. Garces, P. Kuhlmann, H. Eissfeller (1992) *J. Phys. A* **25**:L1335

[56] J. Hertz, A. Krogh, R.G. Palmer (1991) *Introduction to the Theory of Neural Computation* (Addison–Wesley, Reading, MA)

[57] S. Bös, W. Kinzel, M. Opper (1993) *Phys. Rev. E* **47**:1384

[58] M. Biehl, A. Mietzner (1993) *Europhys. Lett.* **24**:421

[59] T. Kohonen (1988) *Self Organisation and Associative Memory* (Springer-Verlag, Berlin)

[60] M. Opper (1995) in preparation

6

Bayesian Methods for Backpropagation Networks

David J.C. MacKay[1]

with 10 figures

Synposis. Bayesian probability theory provides a unifying framework for data modeling. In this framework, the overall aims are to find models that are well matched to the data, and to use these models to make optimal predictions. Neural network learning is interpreted as an *inference* of the most probable parameters for the model, given the training data. The search in model space (i.e., the space of architectures, noise models, preprocessings, regularizers, and weight decay constants) also then can be treated as an inference problem, in which we infer the relative probability of alternative models, given the data. This provides powerful and practical methods for controlling, comparing, and using adaptive network models. This chapter describes numerical techniques based on Gaussian approximations for implementation of these methods.

6.1 Probability Theory and Occam's Razor

Bayesian probability theory provides a unifying framework for data modeling. A Bayesian data modeler's aim is to develop probabilistic models that are well matched to the data, and to make optimal predictions using those models. The Bayesian framework has several advantages.

Probability theory forces us to make explicit all of our modeling assumptions. Bayesian methods are mechanistic: Once a model space has been defined, then, whatever question we wish to pose, the rules of probability theory give a unique answer that consistently takes into account all of the given information. This is in contrast to non-Bayesian statistics, in which one must invent *estimators* of quantities of interest and then choose between those estimators using some criterion measuring their sampling properties; there is no clear principle for deciding which criterion to use to

[1]Cavendish Laboratory, University of Cambridge, Madingley Road, Cambridge, CB3 0HE, United Kingdom (mackay@mrao.cam.ac.uk).

measure the performance of an estimator; nor, for most criteria, is there any systematic procedure for the construction of optimal estimators.

Bayesian inference satisfies the likelihood principle [1]: Our inferences depend only on the probabilities assigned to the data that were received, not on properties of other data sets which might have occurred but did not.

Probabilistic modeling handles uncertainty in a natural manner. There is a unique prescription (marginalization) for incorporating uncertainty about parameters into our predictions of other variables.

Finally, Bayesian model comparison embodies *Occam's razor*, the principle that states a preference for simple models. This point will be expanded on in Sec. 6.1.1.

The preceding advantages of Bayesian modeling do not make all of our troubles go away. The Bayesian is left with the twin tasks of defining an appropriate model space for the data, and implementing the rules of inference numerically.

6.1.1 Occam's Razor

Occam's razor is the principle that states a preference for simple theories. If several explanations are compatible with a set of observations, Occam's razor advises us to buy the least complex explanation. This principle is often advocated for one of two reasons: The first is aesthetic ["A theory with mathematical beauty is more likely to be correct than an ugly one that fits some experimental data" (Paul Dirac)]; the second reason is the supposed empirical success of Occam's razor. Here we discuss a different justification for Occam's razor, namely,

> Coherent inference embodies Occam's razor automatically and quantitatively.

To explain this statement, we first must introduce the language in which inferences can be expressed; this is the language of *probabilities*. All coherent beliefs and predictions can be mapped onto probabilities. We will use the following notation for conditional probabilities: $P(A|B,\mathcal{H})$ is pronounced "the probability of A, given B and $\mathcal{H}$." The statements B and $\mathcal{H}$ list the conditional assumptions on which this measure of plausibility is based. For example, if A is "it will rain today," and B is "the barometer is rising," then the quantity $P(A|B,\mathcal{H})$ is a number between 0 and 1 that expresses how probable we would think "rain today" is, given that the barometer is rising, and given the overall assumptions $\mathcal{H}$ that define our model of the weather. This conditional probability is related to the joint probability of A *and* B by $P(A|B,\mathcal{H}) = P(A,B|\mathcal{H})/P(B|\mathcal{H})$. Note that the conditioning notation does not imply causation. $P(A|B)$ does not mean "the probability that A is caused by B". Rather, it measures the plausibility of proposition A, assuming that the information in proposition

B is true. With apologies to pure mathematicians, we will use the same notation for probabilities of discrete variables and for probability densities over real variables.

Having enumerated a complete list of these conditional degrees of belief, we then can use the rules of probability to evaluate how our beliefs and predictions should change when we gain new information, i.e., as we change the conditioning statements to the right of our "|" symbol. For example, the probability $P(B|A, \mathcal{H})$ measures how plausible it is that the barometer is rising, given that today is a rainy day; this probability can be obtained by Bayes' rule, $P(A|B, \mathcal{H}) = P(B|A, \mathcal{H})P(A|\mathcal{H})/P(B|\mathcal{H})$. Here, our overall model of the weather, $\mathcal{H}$, is a conditioning statement on the right-hand side of all the probabilities. All inferences are conditional on subjective assumptions. Bayesian methods force us to make these tacit assumptions explicit, and then provide rules for reasoning consistently given those assumptions.

We evaluate the plausibility of two alternative theories $\mathcal{H}_1$ and $\mathcal{H}_2$ in light of data D as follows: Using Bayes' rule, we relate the plausibility of model $\mathcal{H}_1$ given the data $P(\mathcal{H}_1|D)$ to the predictions made by the model about the data $P(D|\mathcal{H}_1)$ and the prior plausibility of $\mathcal{H}_1$, $P(\mathcal{H}_1)$. This gives the following probability ratio between theory $\mathcal{H}_1$ and theory $\mathcal{H}_2$:

$$\frac{P(\mathcal{H}_1|D)}{P(\mathcal{H}_2|D)} = \frac{P(\mathcal{H}_1)}{P(\mathcal{H}_2)} \frac{P(D|\mathcal{H}_1)}{P(D|\mathcal{H}_2)}. \tag{6.1}$$

The first ratio $(P(\mathcal{H}_1)/P(\mathcal{H}_2))$ on the right-hand side measures how much our initial beliefs favored $\mathcal{H}_1$ over $\mathcal{H}_2$. The second ratio expresses how well the observed data were predicted by $\mathcal{H}_1$, compared to $\mathcal{H}_2$.

How does this relate to Occam's razor when $\mathcal{H}_1$ is a simpler model than $\mathcal{H}_2$? The first ratio $(P(\mathcal{H}_1)/P(\mathcal{H}_2))$ gives us the opportunity, if we wish, to insert a prior bias in favor of $\mathcal{H}_1$ on aesthetic grounds, or on the basis of experience. This would correspond to the motivations for Occam's razor discussed in the first paragraph. But this is not necessary: The second ratio, the data-dependent factor, embodies Occam's razor *automatically.* Simple models tend to make precise predictions. Complex models, by their nature, are capable of making a greater variety of predictions (Fig. 6.1). So, if $\mathcal{H}_2$ is a more complex model, it must spread its predictive probability $P(D|\mathcal{H}_2)$ more thinly over the data space than $\mathcal{H}_1$. Thus, in the case where the data are compatible with both theories, the simpler $\mathcal{H}_1$ will turn out to be more probable than $\mathcal{H}_2$, without our having to express any subjective dislike for complex models. Our subjective prior just needs to assign equal prior probabilities to the possibilities of simplicity and complexity. Probability theory then allows the observed data to express their opinion.

Let us turn to a simple example. Here is a sequence of numbers:

$$2,\ 4,\ 6,\ 8 \tag{6.2}$$

The task is to predict what the next two numbers are likely to be, and infer what the underlying process probably was that gave rise to this sequence.

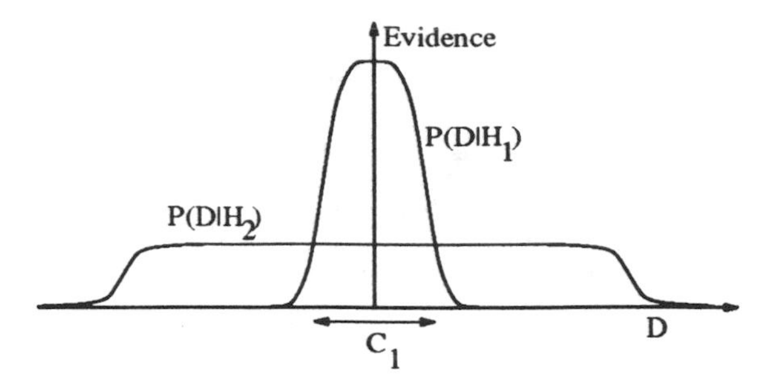

Fig. 6.1. Why Bayesian inference embodies Occam's razor. This figure gives the basic intuition for why complex models are penalized. The horizontal axis represents the space of possible data sets D. Bayes' rule rewards models in proportion to how much they *predicted* the data that occurred. These predictions are quantified by a normalized probability distribution on D. In this chapter, this probability of the data given model $\mathcal{H}_i$, $P(D|\mathcal{H}_i)$, is called the *evidence* for $\mathcal{H}_i$. A simple model $\mathcal{H}_1$ makes only a limited range of predictions, shown by $P(D|\mathcal{H}_1)$; a more powerful model $\mathcal{H}_2$, which has, for example, more free parameters than $\mathcal{H}_1$, is able to predict a greater variety of data sets. This means, however, that $\mathcal{H}_2$ does not predict the data sets in region C_1 as strongly as $\mathcal{H}_1$. Suppose that equal prior probabilities have been assigned to the two models. Then, if the data set falls in region C_1, the *less powerful* model $\mathcal{H}_1$ will be the *more probable* model.

We assume that it is agreed that a plausible prediction and explanation are "10, 12" and "add 2 to the previous number."

What about the alternative answer, "8.91, 8.67" with the underlying rule being, "get the next number from the previous number, x, by evaluating $-x^3/44 + 3/11x^2 + 34/11$"? We assume that this prediction seems rather less plausible. But the second rule fits the data (2, 4, 6, 8) just as well as the rule "add 2." So why should we find it less plausible? Let us give labels to the two general theories:

$\mathcal{H}_a$ — The sequence is an *arithmetic* progression, "add n," where n is an integer.

$\mathcal{H}_c$ — The sequence is generated by a *cubic* function of the form $x \rightarrow cx^3 + dx^2 + e$, where c, d, and e are fractions.

One reason for finding the second explanation, $\mathcal{H}_c$, less plausible might be that arithmetic progressions are more frequently encountered than cubic functions. This would put a bias in the prior probability ratio $P(\mathcal{H}_a)/P(\mathcal{H}_c)$ in Eq. (6.1). But let us give the two theories equal prior probabilities, and concentrate on what the data have to say. How well did each theory predict the data?

To obtain $P(D|\mathcal{H}_a)$, we must specify the probability distribution that each model assigns to its parameters. First, $\mathcal{H}_a$ depends on the added integer n and the first number in the sequence. Let us say that each of these numbers could have been anywhere between -50 and 50. Then, since only the pair of values $\{n = 2, \text{ first number}= 2\}$ give rise to the observed data $D = (2, 4, 6, 8)$, the probability of the data, given $\mathcal{H}_a$, is

$$P(D|\mathcal{H}_a) = \frac{1}{101}\frac{1}{101} = 0.00010. \tag{6.3}$$

To evaluate $P(D|\mathcal{H}_c)$, we must similarly say what values the fractions c, d, and e might take on. A reasonable assignment might be that, for each fraction, the numerator is a number anywhere between -50 and 50, and the denominator is a number between 1 and 50. As for the initial value in the sequence, let us leave its probability distribution the same as in $\mathcal{H}_a$. Then, including a factor of 4 in the probability of $d = 3/11$, since this fraction also can be expressed as 6/22, 9/33, and 12/44, we find that the probability of the observed data, given $\mathcal{H}_c$, is

$$P(D|\mathcal{H}_c) = \left(\frac{1}{101}\right)\left(\frac{1}{101}\frac{1}{50}\right)\left(\frac{4}{101}\frac{1}{50}\right)\left(\frac{1}{101}\frac{1}{50}\right) \tag{6.4}$$

$$= 0.0000000000031. \tag{6.5}$$

Thus, even if our prior probabilities for $\mathcal{H}_a$ and $\mathcal{H}_c$ are equal, the odds, $P(D|\mathcal{H}_a) : P(D|\mathcal{H}_c)$, in favor of $\mathcal{H}_a$ over $\mathcal{H}_c$, given the sequence $D = (2, 4, 6, 8)$, are about three hundred million to one.

This answer depends on several subjective assumptions, in particular, the probability assigned to the free parameters n, c, d, and e of each theory. Bayesians make no apologies for this: There is no such thing as inference or prediction without assumptions. However, the quantitative details of the prior probabilities have no effect on the qualitative Occam's razor effect; the complex theory $\mathcal{H}_c$ always suffers an "Occam factor" because it has more parameters, and so can predict a greater variety of data sets (Fig. 6.1). This was only a small example, and there were only four data points; as we move to larger and more sophisticated problems, the magnitude of the Occam factors typically becomes larger, and the degree to which our inferences are influenced by the quantitative details of our subjective assumptions becomes even smaller.

6.1.2 Bayesian Methods and Data Analysis

Let us now relate the discussion above to real problems in data analysis. There are countless problems in science, statistics, and technology which require that, given a limited data set, preferences be assigned to alternative models of differing complexities. For example, two alternative hypotheses accounting for planetary motion are Mr. Inquisition's geocentric model based on "epicycles," and Mr. Copernicus's simpler model of the solar system. The epicyclic model fits data on planetary motion at least as well as the Copernican model, but it does so using more parameters. Coincidentally for Mr. Inquisition, two of the extra epicyclic parameters for every planet are found to be identical to the period and radius of the sun's "cycle around the earth." Intuitively, we find Mr. Copernicus's theory to be more probable. We now explain in more detail how Mr. Inquisition's excess parameters are penalized automatically under probability theory.

6.1.3 The Mechanism of the Bayesian Occam's Razor: The Evidence and the Occam Factor

Two levels of *inference* often can be distinguished in the task of data modeling. At the first level of inference, we assume that a particular model is true, and we fit that model to the data. Typically, a model includes some free parameters; fitting the model to the data involves inferring what values those parameters should probably take, given the data. The results of this inference often are summarized by the most probable parameter values, and error bars on those parameters. This analysis is repeated for each model. The second level of inference is the task of model comparison. Here we wish to compare the models in light of the data, and assign some sort of preference or ranking to the alternatives.[2]

Bayesian methods consistently and quantitatively are able to solve both of the inference tasks. There is a popular myth that states that Bayesian methods only differ from orthodox (also known as "frequentist" or "sampling-theoretical") statistical methods by the inclusion of subjective priors which are arbitrary and difficult to assign, and usually do not make much difference to the conclusions. It is true that, at the first level of inference, a Bayesian's results often will differ little from the outcome of an orthodox attack. What is not widely appreciated is how Bayes performs the second level of inference; this section therefore will focus on Bayesian model comparison. This emphasis should not be misconstrued as implying a belief that one ought to use the Bayesian rankings to "choose" a single best model. What we do with the Bayesian posterior probabilities is another issue. If we wish to make predictions, for example, then we should integrate over the alternative models, weighted by their posterior probabilities (Sec. 6.5).

Model comparison is a difficult task because it is not possible simply to choose the model that fits the data best: more complex models can always fit the data better, so the maximum likelihood model choice would lead us inevitably to implausible, overparameterized models which generalize poorly. Occam's razor is needed.

Let us write down Bayes' rule for the two levels of inference described

---

[2]Note that both levels of *inference* are distinct from *decision theory.* The goal of inference is, given a defined hypothesis space and a particular data set, to assign probabilities to hypotheses. Decision theory typically chooses between alternative *actions* on the basis of these probabilities so as to minimize the expectation of a "loss function." This chapter concerns inference alone, and no loss functions are involved.

above, so as to see explicitly how Bayesian model comparison works. Each model $\mathcal{H}_i$ is assumed to have a vector of parameters $\mathbf{w}$. A model is defined by a collection of probability distributions: a "prior" distribution $P(\mathbf{w}|\mathcal{H}_i)$, which states what values the model's parameters might plausibly take, and a set of probability distributions, one for each value of $\mathbf{w}$, which defines the predictions $P(D|\mathbf{w}, \mathcal{H}_i)$ that the model makes about the data D.

1. **Model fitting.** At the first level of inference, we assume that one model, say, the ith, is true, and we infer what the model's parameters $\mathbf{w}$ might be given the data D. Using Bayes' rule, the *posterior probability* of the parameters $\mathbf{w}$ is

$$P(\mathbf{w}|D, \mathcal{H}_i) = \frac{P(D|\mathbf{w}, \mathcal{H}_i)P(\mathbf{w}|\mathcal{H}_i)}{P(D|\mathcal{H}_i)}, \tag{6.6}$$

that is,

$$\text{Posterior} = \frac{\text{Likelihood} \times \text{Prior}}{\text{Evidence}}.$$

The normalizing constant $P(D|\mathcal{H}_i)$ is commonly ignored since it is irrelevant to the first level of inference, i.e., the choice of $\mathbf{w}$; but it becomes important in the second level of inference, and we name it the *evidence* for $\mathcal{H}_i$. It is common practice to use gradient-based methods to find the maximum of the posterior, which defines the most probable value for the parameters, $\mathbf{w}_{\text{MP}}$; it is then usual to summarize the posterior distribution by the value of $\mathbf{w}_{\text{MP}}$ and error bars on these best-fit parameters. The error bars are obtained from the curvature of the posterior; evaluating the Hessian at $\mathbf{w}_{\text{MP}}$, $\mathbf{A} = -\nabla\nabla \log P(\mathbf{w}|D, \mathcal{H}_i)$, and Taylor-expanding the log posterior with $\Delta\mathbf{w} = \mathbf{w} - \mathbf{w}_{\text{MP}}$:

$$P(\mathbf{w}|D, \mathcal{H}_i) \simeq P(\mathbf{w}_{\text{MP}}|D, \mathcal{H}_i) \exp\left(-\tfrac{1}{2}\Delta\mathbf{w}^{\text{T}}\mathbf{A}\Delta\mathbf{w}\right), \tag{6.7}$$

we see that the posterior can be locally approximated as a Gaussian with a covariance matrix (equivalent to error bars) $\mathbf{A}^{-1}$. Whether this approximation is good or not will depend on the problem we are solving. The maximum and mean of the posterior distribution have no fundamental status in Bayesian inference — they both can be arbitrarily changed by nonlinear reparameterizations. Maximization of a posterior probability is only useful if an approximation like Eq. (6.7) gives a good summary of the distribution.

2. **Model comparison.** At the second level of inference, we wish to infer which model is most plausible given the data. The posterior probability of each model is

$$P(\mathcal{H}_i|D) \propto P(D|\mathcal{H}_i)P(\mathcal{H}_i). \tag{6.8}$$

Notice that the data-dependent term $P(D|\mathcal{H}_i)$ is the evidence for $\mathcal{H}_i$, which appeared as the normalizing constant in Eq. (6.6). The second term, $P(\mathcal{H}_i)$, is the subjective prior over our hypothesis space, which expresses how plausible we thought the alternative models were before the data arrived. Assuming that we choose to assign equal priors $P(\mathcal{H}_i)$ to the alternative models, *models $\mathcal{H}_i$ are ranked by evaluating the evidence.* Equation (6.8) has not been normalized because in the data modeling process we may develop new models after the data have arrived, when an inadequacy of the first models is detected, for example. Inference is open-ended: we continually seek more probable models to account for the data we gather.

To reiterate the key concept: To assign a preference to alternative models $\mathcal{H}_i$, a Bayesian evaluates the evidence $P(D|\mathcal{H}_i)$. This concept is very general: The evidence can be evaluated for parametric and nonparametric models alike; whatever our data modeling task — a regression problem, a classification problem, or a density estimation problem — the Bayesian evidence is a transportable quantity for comparing alternative models. In all of these cases the evidence naturally embodies Occam's razor.

Evaluating the Evidence

Let us now study the evidence more closely to gain insight into how the Bayesian Occam's razor works. The evidence is the normalizing constant for Eq. (6.6):

$$P(D\,|\mathcal{H}_i) = \int P(D|\mathbf{w}, \mathcal{H}_i) P(\mathbf{w}|\mathcal{H}_i)\, d\mathbf{w}. \tag{6.9}$$

For many problems, including interpolation, it is common for the posterior $P(\mathbf{w}|D, \mathcal{H}_i) \propto P(D|\mathbf{w}, \mathcal{H}_i)P(\mathbf{w}|\mathcal{H}_i)$ to have a strong peak at the most probable parameters $\mathbf{w}_{\text{MP}}$ (Fig. 6.2). Then, taking for simplicity the one-dimensional case, the evidence can be approximated by the height of the peak of the integrand $P(D|\mathbf{w}, \mathcal{H}_i)P(\mathbf{w}|\mathcal{H}_i)$ times its width, $\sigma_{w|D}$:

$$\begin{array}{ccccc} P(D\,|\mathcal{H}_i) & \simeq & \underbrace{P(D\,|\mathbf{w}_{\text{MP}}, \mathcal{H}_i)} & \underbrace{P(\mathbf{w}_{\text{MP}}|\mathcal{H}_i)\,\sigma_{w|D}} . \\ \text{Evidence} & \simeq & \text{Best fit likelihood} & \times\text{Occam factor} \end{array} \tag{6.10}$$

Thus, the evidence is found by taking the best-fit likelihood that the model can achieve and multiplying it by an "Occam factor" [2], which is a term with magnitude less than 1 that penalizes $\mathcal{H}_i$ for having the parameter $\mathbf{w}$.

Interpretation of the Occam Factor

The quantity $\sigma_{w|D}$ is the posterior uncertainty in $\mathbf{w}$. Suppose for simplicity that the prior $P(\mathbf{w}|\mathcal{H}_i)$ is uniform on some large interval σ_w, representing

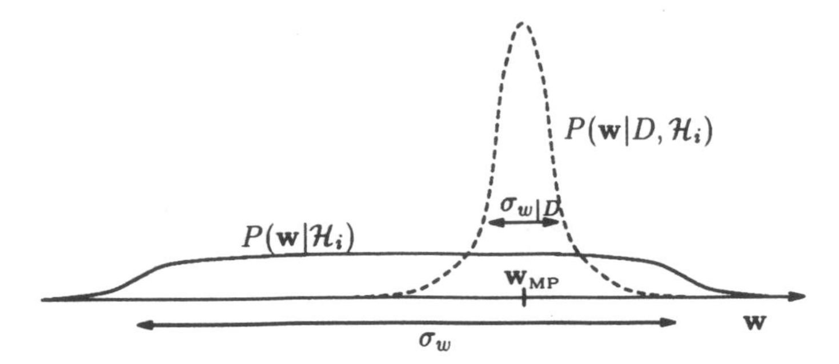

Fig. 6.2. The Occam factor. This figure shows the quantities that determine the Occam factor for a hypothesis $\mathcal{H}_i$ having a single parameter $\mathbf{w}$. The prior distribution (solid line) for the parameter has width σ_w. The posterior distribution (dashed line) has a single peak at $\mathbf{w}_{\text{MP}}$ with characteristic width $\sigma_{w|D}$. The Occam factor is $(\sigma_{w|D}/\sigma_w)$.

the range of values of $\mathbf{w}$ that $\mathcal{H}_i$ thought possible before the data arrived (Fig. 6.2). Then, $P(\mathbf{w}_{\text{MP}}|\mathcal{H}_i) = 1/\sigma_w$, and

$$\text{Occam factor} = \frac{\sigma_{w|D}}{\sigma_w}, \tag{6.11}$$

i.e., *the Occam factor is equal to the ratio of the posterior accessible volume of $\mathcal{H}_i$'s parameter space to the prior accessible volume,* or the factor by which $\mathcal{H}_i$'s hypothesis space collapses when the data arrive [2, 3]. The model $\mathcal{H}_i$ can be viewed as consisting of a certain number of exclusive submodels, of which only one survives when the data arrive. The Occam factor is the inverse of that number. The logarithm of the Occam factor can be interpreted as the amount of information gained about the model when the data arrive.

A complex model having many parameters, each of which is free to vary over a large range σ_w, typically will be penalized by a larger Occam factor than a simpler model. The Occam factor also penalizes models that have to be finely tuned to fit the data, and favors models for which the required precision of the parameters $\sigma_{w|D}$ is coarse. The Occam factor is thus a measure of complexity of the model but, unlike the VC dimension or algorithmic complexity, it relates to the complexity of the predictions that the model makes in data space. This depends not only on the number of parameters in the model, but also on the prior probability that the model assigns to them. Which model achieves the greatest evidence is determined by a trade-off between minimizing this natural complexity measure and minimizing the data misfit.

Figure 6.3 displays an entire hypothesis space so as to illustrate the various probabilities in the analysis. There are three models, $\mathcal{H}_1, \mathcal{H}_2$, and $\mathcal{H}_3$, which have equal prior probabilities. Each model has one parameter

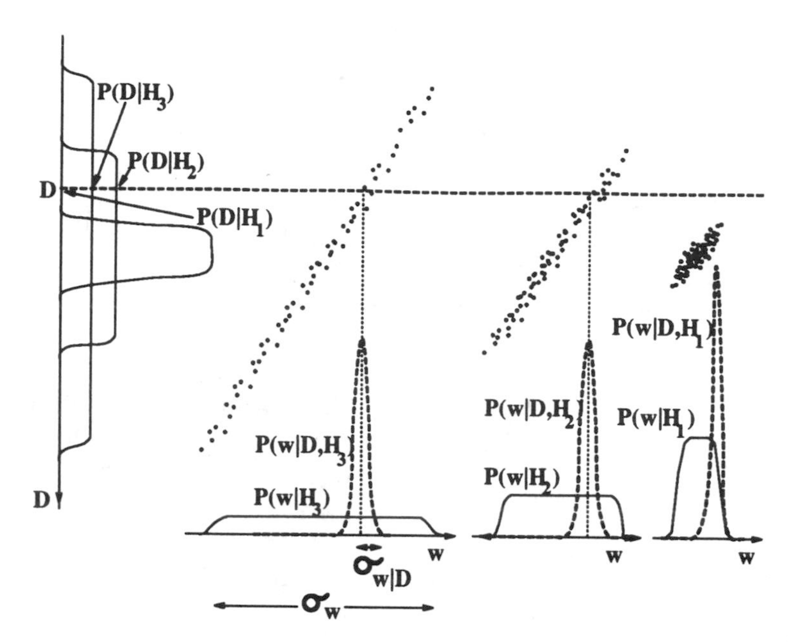

Fig. 6.3. A *hypothesis space* consisting of three exclusive models, each having one parameter **w**, and a one-dimensional data set D. The dashed horizontal line shows a particular observed data set. The dashed curves below show the posterior probability of **w** for each model given this data set (cf. Fig. 6.1). The evidence for the different models is obtained by marginalizing onto the D axis at the left-hand side (cf. Fig. 6.2).

w (each shown on a horizontal axes) but assigns a different prior range σ_w to that parameter. $\mathcal{H}_3$ is the most flexible, i.e., the most complex model, assigning the broadest prior range. A one-dimensional data space is shown by the vertical axis. Each model assigns a joint probability distribution $P(\mathcal{D}, \mathbf{w}|\mathcal{H}_i)$ to the data and the parameters, illustrated by a cloud of dots. These dots represent random samples from the full probability distribution. The total number of dots in each of the three model subspaces is the same, because we assigned equal priors to the models.

When a particular data set D is received (horizontal line), we infer the posterior distribution of **w** for a model ($\mathcal{H}_3$, say) by reading out the density along that horizontal line and normalizing. The posterior probability $P(\mathbf{w}|D, \mathcal{H}_3)$ is shown by the dotted curve at the bottom. Also shown is the prior distribution $P(\mathbf{w}|\mathcal{H}_3)$ (cf. Fig. 6.2).

We obtain Fig. 6.1 by marginalizing the joint distributions $P(D, \mathbf{w}|\mathcal{H}_i)$ onto the D axis at the left-hand side. This procedure gives the predictions of each model in data space. For the data set D shown by the dotted horizontal line, the evidence $P(D|\mathcal{H}_3)$ for the more flexible model $\mathcal{H}_3$ has a smaller value than the evidence for H_2. This is because $\mathcal{H}_3$ placed less predictive probability (fewer dots) on that line. Looking back at the distributions over $\mathbf{w}$, $\mathcal{H}_3$ has smaller evidence because the Occam factor $\sigma_{w|D}/\sigma_w$ is smaller for $\mathcal{H}_3$ than for $\mathcal{H}_2$. The simplest model $\mathcal{H}_1$ has the smallest evidence of all, because the best fit that it can achieve to the data D is very poor. Given this data set, the most probable model is $\mathcal{H}_2$.

Occam Factor for Several Parameters

If $\mathbf{w}$ is k-dimensional, and if the posterior is well approximated by a Gaussian, then the Occam factor is obtained from the determinant of the corresponding covariance matrix [c.f. Eq. (6.10)]:

$$\underset{\text{Evidence}}{P(D\,|\mathcal{H}_i)} \;\simeq\; \underbrace{P(D\,|\mathbf{w}_{\text{MP}}, H_i)}_{\text{Best-fit likelihood}} \; \underbrace{P(\mathbf{w}_{\text{MP}}|\mathcal{H}_i)\,(2\pi)^{k/2}\text{det}^{-1/2}\mathbf{A}}_{\times\text{Occam factor}}, \tag{6.12}$$

where $\mathbf{A} = -\nabla\nabla \log P(\mathbf{w}|D, \mathcal{H}_i)$, the Hessian which we evaluated when we calculated the error bars on $\mathbf{w}_{\text{MP}}$ [Eq. (6.7)]. As the number of data collected, N, increases, this Gaussian approximation is expected to become increasingly accurate.

In summary, Bayesian model selection is a simple extension of maximum likelihood model selection: *The evidence is obtained by multiplying the best-fit likelihood by the Occam factor.*

To evaluate the Occam factor, we need only the Hessian $\mathbf{A}$ if the Gaussian approximation is good. Thus, the Bayesian method of model comparison by evaluating the evidence is no more demanding computationally than the task of finding for each model the best-fit parameters and their error bars.

For background reading on Bayesian methods, the following references may be helpful. Bayesian methods are introduced and contrasted with orthodox statistics in [2a, 3, 4]. The Bayesian Occam's razor is demonstrated on model problems in [2, 5]. Useful textbooks are [1, 6].

Bayesian Methods Meet Neural Networks

The two ideas of neural network modeling and Bayesian statistics might at first glance seem to be uneasy bedfellows. Neural networks are nonlinear parallel computational devices inspired by the structure of the brain. *Backpropagation networks* are able to learn, by example, to solve prediction and classification problems. Such a neural network is typically viewed as a black box that slaps together, by hook or by crook, an incomprehensible solution to a poorly understood problem. In contrast, Bayesian statistics are characterized by an insistence on coherent inference based on clearly defined axioms; in Bayesian circles, an "ad hockery" is a capital offense. Thus, Bayesian statistics and neural networks might seem to occupy opposite extremes of the data modeling spectrum.

However, there is a common theme uniting the two. Both fields aim to create models that are well matched to the data. Neural networks can be viewed as more flexible versions of traditional regression techniques. Because they are more flexible (nonlinear), they are able to fit the data better and model regularities in the data that linear models cannot capture. The problem with neural networks is that an overflexible network might be duped by stray correlations in the data into "discovering" nonexistent structures. This is where Bayesian methods play a complementary role. Using Bayesian probability theory, one can automatically infer how flexible a model is warranted by the data; the Bayesian Occam's razor automatically suppresses the tendency to discover spurious structures in data. The philosophy advocated here is to use flexible models, like neural networks, and then control the complexity of these models in light of the data using Bayesian methods.

Occam's razor is needed in neural networks for the reason illustrated in Fig. 6.4(A). Consider a control parameter that influences the complexity of a model, for example, a regularization constant (weight decay parameter). As the control parameter is varied to increase the complexity of the model [from left to right across Fig. 6.4(A)], the best fit to the *Training* data that the model can achieve becomes increasingly good. However, the empirical performance of the model, the *Test error*, has a minimum as a function of the control parameters. An overcomplex model overfits the data and generalizes poorly. Finding values for model control parameters that are well matched to the data is therefore an important and nontrivial problem.

A central message of this chapter is illustrated in Fig. 6.4(B). When we evaluate the *posterior probability* distribution of the control parameters, we find the Bayesian Occam's razor at work. The probability of a model given the data is not the same thing as the best quality of fit that the model can achieve. Overcomplex models are less probable because they predict the data less strongly. Thus, the "evidence" $P(\text{Data}|\text{Control Parameters})$ can be used as an objective function for optimization of model control parameters.

Bayesian optimization of model control parameters has four important advantages: (1) no validation set is involved, so all of the training data can be devoted to both model fitting and model comparison; (2) regularization constants can be optimized on-line, i.e., simultaneously with the optimization of ordinary model parameters; (3) the Bayesian objective function is not noisy, in contrast to a cross-validation measure; and (4) the gradient of the evidence with respect to the control parameters can be evaluated, making it possible to simultaneously optimize a large number of control parameters.

6.2 Neural Networks as Probabilistic Models

A supervised neural network is a nonlinear parameterized mapping from an input $\mathbf{x}$ to an output $\mathbf{y} = \mathbf{y}(\mathbf{x}; \mathbf{w}, \mathcal{A})$. The output is a continuous function

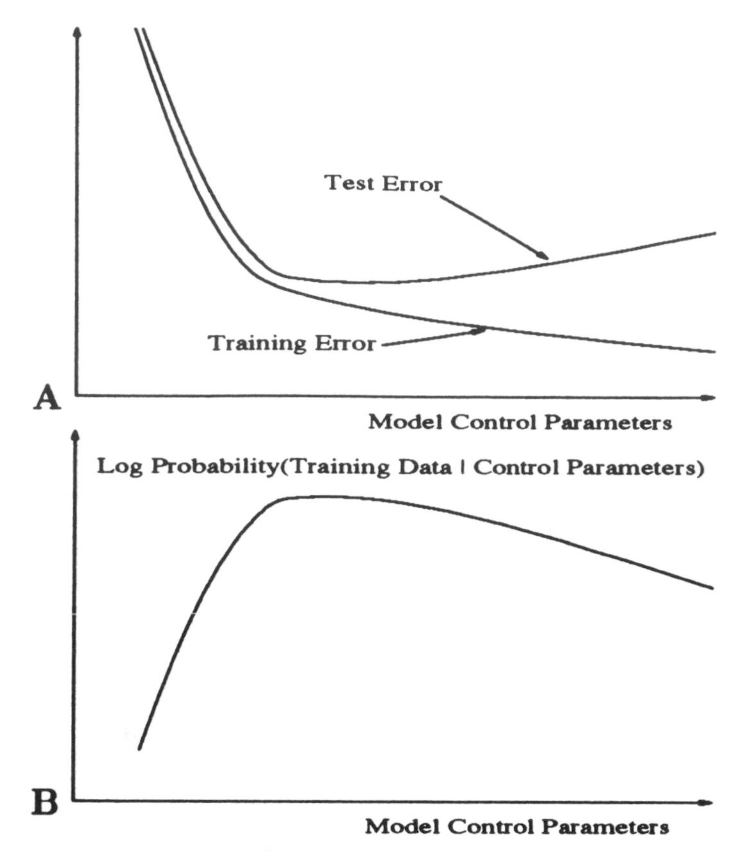

Fig. 6.4. Optimization of model complexity.

of the input and of the parameters $\mathbf{w}$; the architecture of the net, i.e., the functional form of the mapping, is denoted by $\mathcal{A}$. Such networks can be "trained" to perform regression and classification tasks.

6.2.1 Regression Networks

In the case of a regression problem, the mapping for a network with one hidden layer may have the form:

$$\text{Hidden layer:} \quad a_j^{(1)} = \sum_l w_{jl}^{(1)} x_l + \theta_j^{(1)}; \qquad h_j = f^{(1)}(a_j^{(1)}) \tag{6.13}$$

$$\text{Output layer:} \quad a_i^{(2)} = \sum_j w_{ij}^{(2)} h_j + \theta_i^{(2)}; \qquad y_i = f^{(2)}(a_i^{(2)}), \tag{6.14}$$

where, for example, $f^{(1)}(a) = \tanh(a)$, and $f^{(2)}(a) = a$. The "weights" w and "biases" θ together make up the parameter vector $\mathbf{w}$. The nonlinear "sigmoid" function $f^{(1)}$ at the hidden layer gives the neural network greater computational flexibility than a standard linear regression model.

This network is trained using a data set $D = \{\mathbf{x}^{(m)}, \mathbf{t}^{(m)}\}$ by adjusting $\mathbf{w}$ so as to minimize an error function, e.g.,

$$E_D(\mathbf{w}) = \frac{1}{2}\sum_m \sum_i \left(t_i^{(m)} - y_i(\mathbf{x}^{(m)}; \mathbf{w})\right)^2 . \tag{6.15}$$

This minimization is based on repeated evaluation of the gradient of E_D using *backpropagation* (the chain rule) [7]. Often, regularization (also known as "weight decay") is included, modifying the objective function to:

$$M(\mathbf{w}) = \beta E_D + \alpha E_W, \tag{6.16}$$

where, for example, $E_W = \frac{1}{2}\sum_i w_i^2$. This additional term favors small values of $\mathbf{w}$ and decreases the tendency of a model to "overfit" noise in the training data.

6.2.2 Neural Network Learning as Inference

The neural network learning process above can be given the following probabilistic interpretation. The error function is interpreted as minus the log likelihood for a noise model:

$$P(D|\mathbf{w}, \beta, \mathcal{H}) = \frac{1}{Z_D(\beta)} \exp(-\beta E_D). \tag{6.17}$$

Thus, the use of the sum-squared error E_D [Eq. (6.15)] corresponds to an assumption of Gaussian noise on the target variables, and the parameter β defines a noise level $\sigma_\nu^2 = 1/\beta$.

Similarly, the regularizer is interpreted in terms of a log prior probability distribution over the parameters:

$$P(\mathbf{w}|\alpha, \mathcal{H}) = \frac{1}{Z_W(\alpha)} \exp(-\alpha E_W). \tag{6.18}$$

If E_W is quadratic as defined above, then the corresponding prior distribution is a Gaussian with variance $\sigma_w^2 = 1/\alpha$. The probabilistic model $\mathcal{H}$ specifies the functional form $\mathcal{A}$ of the network, the likelihood [Eq. (6.17)], and the prior [Eq. (6.18)].

The objective function $M(\mathbf{w})$ then corresponds to the *inference* of the parameters $\mathbf{w}$ given the data:

$$P(\mathbf{w}|D, \alpha, \beta, \mathcal{H}) = \frac{P(D|\mathbf{w}, \beta, \mathcal{H}) P(\mathbf{w}|\alpha, \mathcal{H})}{P(D|\alpha, \beta, \mathcal{H})} \tag{6.19}$$

$$= \frac{1}{Z_M} \exp(-M(\mathbf{w})). \tag{6.20}$$

The $\mathbf{w}$ found by (locally) minimizing $M(\mathbf{w})$ is then interpreted as the (locally) most probable parameter vector, $\mathbf{w}_{\text{MP}}$.

Why is it natural to interpret the error functions as *log* probabilities? Error functions are usually additive. For example, E_D is a *sum* of squared errors. Probabilities, on the other hand, are multiplicative: For independent events A and B, the joint probability is $P(A,B) = P(A)P(B)$. The logarithmic mapping maintains this correspondence.

The interpretation of $M(\mathbf{w})$ as a log probability adds little new at this stage. But new tools will emerge when we proceed to other inferences. First, though, let us establish the probabilistic interpretation of classification networks, to which the same tools apply.

6.2.3 Binary Classification Networks

If the targets t in a data set are binary classification labels (0,1), it is natural to use a neural network whose output $y(\mathbf{x};\mathbf{w},\mathcal{A})$ is bounded between 0 and 1, and is interpreted as a probability $P(t=1|\mathbf{x},\mathbf{w},\mathcal{A})$. For example, a network with one hidden layer could be described by Eqs. (6.13) and (6.14), with $f^{(2)}(a) = 1/(1+e^{-a})$. The error function βE_D is replaced by the log likelihood:

$$G(\mathbf{w}) = \sum_m t^{(m)} \log y(\mathbf{x}^{(m)};\mathbf{w}) + (1-t^{(m)}) \log(1-y(\mathbf{x}^{(m)};\mathbf{w})). \tag{6.21}$$

The total objective function is then $M = -G + \alpha E_W$. Note that this includes no parameter β.

6.2.4 Multiclass Classification Networks

For a multiclass classification problem, we can represent the targets by a vector, $\mathbf{t}$, in which a single element is set to 1, indicating the correct class, and all other elements are set to 0. In this case, it is appropriate to use a "softmax" network [8] having coupled outputs which sum to 1 and are interpreted as class probabilities $y_i = P(t_i=1|\mathbf{x},\mathbf{w},\mathcal{A})$. The last part of Eq. (6.14) is replaced by:

$$y_i = \frac{e^{a_i}}{\sum_{i'} e^{a_{i'}}}. \tag{6.22}$$

The log likelihood in this case is

$$G = \sum_m \sum_i t_i \log y_i(\mathbf{x}^{(m)};\mathbf{w}). \tag{6.23}$$

As in the case of the regression network, the minimization of the objective function $M(\mathbf{w}) = -G + \alpha E_W$ corresponds to an inference of the form in Eq. (6.20). Let us now study the variety of useful results that can be built on this interpretation. The results will refer to regression models; the corresponding results for classification models are obtained by replacing βE_D by $-G$, and $Z_D(\beta)$ by 1.

6.2.5 IMPLEMENTATION

Bayesian inference for data-modeling problems may be implemented by analytical methods, by Monte Carlo sampling, or by deterministic methods employing Gaussian approximations. For neural networks, there are few analytic methods. Sophisticated Monte Carlo methods that make use of gradient information have been applied to some model problems [9]. The methods reviewed here are based on Gaussian approximations to the posterior distribution.

6.3 Setting Regularization Constants α and β

The control parameters α and β determine the complexity of the model. The term model here refers to a triple: the network architecture; the form of the prior on the parameters; and the form of the noise model. Different values for the hyperparameters α and β define different submodels. To infer α and β given the data, we simply apply the rules of probability theory:

$$P(\alpha, \beta | D, \mathcal{H}) = \frac{P(D|\alpha, \beta, \mathcal{H}) P(\alpha, \beta | \mathcal{H})}{P(D|\mathcal{H})}. \tag{6.24}$$

The data-dependent factor $P(D|\alpha, \beta, \mathcal{H})$ is the normalizing constant from our previous inference [Eq. (6.19)]; we call this factor the *evidence* for α and β.

Assuming that we have only weak prior knowledge about the noise level and the smoothness of the interpolant, the evidence framework optimizes the constants α and β by finding the maximum of the evidence for α and β. If we can approximate the posterior probability distribution in Eq. (6.20) by a single Gaussian,

$$P(\mathbf{w}|D, \alpha, \beta, \mathcal{H}) \simeq \frac{1}{Z'_M} \exp\left(-M(\mathbf{w}_{\text{MP}}) - \frac{1}{2}(\mathbf{w} - \mathbf{w}_{\text{MP}})^{\text{T}} \mathbf{A} (\mathbf{w} - \mathbf{w}_{\text{MP}})\right), \tag{6.25}$$

where $\mathbf{A} = -\nabla\nabla \log P(\mathbf{w}|D, \mathcal{H})$, then the evidence for α and β can be written as

$$\log P(D|\alpha, \beta, \mathcal{H}) = \log \frac{Z'_M}{Z_W(\alpha) Z_D(\beta)} \tag{6.26}$$

$$= -M(\mathbf{w}_{\text{MP}}) - \frac{1}{2} \log \det \mathbf{A} - \log Z_W(\alpha) - \log Z_D(\beta) + \frac{k}{2} \log 2\pi, \tag{6.27}$$

where k is the number of parameters in $\mathbf{w}$. The terms $-\frac{1}{2} \log \det \mathbf{A} - \log Z_W(\alpha)$ constitute the log of a volume factor that penalizes small values of α: The ratio $(2\pi)^{k/2} \det^{-1/2} \mathbf{A} / Z_W(\alpha)$ is the ratio of the posterior accessible volume in parameter space to the prior accessible volume. The maximum of the evidence has some elegant properties which allow it to be located efficiently by on-line reestimation techniques. Technically, there may be multiple evidence maxima, but this is not common when the model space is well matched to the data. As is shown in [10, 5], the maximum evidence $\alpha = \alpha_{\text{MP}}$ satisfies the following self-consistent equation:

$$1/\alpha_{\text{MP}} = \sum_i w_i^{\text{MP}\,2}/\gamma, \tag{6.28}$$

where $\mathbf{w}^{\text{MP}}$ is the parameter vector that minimizes the objective function $M = \beta E_D + \alpha E_W$, and γ is the "number of well-determined parameters," given by

$$\gamma = k - \alpha \text{Trace}(\mathbf{A}^{-1}). \tag{6.29}$$

Here, k is the total number of parameters, and the matrix $\mathbf{A}^{-1}$ measures the size of the error bars on the parameters $\mathbf{w}$ [Eq. (6.7)]. Thus, $\gamma \rightarrow k$ when the parameters are all well determined in relation to their prior range, which is defined by α. The quantity γ always lies between 0 and k. Recalling that α corresponds to the variance $\sigma_w^2 = 1/\alpha$ of the assumed distribution for $\{w_i\}$, Eq. (6.28) specifies an intuitive condition for matching the prior to the data: The variance is estimated by $\sigma_w^2 = \langle w^2 \rangle$, where the average is over the γ effective well-determined parameters; the other $k - \gamma$ effective parameters having been set to 0 by the prior.

Similarly, in a regression problem with a Gaussian noise model, the maximum evidence value of β satisfies:

$$1/\beta_{\text{MP}} = 2E_D/(N - \gamma). \tag{6.30}$$

Since $2E_D$ is the sum of squared residuals, this expression can be recognized as a variance estimator with the number of degrees of freedom set to γ.

Equations (6.28) and (6.30) can be used as reestimation formulas for α and β. The computational overhead for these Bayesian calculations is not severe: It is only necessary to evaluate properties of the error bar matrix, $\mathbf{A}^{-1}$. This matrix may be evaluated explicitly [11, 12, 13, 14], which does not take significant time when the number of parameters is small (a few hundred). For large problems, these calculations can be performed more efficiently using algorithms that evaluate products $\mathbf{Av}$ without explicitly evaluating $\mathbf{A}$ [15, 16].

Thodberg [12] combines Eqs. (6.28) and (6.30) into a single reestimation formula for the ratio α/β. This ratio is all that matters if only the best-fit parameters are of interest. An advantage of keeping α and β distinct, however, is that knowledge from other sources (bounds on the value of the noise level, for example) can be explicitly incorporated. Also, if we move to noise models more sophisticated than a Gaussian, a separation of these two control parameters is essential. Finally, if we wish to construct error bars, or generate a sample from the posterior parameter distribution for use in a Monte Carlo estimation procedure, the separate values of α and β become relevant.

6.3.1 Relationship to Ideal Hierarchical Bayesian Modeling

Bayesian probability theory has been used above to *optimize* the hyperparameters α and β. This procedure is known in some circles as *generalized maximum likelihood.* Ideally, we would *integrate over* these nuisance parameters in order to obtain the posterior distribution over the parameters $P(\mathbf{w}|D,\mathcal{H})$ and the predictive distributions $P(\mathbf{t}^{(N+1)}|D,\mathcal{H})$; however, if a hyperparameter is well determined by the data, integrating over it is very much like estimating the hyperparameter from the data and then using that estimate in our equations [17, 2, 18]. The intuition is that if, in the predictive distribution

$$P(\mathbf{t}^{(N+1)}|D,\mathcal{H}) = \int d\alpha\, P(\mathbf{t}^{(N+1)}|D,\alpha,\mathcal{H})P(\alpha|D,\mathcal{H}), \tag{6.31}$$

the posterior $P(\alpha|D,\mathcal{H})$ is sharply peaked at $\alpha = \alpha_{\text{MP}}$ with width $\sigma_{\log\alpha|D}$, and if the distribution $P(\mathbf{t}^{(N+1)}|D,\alpha,\mathcal{H})$ varies slowly with $\log\alpha$ on a scale of $\sigma_{\log\alpha|D}$, then $P(\alpha|D,\mathcal{H})$ is effectively a delta-function, so that:

$$P(\mathbf{t}^{(N+1)}|D,\mathcal{H}) \simeq P(\mathbf{t}^{(N+1)}|D,\alpha_{\text{MP}},\mathcal{H}). \tag{6.32}$$

Now the error bars on $\log\alpha$ and $\log\beta$, found by differentiating $\log P(D|\alpha,\beta,\mathcal{H})$ twice, are [5]:

$$\sigma^2_{\log\alpha|D} \simeq 2/\gamma; \quad \sigma^2_{\log\beta|D} \simeq 2/(N-\gamma). \tag{6.33}$$

Thus, the error introduced by optimizing α and β is expected to be small for $\gamma \gg 1$ and $N-\gamma \gg 1$. How large γ needs to be depends on the problem; but for many neural network problems, a value of γ greater than 3 may suffice, since the predictions of an optimized network are often insensitive to an e-fold change in α.

It is often possible to integrate over α and β early in the calculation, obtaining a true prior and a true likelihood. Some authors have recommended this procedure [19, 20], but it is counterproductive as far as practical manipulation is concerned [18]: the resulting true posterior is a skew-peaked distribution, and, apart from Monte Carlo methods, there are currently no computational techniques that can cope directly with such distributions.

Later, a correction term will be given which approximates the integration over α and β when predictions are made, i.e., as a last step in the calculations.

6.3.2 Multiple Regularization Constants

For simplicity, it so far has been assumed that there is only a single class of weights, which are modeled as coming from a single Gaussian prior with $\sigma_w^2 = 1/\alpha$. However, in dimensional terms, weights usually fall into three or more distinct groups, which for consistency should not be modeled as coming from a single prior. It therefore is desirable to divide the parameters into several classes c with independent scales α_c. Assuming a Gaussian prior for each class, we can define $E_{W(c)} = \sum_{i \in c} w_i^2/2$, and assign a Gaussian prior:

$$P(\{w_i\}|\alpha_c, \mathcal{H}) = \frac{1}{\prod Z_{W(c)}} \exp\left(-\sum_c \alpha_c E_{W(c)}\right). \tag{6.34}$$

This gives a weight decay scheme with a different decay rate α_c for each class. It often is found that network performance can be enhanced by this division of weights into different classes. The automatic relevance determination model (Sec. 6.7) uses this prior.

The evidence framework optimizes the decay constants by finding their most probable value, i.e., the maximum over $\{\alpha_c\}$ of $P(D|\{\alpha_c\}, \mathcal{H})$, and, as before, the maximum evidence $\{\alpha_c\}$ satisfy the following self-consistent equations:

$$1/\alpha_c^{\text{MP}} = \sum_{i \in c} w_i^{\text{MP}\,2}/\gamma_c, \tag{6.35}$$

where $\mathbf{w}^{\text{MP}}$ is the parameter vector that minimizes the objective function $M = \beta E_D + \sum_c \alpha_c E_{W(c)}$, and γ_c is the number of well-determined parameters in class c, $\gamma_c = k_c - \alpha_c \text{Trace}_c(\mathbf{A}^{-1})$, where k_c is the number of parameters in class c, and the trace is over those parameters only.

For simplicity, the following discussion will assume once more that there is only a single parameter α.

6.4 Model Comparison

The evidence framework divides our inferences into distinct "levels of inference," of which we now have completed the first two:

- **Level 1:** Infer the parameters $\mathbf{w}$ for given values of α, β:

$$P(\mathbf{w}|D, \alpha, \beta, \mathcal{H}) = \frac{P(D|\mathbf{w}, \alpha, \beta, \mathcal{H})P(\mathbf{w}|\alpha, \beta, \mathcal{H})}{P(D|\alpha, \beta, \mathcal{H})}. \tag{6.36}$$

- **Level 2a:** Infer α, β:

$$P(\alpha, \beta|D, \mathcal{H}) = \frac{P(D|\alpha, \beta, \mathcal{H})P(\alpha, \beta|\mathcal{H})}{P(D|\mathcal{H})}. \tag{6.37}$$

- **Level 2b:** Compare models:

$$P(\mathcal{H}|D) \propto P(D|\mathcal{H})P(\mathcal{H}). \tag{6.38}$$

There is a pattern in these three applications of Bayes rule: At each of the higher levels 2a and 2b, the data-dependent factor (e.g., in level 2a, $P(D|\alpha, \beta, \mathcal{H})$) is precisely the normalizing constant (the "evidence") from the preceding level of inference. This pattern of inference continues when we compare different models $\mathcal{H}$, which might use different architectures, preprocessings, regularizers, or noise models. Alternative models are ranked by evaluating $P(D|\mathcal{H})$, the normalizing constant of inference in Eq. (6.37).

In the preceding section we reached level 2a by using a Gaussian approximation to $P(\mathbf{w}|D, \alpha, \beta, \mathcal{H})$. We now evaluate the evidence for $\mathcal{H}$. Using a Gaussian approximation for $P(\log \alpha, \log \beta | D, \mathcal{H})$, and neglecting the slight correlations in this posterior, we obtain the estimate

$$P(D|\mathcal{H}) \simeq P(D|\alpha_{\text{MP}}, \beta_{\text{MP}}, \mathcal{H}) P(\log \alpha_{\text{MP}}, \log \beta_{\text{MP}}|\mathcal{H})\, 2\pi \sigma_{\log \alpha|D} \sigma_{\log \beta|D}, \tag{6.39}$$

where $P(D|\alpha_{\text{MP}}, \beta_{\text{MP}}, \mathcal{H})$ is obtained from Eq. (6.27), and the error bars on $\log \alpha$ and $\log \beta$ are as given in Eq. (6.33). This Gaussian approximation over α and β holds good for $\gamma \gg 1$ and $N - \gamma \gg 1$ [18].

6.4.1 Multimodal Distributions

The preceding exposition falls into difficulty if the posterior distribution $P(\mathbf{w}|D, \alpha, \beta, \mathcal{H})$ is significantly multimodal; this is usually the case for multilayer neural networks. However, we can persist with the use of Gaussian approximations if we introduce two modifications.

First, we recognize that a typical optimum $\mathbf{w}_{\text{MP}}$ will be related to a number of equivalent optima by symmetry operations, such as the interchange of hidden units and the inversion of signs of weights. When evaluating the evidence using a local Gaussian approximation, a symmetry factor should be included in Eq. (6.26) to take into account these equivalent islands of probability mass. In the case of a net with one hidden layer of H units, the appropriate permutation factor is $H!2^H$, for general $\mathbf{w}_{\text{MP}}$.

Second, there are multiple optima which are not related to each other by model symmetries. We modify the above framework by changing our goals; specifically, we view each of the local probability peaks as a distinct model. Instead of inferring the posterior over α, β for the entire model $\mathcal{H}$, we allow each local peak of the posterior to choose its own optimal value for these parameters. Similarly, instead of evaluating the evidence for the entire model $\mathcal{H}$, we aim to calculate the posterior probability mass in each local optimum. This seems natural, since a typical implementation of the model will involve setting the parameter vector to a particular value or a small set of values. Thus, we do not care about the probability of an entire model; what matters is the probability of the local solutions we find.

The same method of chopping up a complex model space is used in the unsupervised classification system, AutoClass [21].

Henceforth, the term "model" will refer to a pair $\{\mathcal{H}, S_{\mathbf{w}^*}\}$, where $\mathcal{H}$ denotes the model specification and $S_{\mathbf{w}^*}$ specifies a solution neighborhood around an optimum $\mathbf{w}^*$. Adopting this shift in objective, the Gaussian integrals above can be used without alteration to set α and β and to compare alternative solutions, assuming that the posterior probability consists of well-separated islands in parameter space that are roughly Gaussian.

For general α and β, the Gaussian approximation over $\mathbf{w}$ will not be accurate; however, we only need it to be accurate for the small range of α and β close to their most probable values. For sufficiently large amounts of data compared to the number of parameters, this approximation is expected to hold. Practical experience indicates that this is a useful approximation for many real problems.

6.5 Error Bars and Predictions

Having progressed up the three levels of modeling, the next inference task is to make predictions with our adapted model. It is common practice simply to use the most probable values of $\mathcal{H}, \mathbf{w}$, etc., when making predictions, but this is not optimal. Bayesian prediction of a new datum $\mathbf{t}^{(N+1)}$ involves *marginalizing* over all of these levels of uncertainty:

$$P(\mathbf{t}^{(N+1)}|D) = \sum_{\mathcal{H}} \int d\alpha\, d\beta \int d^k\mathbf{w}\, P(\mathbf{t}^{(N+1)}|\mathbf{w}, \alpha, \beta, \mathcal{H}) P(\mathbf{w}, \alpha, \beta, \mathcal{H}|D). \tag{6.40}$$

The evaluation of the distribution $P(\mathbf{t}^{(N+1)}|\mathbf{w}, \alpha, \beta, \mathcal{H})$ for specified model parameters $\mathbf{w}$ is generally straightforward, requiring a single forward pass through the network. Typically, marginalization over $\mathbf{w}$ and $\mathcal{H}$ affects the predictive distribution significantly, but integration over α and β has a lesser effect.

6.5.1 IMPLEMENTATION

Marginalization sometimes can be done analytically. When this fails, Monte Carlo methods [9] may be used. The average of a function of an uncertain parameter $\mathbf{q}$, $t(\mathbf{q})$, under the posterior over $\mathbf{q}$, can be estimated with tolerable error by obtaining a small number of samples from the posterior distribution for $\mathbf{q}$ and then evaluating the mean value of t. The variance of this estimator is independent of the dimensionality of $\mathbf{q}$ and scales inversely with the sample size. A cheap and cheerful way of obtaining such samples is described later in Sec. (6.9). Here, methods based on Gaussian approximations are described.

6.5.2 Error Bars in Regression

Integrating first over $\mathbf{w}$ for fixed α and β, the predictive distribution is

$$P(\mathbf{t}^{(N+1)}|D,\alpha,\beta,\mathcal{H}) = \int d^k\mathbf{w}\, P(\mathbf{t}^{(N+1)}|\mathbf{w},\beta,\mathcal{H})P(\mathbf{w}|D,\alpha,\beta,\mathcal{H}). \quad (6.41)$$

If a Gaussian approximation is made for the posterior $P(\mathbf{w}|D,\alpha,\beta,\mathcal{H})$, if the noise model is Gaussian, and if a local linearization of the output is made as a function of the parameters,

$$y(\mathbf{x}^{N+1},\mathbf{w}) \simeq y(\mathbf{x}^{N+1};\mathbf{w}_{\mathrm{MP}}) + \mathbf{g}\cdot(\mathbf{w}-\mathbf{w}_{\mathrm{MP}}), \quad (6.42)$$

with $\mathbf{g} = \partial y/\partial \mathbf{w}$, then the predictive distribution in Eq. (6.41) is a straightforward Gaussian integral. This distribution has mean $y(\mathbf{x}^{N+1},\mathbf{w}_{\mathrm{MP}})$ and variance $\sigma^2_{t|\alpha,\beta} = \mathbf{g}^{\mathrm{T}}\mathbf{A}^{-1}\mathbf{g} + \sigma^2_\nu$, where $\mathbf{A} = \nabla\nabla \log P(\mathbf{w}|D,\alpha,\beta,\mathcal{H})$.

Integration over the regularization constants α and β contributes an additional variance in only one direction; to leading order in γ^{-1}, $P(\mathbf{t}^{(N+1)}|D,\mathcal{H})$ is normal, with variance [18]:

$$\sigma^2_t = \mathbf{g}^{\mathrm{T}}\left(\mathbf{A}^{-1} + (\sigma^2_{\log\alpha|D} + \sigma^2_{\log\beta|D})\mathbf{w}'_{\mathrm{MP}}{\mathbf{w}'_{\mathrm{MP}}}^{\mathrm{T}}\right)\mathbf{g} + \sigma^2_\nu, \quad (6.43)$$

where $\mathbf{w}'_{\mathrm{MP}} \equiv \partial\mathbf{w}_{\mathrm{MP}|\alpha}/\partial(\log\alpha) = \alpha\mathbf{A}^{-1}\mathbf{w}_{\mathrm{MP}}$, and $\sigma^2_{\log\alpha|D} = 2/\gamma$ and $\sigma^2_{\log\beta|D} = 2/N-\gamma$.

6.5.3 Integrating Over Models: Committees

If we have multiple regression models $\mathcal{H}$, then our predictive distribution is obtained by summing together the predictive distribution of each model, weighted by its posterior probability. If a single prediction is required and the loss function is quadratic, the optimal prediction is a weighted mean of the models' predictions $y(\mathbf{x}^{N+1};\mathbf{w}_{\mathrm{MP}},\mathcal{H})$. The weighting coefficients are the posterior probabilities, which are obtained from the evidences $P(D|\mathcal{H})$. If we cannot evaluate these accurately, then alternative pragmatic prescriptions for the weighting coefficients exist [12, 38, 22].

6.5.4 Error Bars in Classification

In the case of linearized regression discussed above, the mean of the predictive distribution in Eq. (6.41) was identical to the prediction of the mean, $\mathbf{w}_{\mathrm{MP}}$. This is not the case in classification problems. The best-fit parameters give overconfident predictions. A non-Bayesian approach to this problem is to downweight all predictions uniformly, by an empirically determined factor [23]. But a Bayesian viewpoint helps us to understand the cause of the problem, and provides a straightforward solution that is demonstrably superior to this ad hoc procedure.

This issue is illustrated for a simple two-class problem in Fig. 6.5. Figure 6.5(a) shows a binary data set, which, in Fig. 6.5(b) is modeled with a linear logistic function. The best-fit parameter values give predictions which are shown by three contours. Are these reasonable predictions? Consider new data arriving at points A and B. The best-fit model assigns both of these examples probability 0.9 of being in class 1. But intuitively we might be inclined to assign a less confident probability (closer to 0.5) at B than at A, since point B is far from the training data.

Precisely this result is obtained by marginalizing over the parameters, whose posterior probability distribution is depicted in Fig. 6.5(c). Two random samples from the posterior define two different classification surfaces, which are illustrated in Figs. 6.5(d) and (e). The point B is classified differently by these different plausible classifiers, whereas the classification of A is relatively stable. We obtain the Bayesian predictions [Fig. 6.5(f)] by averaging together the predictions of the plausible classifiers. The resulting 0.5 contour remains similar to that for the best-fit parameters. However, the width of the decision boundary increases as we move away from the data, in full accordance with intuition.

The Bayesian approach is superior because the best-fit model's predictions are *selectively* downweighted to a different degree for each test case. The consequence is that a Bayesian classifier is better able to identify the points where the classification is uncertain. This pleasing behavior results simply from a mechanical application of the rules of probability.

For a binary classifier, a numerical approximation to the integral over a Gaussian posterior distribution is given in [24]. An equivalent approximation for a multiclass classifier has not yet been implemented.

This marginalization also can be done by Monte Carlo methods. A disadvantage of a straightforward Monte Carlo approach would be that it is a poor way of estimating the probability of an improbable event, i.e., a $P(t|D,\mathcal{H})$ that is very close to 0, if the improbable event is most likely to occur in conjunction with improbable parameter values. In such cases one might instead temporarily add the event in question to the data set and evaluate the evidence $P(D,t^{(N+1)}|\mathcal{H})$. The desired probability is given by comparing this with either the previous evidence $P(D|\mathcal{H})$ or the evidence for the complementary virtual data set $P(D,\overline{t^{(N+1)}}|\mathcal{H})$.

6.6 Pruning

The evidence can serve as a guide for *pruning*, i.e., changing the model by setting selected parameters to 0. Thodberg [12] has done this in the straightforward way: Each parameter in the network is tentatively pruned; then the new model is optimized, and the evidence is evaluated to decide whether to accept the pruning.

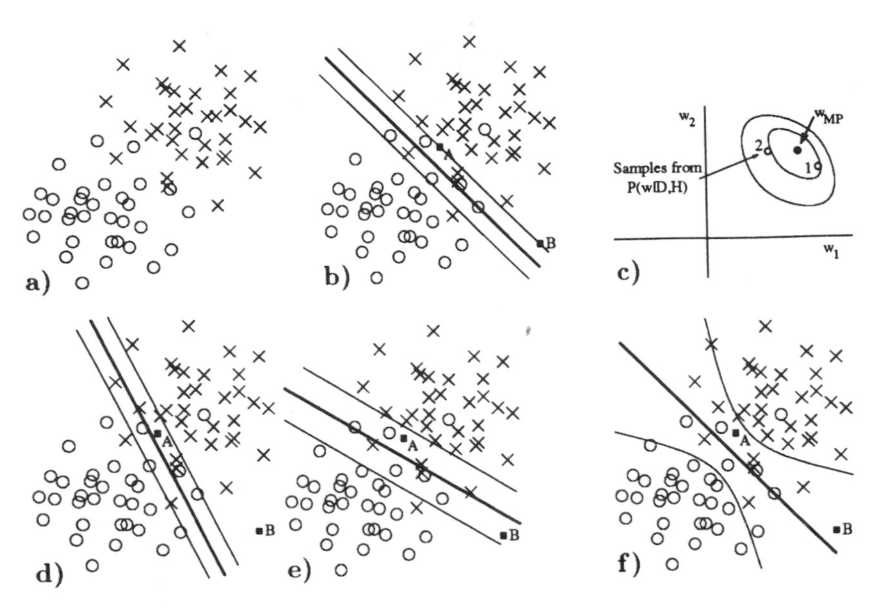

Fig. 6.5. Integrating over error bars in a classifier. (a) A binary data set. The two classes are denoted by the point styles ×=1, o=0. (b) The data are modeled with a linear logistic function. Here, the best-fit model is shown by its 0.1, 0.5, and 0.9 predictive contours. The best-fit model assigns probability 0.9 of being in class 1 to both inputs A and B. (c) The posterior probability distribution of the model parameters, $P(\mathbf{w}|D, \mathcal{H})$ (schematic; the third parameter, the bias, is not shown). The parameters are not perfectly determined by the data. Two typical samples from the posterior are indicated by the points labeled 1 and 2. The following two panels show the corresponding classification contours. (d) Sample 1. (e) Sample 2. Notice how the point B is classified differently by these different plausible classifiers, whereas the classification of A is relatively stable. (f) We obtain the Bayesian predictions by integrating over the posterior distribution of $\mathbf{w}$. The width of the decision boundary increases as we move away from the data (point B). See text for further discussion.

Here, an alternative procedure using the Gaussian approximation is described. Whether pruning is in fact a good idea is questioned later in Secs. 6.7 and 6.10.

Suppose that a model's parameters have prior and posterior distributions which are exactly Gaussian (i.e., assume that the model is locally linear, and that both α and β are well determined):

$$\begin{aligned} P(\mathbf{w}|\mathcal{H}) &= \frac{1}{Z_W} \exp\left(-\alpha \tfrac{1}{2} \mathbf{w}^{\mathrm{T}} \mathbf{I} \mathbf{w}\right) \\ P(\mathbf{w}|D, \mathcal{H}) &= \frac{1}{Z_M} \exp\left(-M_{\mathrm{MP}} - \tfrac{1}{2} \Delta\mathbf{w}^{\mathrm{T}} \mathbf{A} \Delta\mathbf{w}\right), \end{aligned}$$

where $\Delta\mathbf{w} = \mathbf{w} - \mathbf{w}_{\mathrm{MP}}$. For brevity, α and β are omitted here from the conditioning propositions. The evidence for $\mathcal{H}$ is

$$\log P(D|\mathcal{H}) = -M_{\text{MP}} - \tfrac{1}{2}\log\det \mathbf{A} + \tfrac{1}{2}\log\det \alpha\mathbf{I} + \text{const.} \tag{6.44}$$

We are interested in evaluating the difference in evidence between this model $\mathcal{H}$ and an alternative model $\mathcal{H}_{\bar{s}}$, where the subscript $\bar{s}$ denotes the setting to 0 of parameter s. The remaining parameters of $\mathcal{H}_{\bar{s}}$ still have a Gaussian distribution but are confined to the constraint surface $\mathbf{w}\cdot\mathbf{e}_s = 0$, where $\mathbf{e}_s$ is the unit vector in the direction of the deleted parameter.

We can evaluate the difference in evidence between $\mathcal{H}$ and $\mathcal{H}_{\bar{s}}$ by: (1) finding the location of the new optimum $\mathbf{w}_{\bar{s}}^{\text{MP}}$ and evaluating the change in M_{MP}, $\Delta M_{\text{MP}} = -\frac{1}{2}\Delta\mathbf{w}^{\text{T}}\mathbf{A}\Delta\mathbf{w}$ there; and (2) evaluating the change in log determinant of the distribution.

The first task is accomplished by introducing a Lagrange multiplier. We find:

$$\mathbf{w}_{\bar{s}}^{\text{MP}} = \mathbf{w}_{\text{MP}} - \frac{w_s}{\sigma_s^2}\mathbf{A}^{-1}\mathbf{e}_s; \qquad \Delta M_{\text{MP}} = \frac{w_s^2}{2\sigma_s^2}, \tag{6.45}$$

where the marginal error bars on parameter w_s are $\sigma_s^2 = \mathbf{e}_s\mathbf{A}^{-1}\mathbf{e}_s = \mathbf{A}_{ss}^{-1}$. The quantity ΔM_{MP} is the saliency term that has been advocated as a guide for "optimal brain damage" [14, 25]. The change in evidence, however, involves a second "Occam factor" term that is simple to calculate. The change in evidence when a single parameter s is deleted is

$$\log P(D|\mathcal{H}) - \log P(D|\mathcal{H}_{\bar{s}}) = \frac{w_s^2}{2\sigma_s^2} + \log\frac{\sigma_s}{\sigma_w}, \tag{6.46}$$

where σ_w^2 is the prior variance for the parameter w_s. This objective function can be used to select which parameter to delete. It also tells us to stop pruning (or, to be precise, that pruning is yielding a less probable model) once it is positive, for all parameters, s.

An equivalent expression can be worked out for the case of simultaneous pruning of multiple parameters. Consider the pruning of k_s parameters. We obtain the joint $(k_s \times k_s)$ covariance matrix for the pruned parameters, Σ_s, by reading out the appropriate submatrix of $\mathbf{A}^{-1}$. Then the evidence difference is

$$\log P(D|\mathcal{H}) - \log P(D|\mathcal{H}_{\bar{s}}) = \frac{1}{2}\mathbf{w}_s\Sigma_s^{-1}\mathbf{w}_s + \log\frac{\det^{1/2}\Sigma_s}{\prod_1^{k_s}\sigma_w}. \tag{6.47}$$

Thus the Bayesian formulas incorporate additional volume terms not included in the "brain surgery" literature. It is not clear whether these terms would make a big difference in practice. In our opinion, the pruning technique now is superseded by the use of more sophisticated regularizers, as discussed in Sec. 6.7.

6.7 Automatic Relevance Determination

The automatic relevance determination (ARD) model [26] can be implemented with the methods described in the previous sections.

Suppose that in a regression problem there are many input variables, of which some are irrelevant to the prediction of the output variable. Because a finite data set will show random correlations between the irrelevant inputs and the output, any conventional neural network (even with weight decay) will fail to set the coefficients for these junk inputs to 0. Thus, the irrelevant variables will hurt the model's performance, particularly when the variables are many and the data are few.

What is needed is a model whose prior over the regression parameters embodies the concept of relevance, so that the model effectively is able to infer which variables are relevant and then switch the others off. A simple and "soft" way of doing this is to introduce multiple weight decay constants, one α associated with each input. The decay rates for junk inputs automatically will be inferred to be large, preventing those inputs from causing significant overfitting.

The ARD model uses the prior of Eq. (6.34). For a network having one hidden layer, the weight classes are: one class for each input, consisting of the weights from that input to the hidden layer; one class for the biases to the hidden units; and one class for each output, consisting of its bias and all the weights from the hidden layer. Control of the ARD model can be implemented using Eq. (6.35).

Automatic relevance determination is expected to be a useful alternative to the technique of pruning (Sec. 6.6), which also embodies the concept of relevance, but in a discrete manner. Possible advantages of ARD include the following:

1. Pruning using Bayesian model comparison requires the evaluation of determinants or inverses of large Hessian matrices, which may be ill-conditioned. ARD, on the other hand, can be implemented using evaluations of the trace of the Hessian alone, which is more robust.

2. Compared with a non-Bayesian cross-validation method, ARD simultaneously infers the utility of large numbers of possible input variables. With only a single cross-validation measure, one might have to explicitly prune one variable at a time in order to estimate which variables are useful. In contrast, ARD returns two *vectors* measuring the relevance of all input variables x_i: the regularization constants α_i and the "well determinednesses" γ_i, and it suppresses the irrelevant inputs without further intervention.

3. ARD allows large numbers of input variables of unknown relevance to be left in the model without harm.

Practical problems found in implementing the ARD model using Gaussian approximations are as follows:

1. If irrelevant variables are not explicitly pruned from a large model, then computation times remain wastefully large.

2. The presence of large numbers of irrelevant variables in a model hampers the calculation of the "evidence" for different models. Numerical problems arise with the calculation of determinants of Hessians. This does not interfere with the Bayesian optimization of regularization constants, but it prevents the use of Bayesian model comparison methods.

3. Although the ARD model is intended to embody a soft version of pruning, the approximations of the evidence framework can lead to singularities with an α_c going to ∞ if the signal-to-noise ratio is low; this causes inputs to be irreversibly shut off.

In spite of these reservations, we are confident that the right direction for adaptive modeling methods lies in the replacement of discrete model choices (e.g., pruning) by continuous control parameters (e.g., sophisticated regularizers).

A common concern is whether the extra hyperparameters $\{\alpha_c\}$ might cause overfitting. There is no cause for worry; there are two reasons. First, if we can evaluate the evidence, then we can evaluate objectively whether the new model is more probable, given the data. The extra parameters are penalized by Occam factors so, eventually, if we increased the number of hyperparameters, an evidence maximum would be reached. In fact, the Occam factors for regularization constants are very weak; the error bars on $\log \alpha_c$ scale only as $1/\sqrt{\gamma_c}$. This fact relates to the second reason why the extra parameters $\{\alpha_c\}$ are incapable of causing overfitting of the data: The extra parameters do not make the model capable of fitting more complicated data sets. Only the parameters $\mathbf{w}$ can overfit noise, and the worst overfitting occurs when the regularization constants α_c are all switched to 0. Thus, the extra hyperparameters have no effect on the worst-case capacity of the model. Their effect is a positive one, namely, a damping out of unneeded degrees of freedom in the model. There is a weak probabilistic penalty for the extra parameters, simply because they increase the variety of simple data sets that the model is capable of predicting. A model with only one hyperparameter α is capable of realizing only one "flavor of simplicity," namely, "all parameters w_i are small," as opposed to the complex flavor, "most parameters w_i are big." A model having, say, three hyperparameters $\{\alpha_c\}$, can predict a total of $2^3 = 8$ flavors of simplicity and complexity including the two above.

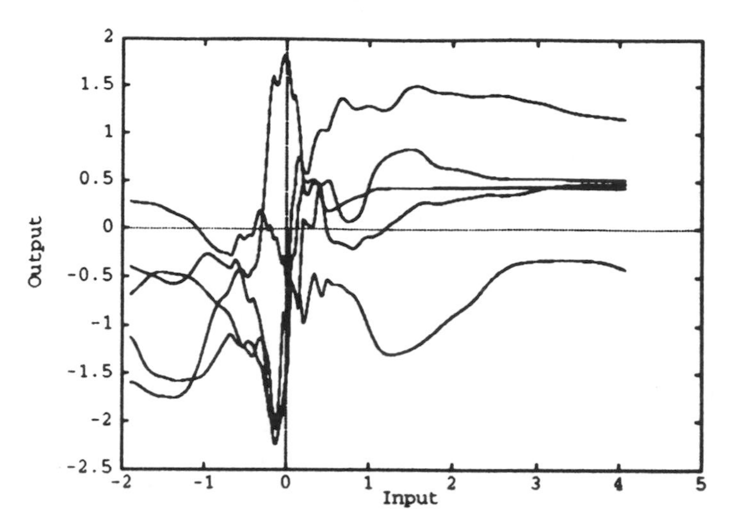

Fig. 6.6. Samples from the prior of a one-input network, with varying number of hidden units. For each curve a different number of hidden units, H, is used: 100, 200, 400, 800, and 1600. The regularization constants for the input weights and hidden unit biases are fixed at $\sigma_{\text{in}}^{w} = 40$ and $\sigma_{\text{bias}}^{w} = 8$. The output weights have $\sigma_{\text{out}}^{w} = 1/\sqrt{H}$ to keep the dynamic range of the function constant.

6.8 Implicit Priors

It is interesting to examine what sort of functions are generated when nets are created by sampling from the prior distributions of Eqs. (6.18) and (6.34). The study of these prior distributions provides guidelines for the expected scaling behavior of regularization constants with the number of hidden units, H. It also identifies which control parameters are responsible for controlling the "complexity" of the function, and which are merely scaling constants. For regression nets with one hidden layer of tanh units and a standard Gaussian prior, we find the following interesting result [27].

In the limit as $H \to \infty$, the complexity of the functions generated by the prior is independent of the number of hidden units. The prior on the input to hidden weights determines the spatial scale (over the inputs) of variations in the function. The prior on the biases of the hidden units determines the characteristic number of fluctuations in the function. The prior on the output weights simply determines the vertical scale of the output, and has no other influence on complexity.

Figures 6.6–6.8 illustrate samples from priors for a one-input–one-output network with a large number of hidden units.

Figure 6.6 illustrates that, as the number of hidden units H is increased, while keeping $\{\sigma_{\text{in}}^{w}, \sigma_{\text{bias}}^{w}, (\sigma_{\text{out}}^{w}\sqrt{H})\}$ fixed, the properties of a random sample from the prior remain stable. (The output weights must get smaller in

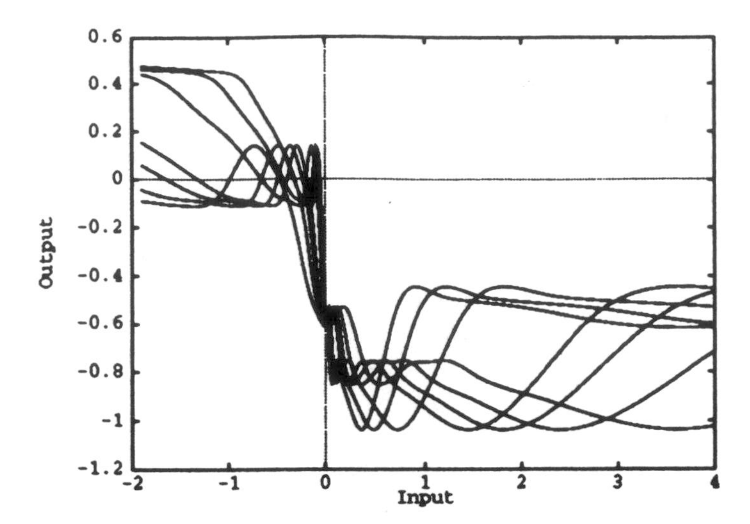

Fig. 6.7. Samples from the prior of a one-input network, with varying σ_{in}^{w}. Varying σ_{in}^{w} alone changes both the characteristic scale length of the oscillations and the overall width of the region in input space in which the action occurs. $\{H, \sigma_{\text{bias}}^{w}, \sigma_{\text{out}}^{w}\} = \{400, 2.0, 0.05\}$. $\sigma_{\text{in}}^{w} = 40, 30, 20, 10, 8, 6, 4$. The smaller the value of σ_{in}^{w}, the less steep the function.

accordance with $\sigma_{\text{out}}^{w} \propto 1/\sqrt{H}$ in order to keep constant the vertical range of the function, which is a sum of H independent random variables with finite variance.)

Figure 6.7 illustrates the effect of varying σ_{in}^{w} alone. Finally, Fig. 6.8 illustrates the effect of varying both σ_{in}^{w} and σ_{bias}^{w}, so as to keep the range of the "action" over the input variable constant. The parameter σ_{bias}^{w} determines the total number of fluctuations in the function.

Progressing to multiple inputs, we obtain Fig. 6.9 by setting the weights into a 2:400:1 net to random values and plotting the output of the net. The picture shows that you can get a "random-looking" function from this model even though the hidden units' activities are based on linear functions of the inputs.

The prior distribution over functions is symmetrical about 0, in both the input space and the output space. It is therefore wise, if this Bayesian model is used, to preprocess the inputs and targets so that 0 is at the expected center of the action.

6.9 Cheap and Cheerful Implementations

The following methods can be used to solve the tasks of automatic optimization of $\{\alpha_c\}$ and β (Sec. 6.3) and calculation of error bars on parameters

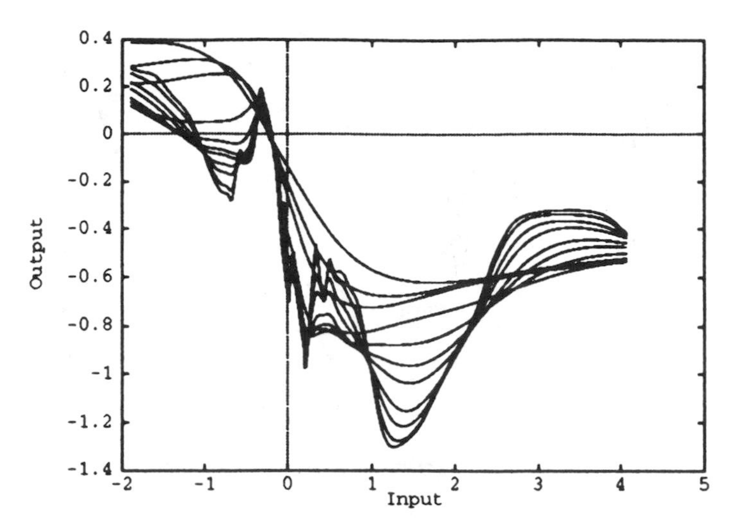

Fig. 6.8. Samples from the prior of a one-input network, with varying σ^w_{bias}. The number of hidden units is kept fixed at 400, with $\sigma^w_{\text{out}} = 0.05$, for all of these samples. The same seed was used, so that all of the weights are simply scaled by the regularization constants as the "movie" progresses. The ratio $\sigma^w_{\text{in}}/\sigma^w_{\text{bias}} = 5.0$ in all cases, so as to keep the action in the range ± 5.0. The constant σ^w_{bias} took the following values: 8, 6, 4, 3, 2, 1.6, 1.2, 0.8, 0.4, 0.3, 0.2. This constant determines the total number of ups and downs in the function. The constant σ^w_{in} determines the input scale on which the ups and downs occur.

and predictions (Sec. 6.5) without calculation of Hessians or sophisticated Monte Carlo methods. These methods depend on the same Gaussian assumptions as does the rest of this chapter; further approximations also are made.

6.9.1 Cheap Approximations for Optimization of α and β

On neglecting the distinction between well-determined and poorly determined parameters, we obtain the following update rules for α and β [cf. Eqs. (6.35) and (6.30)]:

$$\alpha_c := k_c/2E^c_W \qquad \beta := N/2E_D.$$

This easy-to-program procedure is expected to break down when there are a large number of poorly determined parameters.

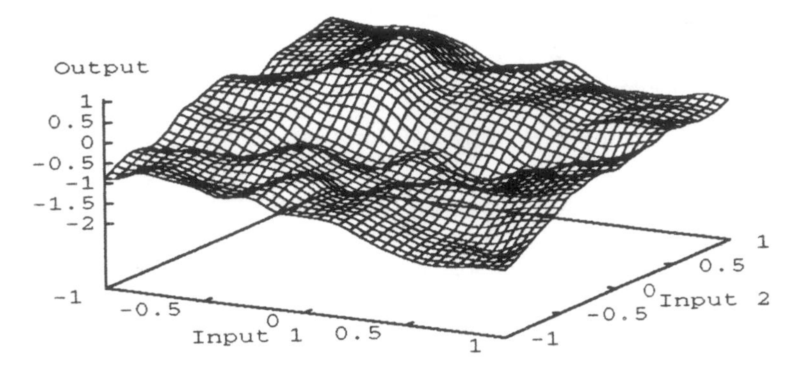

Fig. 6.9. A sample from the prior distribution of a two-input network. $\{H, \sigma_{\text{in}}^{w}, \sigma_{\text{bias}}^{w}, \sigma_{\text{out}}^{w}\} = \{400, 8.0, 8.0, 0.05\}$.

6.9.2 Cheap Generation of Predictive Distributions

A simple way of obtaining random samples from the posterior probability distribution of the parameters follows. This approximate procedure is accurate when the noise really is Gaussian, and when the model can be treated as locally linear.

1. Start with a converged network, with parameters $\mathbf{w}^*$, trained on the true data set $D^* = \{\mathbf{x}^{(m)}, \mathbf{t}^{(m)}\}$. Estimate the Gaussian noise level from the residuals using, for example, $\sigma_\nu^2 = \sum(t - y(\mathbf{w}^*))^2/(N-k)$; alternatively, estimate σ_ν^2 from a test set.

2. Now define a new data set D_1 by adding artificial Gaussian noise of magnitude σ_ν to the outputs in the true data set D^*. Thus, $D_1 = \{\mathbf{x}^{(m)}, \mathbf{t}_1^{(m)}\}$, where $\mathbf{t}_1^{(m)} = \mathbf{t}^{(m)} + \nu$, where $\nu \sim \text{Normal}(0, \sigma_\nu^2)$. No noise is added to the inputs.

3. Next, starting from $\mathbf{w}^*$, train a new network on D_1. Call the converged weight vector $\mathbf{w}_1$. Because the data set will be changed little by the added noise, $\mathbf{w}_1$ will be close to $\mathbf{w}^*$, and this optimization should not take long.

4. Repeat steps 2 and 3 twelve times, generating a new data set D_j from the original data set D^* each time to obtain a new $\mathbf{w}_j$. Save the list of vectors $\mathbf{w}_j$.

5. Separately, use each of $\mathbf{w}_1, \mathbf{w}_2, \ldots \mathbf{w}_{12}$ to make predictions. For example, in the case of time-series continuation, use each $\mathbf{w}_j$ by itself to generate an entire continuation.

> These predictions can be viewed as samples from the model's predictive distribution. They might be summarized by measuring their mean and variance.

In order to get a true sample from the posterior, we also should perturb the prior. For each weight, the mean to which each weight decays, ordinarily 0, should be randomized by sampling from a Gaussian of variance $\sigma_w^2 = 1/\alpha$.

The above method is used in Bayesian image reconstruction [28]. It should be particularly useful for obtaining error bars when neural nets are used to forecast a time series by bootstrapping the network with its own predictions. A full Bayesian treatment of time-series modeling with neural nets has not yet been made.

6.10 Discussion

6.10.1 Applications

The methods of Secs. 6.2 to 6.7 have been successfully applied to several practical problems.

Thodberg has applied these methods to an industrial problem, the inference of pork fat content from spectroscopic data [12]. The evidence framework yields better performance than standard techniques involving cross-validation. This improvement is attributed to the fact that a Bayesian needs no validation set: All of the available data can be used for parameter fitting, for optimization of model complexity, and for model comparison.

The automatic relevance determination model (Sec. 6.7) also has been used to win a recent prediction competition, involving modeling of the energy consumption of a building [22]. Here, the success is attributed to the fact that the evidence framework can be used to simultaneously optimize multiple regularization constants $\{\alpha_c\}$ on-line. Over 20 regularization constants were involved in these networks. The scaling up of these methods to larger neural network problems will be helped by the use of implicit second-order methods [15, 16].

6.10.2 Modeling Insights

An advantage of the Bayesian framework for data modeling (in the eyes of Bayesians) is that it forces one to make explicit the assumptions made in constructing the model. When a poor modeling assumption is identified, the probabilistic viewpoint makes it easy to design coherent modifications to the model.

For example, in [11] the standard weight decay model with only one regularization constant α was applied to a regression network. The evidence for different solutions was found to be poorly correlated with the empirical performance of the solutions. This failure forced home the insight that, for dimensional consistency, at least three different α's are required: one for the input to hidden weights, one for the biases of the hidden units, and one for all of the connections to the outputs. Changing to this model with multiple regularizers produced solutions with slightly improved empirical performance; most importantly, the evidence for these solutions was beautifully correlated with their generalization error.

Here are some examples of other model modifications that are easily motivated from the probabilistic viewpoint. The use of a sum-squared error corresponds to the assumption that the residuals are Gaussian and uncorrelated among the different target variables. In time-series modeling, this may well be a poor model for residuals, which may show local trends. A better model would, for example, assume Gaussian correlations between residuals, such that the data error βE_D is replaced by:

$$\sum_m \left(\beta_0(t_m - y)^2 + \beta_1(t_m - y)(t_{m+1} - y) + \beta_2(t_m - y)(t_{m+2} - y) + \cdots\right). \tag{6.48}$$

This would modify the "backprop" rule, so that the propagated error signal at each frame would be a weighted combination of the residuals at neighboring frames. The network then would experience less of an urge to fit local trends in the data. And, when predictions are made, the model of correlations among residuals would be able to capture the current trend and modify the net's predictions accordingly. The evidence would be used to optimize the correlation model's parameters $\beta_0, \beta_1, \beta_2$, etc.

The Gaussian noise model also might be modified to include the possibility of outliers, using a Bayesian robust noise model [6]. Probability theory allows us to infer from the data how heavy the tails of the noise model ought to be.

Another assumption is that the output noise level is the same for all input vectors. As is discussed in [29, Chapter 6], this assumption can be relaxed by constructing a parameterized model of $\beta(\mathbf{x})$, which can be learned by evidence maximization.

All three of the above examples could be realized as special cases of the following general model, in which the entire set of network parameters is modeled as changing in a correlated way. The general model could be written $P(\{t^{(m)}\}, \{\mathbf{w}^{(m)}\}, w^* | \alpha, \Phi) = \prod_m P(t^{(m)} | \mathbf{w}^{(m)}) P(\{\mathbf{w}^{(m)}\} | \mathbf{w}^*, \Phi) P(\mathbf{w}^* | \alpha)$. Here, the underlying mapping is parameterized by $\mathbf{w}^*$, which is drawn, say, from the ARD prior. The mapping at time m is parameterized by $\mathbf{w}^{(m)}$, which is a random sample from a distribution centered on $\mathbf{w}^*$. Temporal correlations between these samples are defined by the parameters Φ, which might be optimized by evidence maximization. These parameters also model the noise itself. For example, correlated Gaussian residuals are achieved by introducing correlated Gaussian noise into the bias of the output unit. Non-Gaussian noise (long-tailed, or asymmetric) could be created by introducing a Gaussian noise process earlier in the net; a careful choice of hidden unit activation functions could bias the noise distribution in accordance with our prior beliefs. If we make a model in which noise is modeled by fluctuations in the network parameters, it is easy to imagine that an input-dependent noise level could be learned by this model. Finally, a "mixture of experts" is another special case of this model, obtained when the distribution $P(\{\mathbf{w}^{(m)}\}|\mathbf{w}^*, \Phi)$ is a mixture of delta-functions.

A final example of a probabilistic motivation for a model modification lies in image analysis. If we use a neural net for character recognition, say, then we might expect a well-trained net to have input weights that are spatially correlated. It is desirable to incorporate this prior expectation into the model adaptation process, because such priors on parameters damp out unnecessary degrees of freedom and reduce overfitting. One way of creating such a correlation is to preblur the data before feeding it into the network, and use the normal uncorrelated prior on the parameters. This is equivalent to keeping the original inputs and having a correlated prior on the parameters, where the correlations are defined implicitly by the properties of the preblur. This procedure has been used fruitfully in character recognition work [30].

Whenever a modification to a model is conceived, which can be expressed probabilistically, a coherent algorithm incorporating the modification can be mechanically derived. The automatic relevance determination model (Sec. 6.7) is an example of a successful model developed in this way. One often can observe the unanticipated emergence of elegant formulas when the rules of probability theory are applied to a new model.

6.10.3 Relationship to Theories of Generalization

The Bayesian "evidence" framework assesses within a well-defined hypothesis space *how probable* a set of alternative models are. However, what we often want to know is how well each model is expected to generalize. Empirically, the correlation between the evidence and generalization error is surprisingly good [11, 12]. But a theoretical connection linking the two is not yet established. Here, a brief discussion is given of similarities and differences between the evidence and quantities arising in work on prediction of the generalization error.

Relation to GPE

Moody's "Generalized Prediction Error" (GPE) [31] is a generalization of Akaike's "Final Prediction Error" (FPE) to nonlinear regularized models. These are both estimators of generalization error which can be derived without making assumptions about the distribution of residuals between the data and the true interpolant, and without assuming that the true interpolant belongs to some particular class. Both are derived by assuming that the observed distribution over the inputs in the training set gives a good approximation to the distribution of future inputs.

The difference between the FPE and the GPE is that the total number of parameters k in the FPE is replaced by an effective number of parameters, which is in fact identical to the quantity γ arising in the Bayesian analysis in Eq. (6.29). If E_D is one-half the sum-squared error, then the predicted error per data point is

$$\text{GPE} = \left(E_D + \sigma_\nu^2 \gamma\right)/N. \tag{6.49}$$

The added term $\sigma_\nu^2\gamma$ has an intuitive interpretation in terms of overfitting. For every parameter that is well determined by the data, we unavoidably overfit one "direction" of noise. This has two effects: it makes E_D smaller than it "ought to be," by $\sigma_\nu^2/2$, on average; and it means that our predictions vary from the ideal predictions (those that we would make if we had infinite data) so that our prediction error on the same N input points would on average be worse by $\sigma_\nu^2/2$. The sum of these two terms, multiplied by the effective number of well-determined parameters γ, gives the correction term.

Like the log evidence, the GPE has the form of the data error plus a term that penalizes complexity. However, although the same quantity γ arises in the Bayesian analysis, the Bayesian Occam factor does *not* have the same scaling behavior as the GPE term (see the discussion below). And, empirically, the GPE is not always a good predictor of generalization. One reason is that, in the derivation of the GPE, it is effectively assumed that test samples will be drawn only at the $\mathbf{x}$ locations at which we have already received data. The consequences of this false assumption are most serious for overparameterized and overflexible models. An additional distinction between the GPE and the evidence framework is that the GPE is defined for regression problems only; the evidence can be evaluated for regression, classification, and density models.

Relation to the Effective VC Dimension

Recent work on "structural risk minimization" [30] utilizes empirical expressions of the form:

$$E_{\text{gen}} \simeq E_D/N + c_1 \frac{\log(N/\gamma) + c_2}{N/\gamma}, \tag{6.50}$$

where γ is the "effective VC dimension" of the model and is identical to the quantity in Eq. (6.29). The constants c_1 and c_2 are determined by experiment. The structural risk theory currently is intended to be applied only to nested families of classification models (hence the absence of β: E_D is dimensionless, like G) with monotonic effective VC dimension, whereas the evidence can be evaluated for any models. Interestingly, the scaling behavior of this expression (6.50) is identical to the scaling behavior of the log evidence in Eq. (6.27), subject to two assumptions: first, that the value of the regularization constant satisfies Eq. (6.28); and second, that the significant eigenvalues ($\lambda_a > \alpha$) scale as $\lambda_a \sim N\alpha/\gamma$. (This scaling holds, for example, in the family of interpolation models consisting of a sequence of steps of independent heights, in which we vary the number of steps.) Then it can be shown that the scaling of the log evidence is

$$- \log P(D|\alpha, \beta, \mathcal{H}) \sim \beta E_D^{\mathrm{MP}} + \tfrac{1}{2} \left(\gamma \log(N/\gamma) + \gamma\right). \qquad (6.51)$$

[Readers familiar with the Minimum Description Length (MDL) will recognize the dominant $\gamma/2 \log N$ term; MDL and Bayes are equivalent, as is discussed later.] Thus, the scaling behavior of the log evidence is identical to the structural risk minimization expression (6.50), provided that $c_1 = \frac{1}{2}$ and $c_2 = 1$. Isabelle Guyon has confirmed (personal communication) that the empirically determined values for c_1 and c_2 are indeed close to these Bayesian values. It will be interesting to try to understand and develop this relationship.

6.10.4 Contrasts with Conventional Dogma in Learning Theory and Statistics

Representation Theorems

It is popular to prove the utility of a particular model by demonstrating that the model has arbitrary representational power. For example, "neural networks are good interpolation tools because they can implement any smooth function given enough hidden units."

A Bayesian data modeler takes a different attitude (as, to be fair, do other learning theory researchers). The objective of data modeling is to find a model that is well matched to the data. A model that is too flexible, and which could match arbitrary data, will generalize poorly; and in Bayesian terms such a model is improbable compared to simpler models that also fit the data. Probability theory favors a model that is as *inflexible* as possible: just flexible enough to capture the real structure in the data, but no more. The quality of a model is judged solely by how well matched it is, probabilistically, to the data.

Those who appreciate that the universal representational power of a model is not a good thing are often led astray by a second myth, the supposed need to limit the complexity of a model when there is little data.

"The Complexity of the Model Should Be Matched to the Amount of Data"

A popular idea is that, when there is little data, it is good to use a model with few parameters, even if that model is known to be incapable of representing the true function. An attempt is made to match the "capacity" of the model to the number of data points. This sometimes is used as the motivation for "pruning" a neural network [14].

A Bayesian need never do this; the choice of which models to consider is a matter of prior belief, and should not depend on the amount of data collected. It is now common practice for Bayesians to fit models that have more parameters than the number of data points [5, 32]. It is true that probability theory penalizes models that are too complicated for the data. But we should not therefore deliberately construct models that are so simple that they are incompatible with our prior beliefs. In terms of the evidence, it is not possible for a small data set to systematically favor the wrong model [5].

In a domain such as interpolation, our typical prior belief is that the real underlying function is complex and would require an infinite number of parameters to describe it exactly. There will never be enough data to determine all of the parameters of the true model. But this does not mean that we should use a smaller model: that would give us well-determined but *incorrect* predictions! We should use the model we believe in. No harm can come of this. It may be that the resulting predictions are ill-determined; but if the true model's predictions have huge error bars, it is surely crazy to use a simpler model in order to make the error bars smaller! Alternatively, the prior knowledge of smoothness, etc., included in the true model may constrain the ill-determined parameters such that the predictions are quite satisfactory.

The strength of the Bayesian method, therefore, centers on the prior assigned to the parameters. The prominent role of the prior in Bayesian methods often is regarded as a weakness. But any alternative method of controlling the complexity of an interpolant, say, also embodies implicit priors — except that those implicit priors generally do not correspond to our real beliefs. The way forward, therefore, is to develop more sophisticated probabilistic models and better computational methods for using them. Discrete model choices should be replaced by regularized continua of models, with an arbitrarily large number of parameters.

"My Model Is Better than Your Model"

Much of orthodox statistics is concerned with the invention of estimators and the evaluation of certain average case properties of those estimators (such as bias, variance, sufficiency, consistency, power, etc.). These criteria then are used to choose between different estimators; all this without any reference to the actual data that have been observed. Bayesians need not get involved in debates concerning which properties of an estimator one should concentrate on, or which estimator is intrinsically best. There is no best model. Each model corresponds to a probabilistic statement about the domain. One model will be better matched to some data sets, while another model will be better matched to others. The evaluation of the evidence allows us to infer, from the particular observed data set, which in our space of models is the most probable model.

An alternative way of viewing Bayesian modeling is to say that we only have one supermodel, composed of a number of submodels which make different assumptions, which have different complexities, etc. Once a supermodel is defined, our inferences are given by mechanically following the rules of probability; these inferences implicitly involve comparisons of the submodels, embodying the Bayesian Occam's razor.

The subjective task that a Bayesian still has to tackle is the definition of the entire model space. The inventions of a good model space for a problem, and of numerical techniques for inference in that space, are nontrivial tasks requiring great skill. The recommended philosophy [6] is to aim to incorporate every imaginable possibility into the model space: for example, if it is conceivable that a very simple model might be able to explain the data, one should include simple models in the model space; if the noise might have a long-tailed distribution, one should include a hyperparameter which controls the heaviness of the tails of the distribution, such that one value of the hyperparameter gives the null distribution; if an input variable might be irrelevant to a regression, include it in the regression anyway, with a sophisticated regularizer embodying the concept of relevance. The inclusion of remote possibilities in the model space is "safe," because our inferences will home in on the submodels that are best matched to the data. The inclusion in our initial model space of bizarre models that are subsequently ruled out by the data is not expected to influence predictive performance significantly.

6.10.5 Minimum Description Length (MDL)

A complementary view of Bayesian model comparison is obtained by replacing probabilities of events by the lengths in bits of messages that communicate the event without loss to a receiver. Message lengths $L(\mathbf{x})$ correspond to a probabilistic model over events $\mathbf{x}$ via the relations:

$$P(\mathbf{x}) = 2^{-L(\mathbf{x})}, \qquad L(\mathbf{x}) = -\log_2 P(\mathbf{x}). \tag{6.52}$$

Noninteger coding lengths can be handled by the arithmetic coding procedure [33].

The MDL principle [34] states that one should prefer models which can communicate the data in the smallest number of bits. Consider a message that states which model, $\mathcal{H}$, is to be used, and then communicates the data D within that model, to some prearranged precision δD. This produces a message of length $L(D, \mathcal{H}) = L(\mathcal{H}) + L(D|\mathcal{H})$. The lengths $L(\mathcal{H})$ for different $\mathcal{H}$ can be interpreted in terms of an implicit prior $P(\mathcal{H})$ over the alternative models. Similarly, $L(D|\mathcal{H})$ corresponds to a density $P(D|\mathcal{H})$. Thus, a procedure for assigning message lengths can be mapped onto pos-

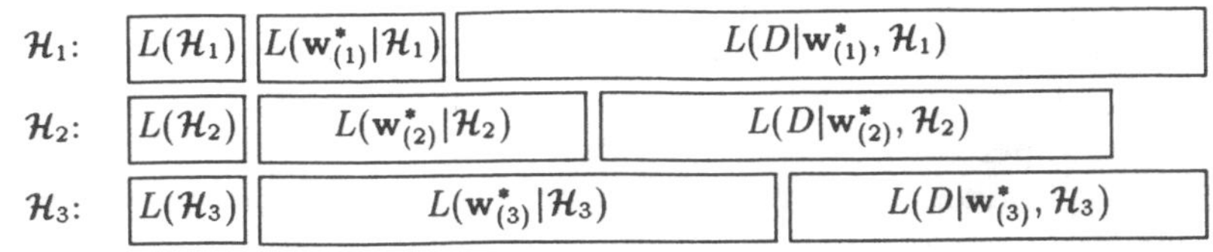

$\mathcal{H}_1$:	$L(\mathcal{H}_1)$	$L(\mathbf{w}^*_{(1)}\|\mathcal{H}_1)$	$L(D\|\mathbf{w}^*_{(1)}, \mathcal{H}_1)$
$\mathcal{H}_2$:	$L(\mathcal{H}_2)$	$L(\mathbf{w}^*_{(2)}\|\mathcal{H}_2)$	$L(D\|\mathbf{w}^*_{(2)}, \mathcal{H}_2)$
$\mathcal{H}_3$:	$L(\mathcal{H}_3)$	$L(\mathbf{w}^*_{(3)}\|\mathcal{H}_3)$	$L(D\|\mathbf{w}^*_{(3)}, \mathcal{H}_3)$

Fig. 6.10. A popular view of model comparison by minimum description length. Each model $\mathcal{H}_i$ communicates the data D by sending the identity of the model, sending the best-fit parameters of the model $\mathbf{w}^*$, and then sending the data relative to those parameters. As we proceed to more complex models, the length of the parameter message increases. On the other hand, the length of the data message decreases, because a complex model is able to fit the data better, making the residuals smaller. In this example, the intermediate model $\mathcal{H}_2$ achieves the optimum trade-off between these two trends.

terior probabilities:

$$\begin{aligned} L(D, \mathcal{H}) &= -\log P(\mathcal{H}) - \log(P(D|\mathcal{H})\delta D) \\ &= -\log P(\mathcal{H}|D) + \text{const.} \end{aligned}$$

In principle, then, MDL always can be interpreted as Bayesian model comparison, and vice versa. However, this simple discussion has not addressed how one would actually evaluate the key data-dependent term $L(D|\mathcal{H})$, which corresponds to the evidence for $\mathcal{H}$. Often, this message is imagined as being subdivided into a "best-fit parameter" block and a data block. This procedure conveys an intuitive picture of model comparison (Fig. 6.10). Models with a small number of parameters have only a short parameter block but do not fit the data well, and so the data message (a list of large residuals) is long. As the number of parameters increases, the parameter block lengthens, and the data message becomes shorter. There is an optimum model complexity ($\mathcal{H}_2$ in the figure) for which the sum is minimized.

This picture is still too simple. We have not specified the precision to which the parameters $\mathbf{w}$ should be sent. This precision has an important effect (unlike the precision δD to which real-valued data D are sent, which, assuming δD is small relative to the noise level, just introduces an additive constant). As we decrease the precision to which $\mathbf{w}$ is sent, the parameter message shortens, but the data message typically lengthens because the truncated parameters do not match the data very well. There is a nontrivial optimal precision. In simple Gaussian cases it is possible to solve for this optimal precision [35], and it is closely related to the posterior error bars on the parameters, $\mathbf{A}^{-1}$, where $\mathbf{A} = -\nabla\nabla \log P(\mathbf{w}|D, \mathcal{H})$. It turns out that the optimal parameter message length is virtually identical to the log of the Occam factor in Eq. (6.12). (The random element involved in parameter truncation means that the encoding is slightly suboptimal.)

With care, therefore, one can replicate Bayesian results in MDL terms. Although some of the earliest work on complex model comparison involved the MDL framework [36], MDL has no apparent advantages over the direct probabilistic approach.

MDL does have its uses as a pedagogical tool. The description length concept is useful for motivating prior probability distributions. Also, different ways of breaking down the task of communicating data using a model can give helpful insights into the modeling process.

On-Line Learning and Cross-Validation

The log evidence can be decomposed as a sum of on-line predictive performances:

$$\log P(D|\mathcal{H}) = \log P(\mathbf{t}^{(1)}|\mathcal{H}) + \log P(\mathbf{t}^{(2)}|\mathbf{t}^{(1)}, \mathcal{H}) + \\ \log P(\mathbf{t}^{(2)}|\mathbf{t}^{(1)}, \mathbf{t}^{(2)}, \mathcal{H}) \cdots + \log P(\mathbf{t}^{(N)}|\mathbf{t}^{(1)} \ldots \mathbf{t}^{(N-1)}, \mathcal{H}).$$

This decomposition emphasizes the difference between the evidence and "leave one out cross-validation" as measures of predictive ability. Cross-validation examines the average value of just the last term, $\log P(\mathbf{t}^{(N)}|t^{(1)} \ldots \mathbf{t}^{(N-1)}, \mathcal{H})$, under random reorderings of the data. The evidence, on the other hand, sums up how well the model predicted all of the data, starting from scratch.

The "Bits Back" Encoding Method

Another MDL thought experiment [37] involves incorporating random bits into our message. The data are communicated using a parameter block and a data block. The parameter vector sent is a random sample from the posterior distribution $P(\mathbf{w}|D, \mathcal{H}) = P(D|\mathbf{w}, \mathcal{H})P(\mathbf{w}|\mathcal{H})/P(D|\mathcal{H})$. This sample $\mathbf{w}$ is sent to an arbitrary small granularity $\delta\mathbf{w}$ using a message length $L(\mathbf{w}|\mathcal{H}) = -\log(P(\mathbf{w}|\mathcal{H})\delta\mathbf{w})$. The data are encoded relative to $\mathbf{w}$ with a message of length $L(D|\mathbf{w}, \mathcal{H}) = -\log(P(D|\mathbf{w}, \mathcal{H})\delta D)$. Once the data message has been received, the random bits used to generate the sample $\mathbf{w}$ from the posterior can be deduced by the receiver. The number of bits so recovered is $-\log(P(\mathbf{w}|D, \mathcal{H})\delta\mathbf{w})$. These recovered bits need not count toward the message length, since we might use some other optimally encoded message as a random bit string, thereby communicating that message at the same time. The net description cost is therefore:

$$\begin{aligned} L(\mathbf{w}|\mathcal{H}) + L(D|\mathbf{w}, \mathcal{H}) - \text{“bits back”} &= -\log \frac{P(\mathbf{w}|\mathcal{H})P(D|\mathbf{w}, \mathcal{H})\delta D}{P(\mathbf{w}|D, \mathcal{H})} \\ &= -\log P(D|\mathcal{H}) - \log \delta D. \end{aligned}$$

Thus, this thought experiment has yielded the optimal description length.

6.10.6 Ensemble Learning

The posterior distribution $P(\mathbf{w}|D,\mathcal{H})$ may be a very complicated density. The methods described in this chapter have assumed that, in local regions that contain significant probability mass, the posterior can be well approximated by a Gaussian found by making a quadratic expansion of $\log P(\mathbf{w}|D,\mathcal{H})$ around a local maximum. (For brevity we omit here the parameters α and β.)

An interesting idea that has been implemented by Hinton and van Camp [37] is to try to improve the quality of this type of approximation by optimizing the entire posterior approximation. We call this *ensemble learning*. Consider a parameterized approximation $Q(\mathbf{w};\theta)$ to the true posterior distribution $P(\mathbf{w}|D,\mathcal{H})$. For example, the parameters θ for a Gaussian distribution would be its mean and covariance matrix. The idea is that a Gaussian fitted somewhere other than the mode of $P(\mathbf{w}|D,\mathcal{H})$ might in some sense be a better approximation to the posterior. One possible measure of the quality of fit of Q to P is the "free energy":

$$F(\theta) = -\int d\mathbf{w}\, Q(\mathbf{w};\theta) \log \frac{P(\mathbf{w}|D,\mathcal{H})}{Q(\mathbf{w};\theta)}. \tag{6.53}$$

It is well known that F has a lower bound of 0 that can be realized only if there are parameters θ such that Q matches P exactly. This measure can be motivated by generalizing the MDL "bits back" thought experiment (Sec. 6.10) with the random sample $\mathbf{w}$ drawn from Q instead of from P [37].

Now the task is to minimize $F(\theta)$. This is in general a challenging task. However, Hinton and van Camp [37] have shown that exact derivatives of F with respect to θ can be obtained for a neural net with one nonlinear hidden layer and a linear output if the Gaussian model $Q(\mathbf{w};\theta)$ is restricted so as to have 0 correlation among the weights.

The weakness of ensemble learning by free energy minimization is that, if the approximating distribution $Q(\mathbf{w};\theta)$ has only a simple form, then the free energy objective function favors distributions that are extremely conservative, placing no probability mass in regions where $P(\mathbf{w})$ is small. For example, if a strongly correlated Gaussian P is modeled by a separable Gaussian Q, then the free energy solution sets the curvature of $\log Q$ to be the same as the diagonal elements of the curvature of $\log P$. This gives an approximating distribution that covers far too small a region of $\mathbf{w}$ space, so that the outcome of ensemble learning would be essentially identical to the outcome of traditional optimization of a point estimate. It is therefore interesting to try to extend the ensemble learning method to more complex models Q.

A possible extension of Hinton's and van Camp's idea is to include in θ an adaptive linear preprocessing of the inputs. Denote the coefficients of this linear mapping from inputs to subinputs by $\mathbf{U}$, and the parameters from the subinputs to the hidden units by $\mathbf{V}$; the effective input weights are given by the product **VU**. A separable Gaussian prior now can be applied to the parameters **V**, so that Hinton's and van Camp's exact derivatives still can be evaluated. Inclusion of the additional parameters **U** in θ defines a richer family of probability distributions $Q(\mathbf{w}; \theta)$ over the effective parameters **w**. It will be interesting to see if these distributions are powerful enough to yield Gaussian approximations superior to those produced by the evidence framework.

Acknowledgments. The author thanks his colleagues at Caltech, the University of Toronto, and the University of Cambridge for invaluable discussions. He also is grateful to Radford Neal for comments on the manuscript.

References

[1] J. Berger (1985) *Statistical Decision Theory and Bayesian Analysis* (Springer-Verlag, New York)

[2] S.F. Gull (1988) Bayesian inductive inference and maximum entropy. In: *Maximum Entropy and Bayesian Methods in Science and Engineering, Vol. 1: Foundations,* G.J. Erickson, C.R. Smith (Eds.) (Kluwer, Dordrecht), pp. 53–74

[2a] E.T. Jaynes (1983) Bayesian intervals versus confidence intervals. In: *E.T. Jaynes. Papers on Probability, Statistics and Statistical Physics*, R.D. Rosencrantz (Ed.) (Kluwer, Dordrecht), p. 151

[3] H. Jeffreys (1939) *Theory of Probability* (Oxford Univ. Press, Oxford, UK)

[4] T.J. Loredo (1990) From Laplace to supernova SN 1987A: Bayesian inference in astrophysics. In: *Maximum Entropy and Bayesian Methods, Dartmouth, U.S.A., 1989*, P. Fougere (Ed.) (Kluwer, Dordrecht), pp. 81–142

[5] D.J.C. MacKay (1992) Bayesian interpolation. *Neural Comput.* **4**(3):415–447

[6] G.E.P. Box, G.C. Tiao (1973) *Bayesian Inference in Statistical Analysis* (Addison-Wesley, Reading, MA)

[7] D.E. Rumelhart, G.E. Hinton, R.J. Williams (1986) Learning representations by back–propagating errors. *Nature* **323**:533–536

[8] J.S. Bridle (1989) Probabilistic interpretation of feedforward classification network outputs, with relationships to statistical pattern recognition. In: *Neuro-Computing: Algorithms, Architectures and Applications* F. Fougelman-Soulie, J. Hérault (Eds.) (Springer-Verlag, New York)

[9] R.M. Neal (1993) Bayesian learning via stochastic dynamics. In: *Advances in Neural Information Processing Systems 5*, C.L. Giles, S.J. Hanson, J.D. Cowan (Eds.) (Morgan Kaufmann, San Mateo, CA), pp. 475–482

[10] S.F. Gull (1989) Developments in maximum entropy data analysis. In: *Maximum Entropy and Bayesian Methods, Cambridge 1988*, J. Skilling (Ed.) (Kluwer, Dordrecht), pp. 53–71

[11] D.J.C. MacKay (1992) A practical Bayesian framework for backpropagation networks. *Neural Comput.* **4**(3):448–472

[12] H.H. Thodberg (1993) Ace of Bayes: Application of neural networks with pruning. Technical Report 1132 E, Danish Meat Research Institute

[13] C.M. Bishop (1992) Exact calculation of the Hessian matrix for the multilayer perceptron. *Neural Comput.*, **4**(4):494–501

[14] B. Hassibi, D.G. Stork (1993) Second order derivatives for network pruning: Optimal brain surgeon. In: *Advances in Neural Information Processing Systems 5*, C.L. Giles, S.J. Hanson, J.D. Cowan (Eds.) (Morgan Kaufmann, San Mateo, CA), pp. 164–171

[15] J. Skilling (1993) Bayesian numerical analysis. In: *Physics and Probability*, W.T. Grandy Jr., P. Milonni (Eds.) (Cambridge University Press, Cambridge, UK)

[16] B. Pearlmutter (1993) *Neural Comput.* to appear

[17] G.L. Bretthorst (1988) *Bayesian Spectrum Analysis and Parameter Estimation.* (Springer-Verlag, New York)

[18] D.J.C. MacKay (1994) Hyperparameters: Optimize, or integrate out? In: *Maximum Entropy and Bayesian Methods, Santa Barbara 1993*, G. Heidbreder (Ed.) (Kluwer, Dordrecht)

[19] W.L. Buntine, A.S. Weigend (1991) Bayesian back–propagation. *Complex Syst.* **5**:603–643

[20] D.H. Wolpert (1993) On the use of evidence in neural networks. In: *Advances in Neural Information Processing Systems 5*, C.L. Giles, S.J. Hanson, J.D. Cowan (Eds.) (Morgan Kaufmann, San Mateo, CA), pp. 539–546

[21] R. Hanson, J. Stutz, P. Cheeseman (1991) Bayesian classification theory. Technical Report FIA–90-12-7-01, NASA Ames

[22] D.J.C. MacKay (1993) Bayesian non-linear modeling for the 1993 energy prediction competition. Technical Report, Cambridge University, Cambridge, UK, in preparation

[23] J.B. Copas (1983) Regression, prediction and shrinkage (with discussion). *J.R. Statist. Soc B* **45**(3):311–354

[24] D.J.C. MacKay (1992) The evidence framework applied to classification networks. *Neural Comput.* **4**(5):698–714

[25] Y. LeCun, J.S. Denker, S.A. Solla (1990) Optimal brain damage. In: *Advances in Neural Information Processing Systems 2*, D.S. Touretzky (Ed.) (Morgan Kaufmann, San Mateo, CA), pp. 598–605

[26] D.J.C. MacKay, R.M. Neal (1993) Automatic relevance determination for neural networks. Technical Report, Cambridge University, Cambridge, UK, in preparation

[27] R.M. Neal (1993) Priors for infinite networks. Technical Report, Univ. of Toronto, Toronto, Canada, in preparation

[28] J. Skilling, D.R.T. Robinson, S.F. Gull (1991) Probabilistic displays. In: *Maximum Entropy and Bayesian Methods, Laramie, 1990*, W.T. Grandy, L. Schick (Eds.) (Kluwer, Dordrecht), pp. 365–368

[29] D.J.C. MacKay (1991) Bayesian Methods for Adaptive Models. PhD thesis, California Institute of Technology, Pasadena

[30] I. Guyon, V.N. Vapnik, B.E. Boser, L.Y. Bottou, S.A. Solla (1991) Structural risk minimization for character recognition. In: *Advances in Neural Information Processing Systems 4*, J.E. Moody, S.J. Hanson, R.P. Lippmann (Eds.) (Morgan Kaufmann, San Mateo, CA), pp. 471–479

[31] J.E. Moody (1992) The *effective* number of parameters: An analysis of generalization and regularization in nonlinear learning systems. In: *Advances in Neural Information Processing Systems 4*, J.E. Moody, S.J. Hanson, R.P. Lippmann (Eds.) (Morgan Kaufmann, San Mateo, CA), pp. 847–854

[32] N. Weir (1991) Applications of maxmimum entropy techniques to HST data. *Proc. ESO/ST–ECF Data Analysis Workshop, April 1991*

[33] I.H. Witten, R.M. Neal, J.G. Cleary (1987) Arithmetic encoding for data compression. *Commun. ACM* **30**(6):520–540

[34] C.S. Wallace, D.M. Boulton (1968) An information measure for classification. *Comput. J.* **11**(2):185–194

[35] C.S. Wallace, P.R. Freeman (1987) Estimation and inference by compact coding. *J.R. Statist. Soc. B* **49**(3):240–265

[36] J.D. Patrick, C.S. Wallace (1982) Stone circle geometries: An information theory approach. In: *Archaeoastronomy in the Old World*, D.C. Heggie (Ed.) (Cambridge Univ. Press, Cambridge, UK), pp. 231–264

[37] G.E. Hinton, D. van Camp (1993) Keeping neural networks simple by minimizing the description length of the weights. *Proc. COLT-93*, to appear

[38] L. Breiman (1992) Stacked regressions. Technical Report 367, Dept. of Stat., Univ. of Cal. Berkeley

7

Penacée: A Neural Net System for Recognizing On-Line Handwriting

I. Guyon,[1] J. Bromley,[2] N. Matić,[3] M. Schenkel,[4] and H. Weissman[5]

with 11 figures

Synopsis. We report on progress in handwriting recognition and signature verification. Our system, which uses pen-trajectory information, is suitable for use in pen-based computers. It has a multimodular architecture whose central trainable module is a time-delay neural network. Results comparing our system and a commercial recognizer are presented. Our best recognizer makes three times less errors on hand-printed word recognition than the commercial one.

7.1 Introduction

This chapter reports on progress in the design of a multimodular system that recognizes on-line handwriting and whose central module is a neural network. By *on-line* we mean that the input to the system is the pen trajectory, sampled at regular time intervals by a touch-sensitive pad. In this chapter, we provide a synthesis of previously published work from our group [1, 2, 3, 4, 5, 6, 7] and a perspective on the on-going research.

There is urgent need for good recognizers to ensure the success of the

[1]AT&T Bell Labs, 955 Craston Road, Berkeley, CA 94708, USA (isabelle@research.att.com).

[2]AT&T Bell Labs, Room 4G-338, Holmdel, NJ 07733, USA (jbromley@research.att.com).

[3]AT&T Bell Labs, presently at Synaptics, 2698 Orchard Parkway, San Jose, CA 95134, USA (nada@synaptics.com).

[4]AT&T Bell Labs and ETH-Zürich, CH-8092 Zürich, Switzerland (schenkel@isi.ethz.ch).

[5]AT&T Bell Labs, presently at 12 Mordehai-Hetez St., Petah-Tikua, Israel (f67361@barilan.bitnet).

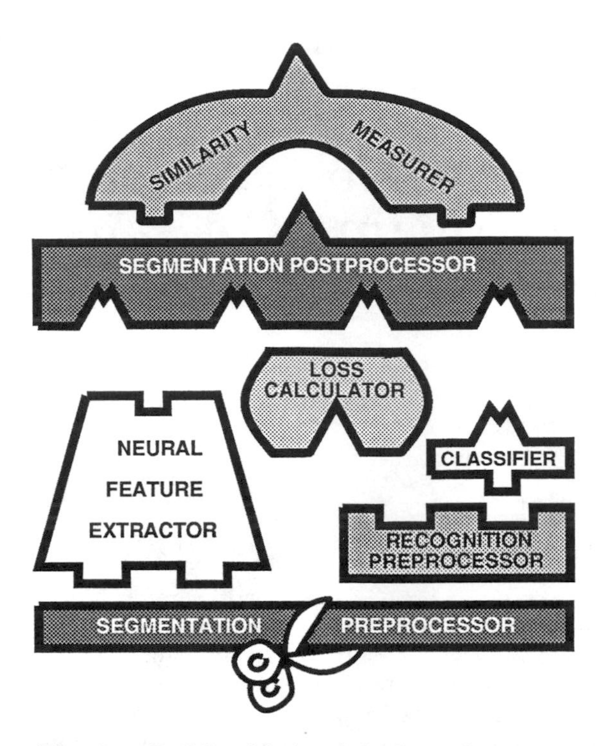

Fig. 7.1. Building blocks of the Penacée system.

first pen-based computers and pen-based personal communicators. In spite of the efforts of many companies and universities, state-of-the-art on-line handwriting recognition accuracy has not yet reached a level that is acceptable to users. We tackle this problem from different angles. First, we address tasks of intermediate difficulty, but of real practical interest, such as the recognition of hand-printed words. Second, we introduce writer adaptation to fine tune the recognizer with examples of a particular user style.

The Penacée system (our Pen panacea...) is composed of several modules (preprocessor, classifier, segmentor, etc.), as is classically done in pattern recognition [8] (see Fig. 7.1). The originality of using neural networks, and perhaps also the main advantage, is that the network itself can be decomposed into two modules [9]: a neural-network–based feature extractor and a classifier.

We make extensive use of our neural feature extractor, which is a trainable module capable of producing a very good and compact feature representation. Our neural network is a time–delay neural network (or TDNN) [10, 11]. It is a convolutional network that has several layers of local feature extractors. It is the one-dimensional version of the network used by the Optical Character Recognition group in our department [12, 13, 14] and is suitable for processing time-varying signals, such as the pen-trajectory.

In Sec. 7.2, we introduce the various modules of our system, and in Sec. 7.3 we present the results of applying it to isolated character recognitions, word recognition, and signature verification.

7.2 Description of the Building Blocks

7.2.1 Recognition Preprocessor

The preprocessor converts the input to the system to a representation that facilitates the recognition process. Preprocessing techniques incorporate human knowledge about the task at hand, such as known invariances and relevant features. In this work, we use a rather crude preprocessing and rely mostly on the neural network to enforce invariances and extract features. Our preprocessing consists of fairly simple normalizations and the extraction of low-level local topological features such as line orientation.

Because the input data is the pen-trajectory, we face the choice of whether or not to use the dynamic information. It is possible to remove the temporal parametrization of the data and represent patterns as pixel images. With such a representation, all of the techniques used for OCR (optical character recognition) are readily applicable. In this work, however, we encode patterns as a sequence of feature vectors, corresponding to the sequence of drawing actions [1, 4].

In Fig. 7.2, we give an example of a preprocessed pattern with sequential encoding. The preprocessing is decomposed into normalization (centering, scaling, deskewing), resampling (to obtain a desired number of regularly spaced points along the trajectory), smoothing (to remove jittering), and feature extraction. In the resulting representation, each point on the trajectory is associated with a feature vector whose components are a subset of x and y coordinates, direction of the trajectory, curvature of the trajectory, speed of the pen, acceleration of the pen, pen-down or -up position (touching or above the writing surface).

Depending on the application, the preprocessing may vary slightly. For instance, it is debatable whether the representation should be invariant under changes in the speed of the pen. For writer-independent character recognition, variations in the speed of the pen are a nuisance. Conversely, for signature verification, the exact dynamics of drawing actions are very precious for the detection of forgeries. Data-collection devices sample the trajectory at regular time intervals (10–12 ms). Some invariance with respect to the speed of the pen is obtained for character recognition applications by resampling to points regularly spaced in arc length, as opposed to regularly spaced in time. For signature verification, the resampling preserves regular time intervals.

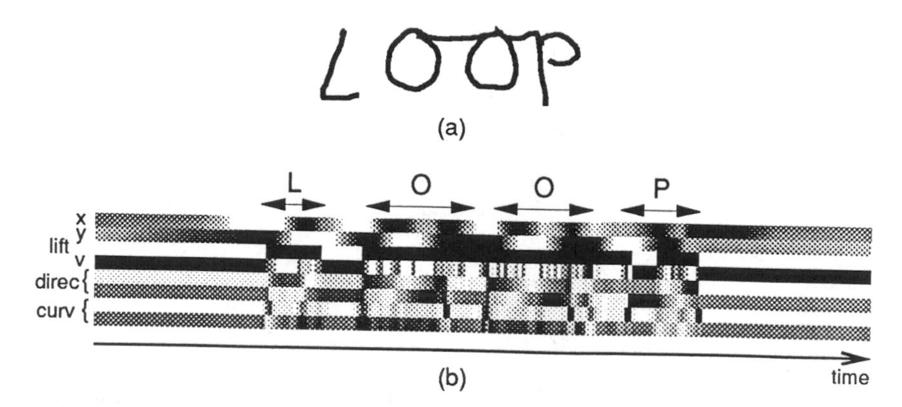

Fig. 7.2. Recognition preprocessing. (a) The original word. (b) The data as presented to the network. The feature vector in this case has eight components encoding x and y coordinates normalized and rescaled, pen-up or -down information, speed of the pen, two components for direction of the pen-trajectory, and two components for its curvature.

7.2.2 Neural Feature Extractor

Our networks consist of several layers of feature extraction followed by a classification layer. The neural feature extractor in Fig. 7.1 is the network up to its classification layer.

We use a convolutional neural network, the time-delay neural network (TDNN). TDNNs first were introduced for speech recognition and are well suited to sequential signal processing [10, 11].

We briefly sketch here the principles of the TDNN (Fig. 7.3). One layer of the network transforms a sequence of feature vectors into another sequence of higher order feature vectors in the following way.

A given neuron detects a particular local topological feature of the pen-trajectory. Its receptive field is restricted to a limited time window. The same neuron is reused along the time axis (Fig. 7.3, the same neuron is replicated in the time direction) to detect the presence or absence of the same feature at different places along the trajectory. By using several different neurons at each time step, the neural network performs the detection of different features (Fig. 7.3, the outputs of different neurons produce a new feature vector in the next layer, at a given time step).

The operations performed by one layer of the network are convolutional in essence (Fig. 7.4). Each neuron k in layer $\ell+1$ has an associated convolution kernel of height m (the number of features in layer $\ell+1$) and of width δ. The coefficient of the kernel are the neuron weights $w_{i,j}^{k,\ell}, i = 0, \ldots (\delta-1)$ and $j = 0, \ldots (m-1)$. The convolution of the states $f_j^{\ell}(t)$ of layer ℓ with kernel k is another sequence of states $f_k^{\ell+1}(t)$ of layer $\ell+1$ corresponding to confidence levels for the presence or absence of a given feature k along the

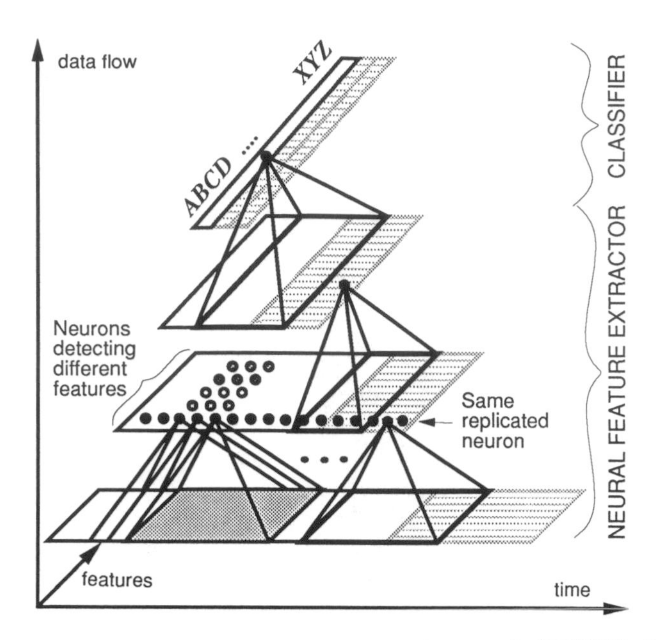

Fig. 7.3. Architecture of the time-delay neural network (TDNN). The connections between layers obey the following rules (not all neurons are represented): (1) neurons are feature detectors with restricted input fields, limited in the time direction; (2) in each layer, a set of neurons scans the output of the previous layer along the time axis and produces higher level feature vectors; and (3) the sequence of feature vectors is subsampled in time at each layer to obtain time contraction. For isolated character recognition, the time contraction is such that characters, which fit into a fixed-size input window, correspond to one output vector, for which the time dimension has been completely eliminated. The figure also shows that the convolution can be extended in time and so that, for a variable-length input, a sequence of output vectors is produced. This is used in Sec. 7.3.2 for the recognition of entire words.

pen-trajectory. State sequences are parametrized by the discrete-time variable t. The confidence levels are squashed by the neuron sigmoid function $g[\cdot]$:

$$f_k^{\ell+1}(t) = g\left[\sum_{i=0}^{\delta-1}\sum_{j=0}^{m-1} w_{i,j}^{k,\ell} \; f_j^{\ell}(t-i)\right] \tag{7.1}$$

with

$$g[x] = \alpha \tanh \beta x, \tag{7.2}$$

where tanh denotes the hyperbolic tangent, $\alpha = 1/\tanh(2/3)$, and $\beta = 2/3$.

The time component of our input representation is gradually eliminated by subsampling the convolution at each layer by a factor of two or three. To partially compensate for the loss of information, the number of features is gradually increased [15]. This is what we call a *bi-pyramidal network*

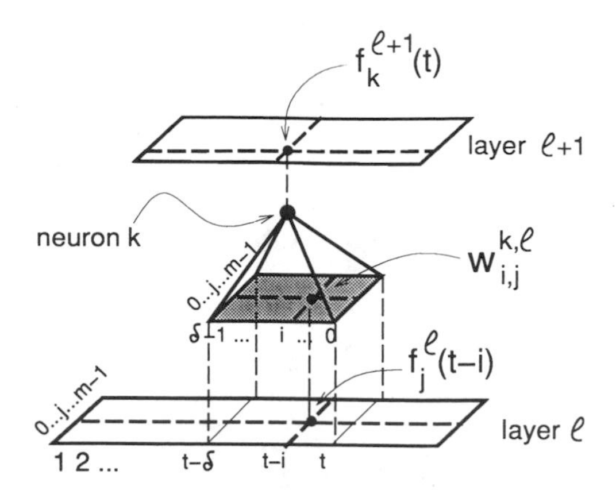

Fig. 7.4. Operations of one layer of the TDNN. Each layer ℓ produces an intermediate representation of the input pattern as a sequence of feature vectors. We denote by $f_j^{\ell}(t)$ the jth component of that vector at time t. The intermediate representation of layer $\ell + 1$ is obtained by sweeping several "neuron feature detectors" over the intermediate representation of layer ℓ. Neuron k produces the sequence $f_k^{\ell+1}(t)$.

architecture. Bi-pyramidal networks progressively convert time information into feature information.

The weights are adjusted during a supervised training session using error backpropagation [16], which performs gradient descent in weight space with a mean-squared-error (MSE) loss function (see Sec. 7.2.6).

Because of the convolutional structure of our network, the same neuron is replicated along the time axis. For the unfolded architecture (Fig. 7.3), one usually talks about "weight sharing" among the various replicas of the same neuron [17]. Weight sharing is enforced during training by averaging the weight updates of the various replicas.

7.2.3 Classifier

The classifier is often part of the neural network itself. On top of the feature extraction layers, the last fully connected layer performs the final classification (Fig. 7.3).

The last layer of the TDNN consists of as many neurons as there are classes, $\{char_1, char_2, \ldots char_C\}$. During training, when a character of class $char_x$ is presented, we impose desired values of -1 to all neurons, except to the neuron associated to the $char_x$ class, which receives a $+1$ desired value. Thus, we bring back our C-class problem to C 2-class problems: Each neuron is trained to separate one class versus all the other ones. After training, the classification is made according to the largest output value. The outputs of the neural network may be used to estimate the *posterior* probabilities $P(char_x|input)$.[6] In that case, the classification is made according to the maximum a posteriori probability.

We make the distinction between neural feature extractor and classifier for two reasons:

- The neural feature extractor can be used independently to provide a compact representation of the data. In applications such as writer adaptation (Sec. 7.3.1) and signature verification (Sec. 7.3.3), patterns are stored for later use in this representation.
- Convolutional networks are good at extracting features, but there may be more suitable classifiers for the task at hand that can be used instead of the last layer of the network [9]. For instance, a K-nearest neighbor classifier or an optimal margin classifier [18] replaces the last layer for writer adaptation in Sec. 7.3.1.

7.2.4 Segmentation Preprocessor and Postprocessor

Our neural network isolated character recognizer is part of a larger system that can perform word recognition. The segmentation techniques that we use to separate a word into characters are closely related to dynamic time warping and hidden Markov models, which are used for speech recognition [19]. Their resemblance with neural networks has been pointed out by several authors [20, 21, 22, 23], but they will not be emphasized here.

We distinguish several levels of difficulty in the segmentation problem (see Fig. 7.5). If characters are entered in boxes, or if they are clearly spaced, the segmentation is trivial and can be handled independently of the recognition process. But, in the absence of boxes, people usually do not space their characters uniformly, which results in many segmentation ambiguities. To address this harder task, our strategy is to let the *segmentation preprocessor* make guesses and provide the recognizer with several likely segmentation hypotheses. The final segmentation and recognition decisions then are taken all at once by the *segmentation postprocessor*, using the recognition scores.

In our nomenclature, the segmentation preprocessor isolates segments of pen-trajectory called *tentative characters*. A *stroke* is an elementary segment between a pen-down and the next pen-up. The segmentation preprocessor may take advantage of such pen-lifts and spaces between strokes and/or minima of the speed of the pen to determine *tentative cuts*. Alterna-

[6]Given our choice for the sigmoid function [Eq. (7.2)] and desired values, the outputs of the network are between -1.7 and $+1.7$. They are rescaled between 0 and 1 to obtain probability estimates.

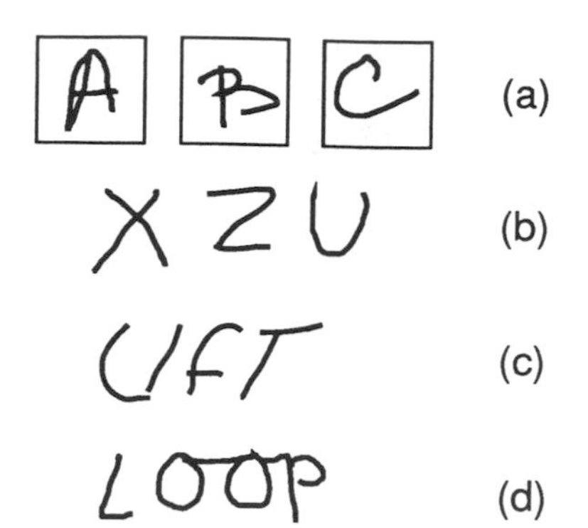

Fig. 7.5. Writing styles handled by the Penacée segmentors. (a) Boxed characters. (b) Clearly spaced characters. (c) Ambiguous spacing but consistent pen-lifts between characters. (d) Connected characters.

tively, the segmentation preprocessor simply can regularly sample windows of the input signal.

Tentative characters usually overlap considerably. The segmentation postprocessor uses an *interpretation graph.* The nodes of this graph contain the tentative character recognition scores, and the transitions between nodes favor particular character chainings. This avoids reusing the same part of a character multiple times and allows the implementation of frequencies of character successions particular to a given language.

We use the recognition scores provided by the neural network to provide estimates of the *posterior* probabilities of the various character interpretations $char_x$ given the input: $P(char_x|input)$. The transition coefficients are estimates of $P(char_y|char_x)$ determined independently of the recognition process. Using (abusively) the assumption of independence between the recognition of the various characters, the segmentation postprocessor computes probability estimates of character sequence interpretations as

$$\begin{aligned} P(char_x, char_y, \ldots |input) &= P(char_x|input) \cdot P(char_y|char_x) \\ &\quad \cdot P(char_y|input) \cdots . \end{aligned} \tag{7.3}$$

The output of the segmentation postprocessor is the best path in the graph, corresponding to the most probable word (or character sequence). Our implementations will be described in Sec. 7.3.2.

7.2.5 Similarity Measurer

So far, we have assumed that the neural network would be used to estimate the posterior probability $P(char_x|input)$. But one may be interested in having the network estimate $P(input_1 \sim input_2)$, that is, the estimate of the similarity between two patterns. Such a network has found applications to classification problems, using a K-nearest neighbor classifier or a kernel classifier [24], and to verification problems such as signature verification [4] (Sec. 7.3.3) and fingerprint verification [25].

Similarity measures are very problem-dependent and usually must be designed by experts. As an alternative, we tailor the similarity measure to the task at hand by training a neural network classifier to separate pairs of similar patterns and pairs of dissimilar ones.

For instance, take two identical neural feature extractors net_1 and net_2. The first one encodes $input_1$ into $output_1$ and the second one $input_2$ into $output_2$. Add on top a *similarity measure module* that computes the dot product between $output_1$ and $output_2$ to obtain the degree of similarity between the two patterns. We named a system with such an architecture a *siamese* neural network [4] (see Fig. 7.2.5).

During training, similar patterns (e.g., two genuine signatures from the same person) are given a large positive desired degree of similarity. Conversely, dissimilar patterns (e.g., a signature and its forgery) are associated with a negative or 0 desired degree of similarity. The network is trained with the backpropagation algorithm by backpropagating errors through the *similarity measure module* (Sec. 7.2.7). The constraint that $net_1 = net_2$ is enforced during training by averaging the weight updates of corresponding weights in net_1 and net_2.

Other *similarity measure modules* can be used instead of the simple dot product, such as a Gaussian similarity measure $\exp(-d^2/\sigma^2)$, where d is the Euclidean distance between *output*1 and *output*2. One also could imagine using an elastic matching module to perform better time alignment. In Sec. 7.3.3, we use a normalized dot product: $\cos(output_1, output_2)$.

7.2.6 Loss Calculator

The loss calculator computes a penalty function or *loss function* which measures the distance of the system from its objective. Combined with an appropriate optimization technique such as gradient descent, the loss function guides the system during training towards our goal. A very commonly used loss function is the square loss:

$$L_{sq} = (output - desired)^2,$$

where *output* is the actual output of the system and *desired* is its corresponding desired value. For instance, for a classification problem with two classes A and B, all elements of class A are assigned a desired value of $+1$ and all elements of class B a desired value of -1. After training, if the output of the system is positive, the input pattern is classified as A, otherwise as B.

The goal of learning is to get best generalization performance. For a classification problem, this means that, after training, the classification error rate on examples not used for training (test set) must be low. It is not clear whether minimizing a loss function using only training examples can yield good generalization. However, both theoretical studies and experimental results [26] show that good generalization can be achieved if the "capacity" of the learning system is matched to the number of training examples. If the capacity is too large, the system can easily learn the training examples, but it usually exhibits poor generalization (overfitting). Conversely, if the capacity is too small, the system is not capable of learning the task at all (underfitting).

The capacity of a neural network is related to the number of free parameters [27]. This is a relatively small number for convolutional networks such as the TDNN, compared to fully connected networks and to networks with local connections but with no weight sharing [17]. More recent work talks rather of an "effective capacity" which incorporates properties of the input space, network architecture, and training algorithm [26]. A simple way of affecting the capacity during training is to modify the loss function by adding, for instance, a weight decay term that pulls the weights to 0.

In our present work, the capacity control is handled by "early stopping." Before training, the weights are initialized with small random values such that the total input of the neurons lies in the linear part of the sigmoid squashing function [Eq. (7.2)]. The initial effective capacity is smaller than what the number of tunable parameters suggests: It is equal to the number of free parameters of the equivalent linear system. During training, as the weights increase, use is made of the nonlinearity of the squashing function and the capacity progressively increases until it reaches an optimum which is detected by cross-validation. We do cross-validation by computing the performance of the system on a small set of patterns that is distinct from the training set and the test set.

Training algorithms usually minimize the average loss over all training examples, with respect to the weights of the network. Such is the case for the original backpropagation algorithm [16], which minimizes the mean-square-error (MSE) or average square loss. This technique is widely applied and well suited to a large variety of problems. An alternative strategy is to minimize the maximum loss over all training patterns ("minimax" training) [28]. This technique is guaranteed to capture the tail of the distribution of input patterns but is very sensitive to outliers [18]. Better generalization than with MSE minimization is obtained with minimax training after careful screening of eventual mislabeled or meaningless patterns that have been inserted in the training data [3] (see Sec. 7.3.1).

7.2.7 Global Optimization Techniques

Once the various modules described in the previous sections are assembled into a system, the problem arises of optimizing all of the parameters (i.e., the weights of the network) with respect to the final objective determined by the loss function. In [29], a general framework has been proposed to achieve the global optimization of a multimodular architecture. We present here a particular example which illustrates the main ideas.

We train our neural network with gradient descent, for which the weight updates are computed as

$$\Delta w = -\epsilon \frac{\partial L}{\partial w}, \tag{7.4}$$

where L is a loss function, w is a weight of the network, and ϵ is a small positive quantity (the gradient step or learning rate). The backpropagation algorithm permits chaining the computation of the partial derivatives from the output to the input of the network [16].

Consider the case of the "siamese" neural network described in Sec. 7.2.5. To compute the weight updates and use the chaining rule of back-propagation, one needs first to compute the partial derivative:

$$\frac{\partial L}{\partial output_i}, \tag{7.5}$$

where i is either 1 or 2. Assume that we use a cosine similarity measure and a squared loss:

$$L(input_1, input_2, \mathbf{w}) = [\cos(output_1, output_2) - desired]^2 \tag{7.6}$$

$$= \left[\left(\frac{output_1 \cdot output_2}{\| output_1 \| \| output_2 \|} \right) - desired \right]^2, \tag{7.7}$$

where $desired = 1$ if the patterns are similar and 0 otherwise. Then, using the condensed notation S for $\cos(output_1, output_2)$, X for $output_1$, Y for $output_2$, and D for $desired$, we obtain:

$$\frac{\partial L}{\partial X} = \frac{\partial L}{\partial S}\frac{\partial S}{\partial X} \tag{7.8}$$

$$= 2(S-D)\frac{1}{\| X \| \| Y \|}\left(Y - S\frac{\| Y \|}{\| X \|}X\right), \tag{7.9}$$

from which all other partial derivatives with respect to the weights of the network can be computed using backpropagation.

7.3 Applications

In this section we present solutions to problems in on-line handwriting recognition and signature verification that involve various combinations of the modules previously described.

We have designed the Penacée system so that ultimately it will handle unconstrained handwriting of any style. We report first the results we have obtained for the writer-independent recognition of digits, uppercase and lowercase letters, written in the boxed mode. Then we explain the

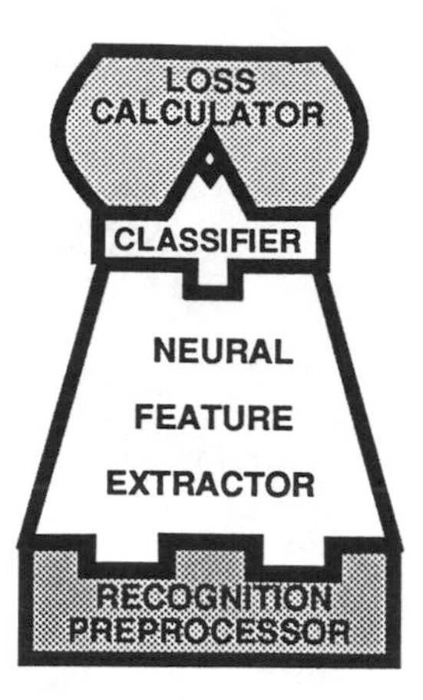

Fig. 7.6. The architecture of the isolated character recognizer.

techniques that we have used to handle the recognition of unsegmented hand-printed uppercase words (run-on mode). We report the results of two methods of segmentation: INSEG (for INput SEGmentation) and OUTSEG (for OUTput SEGmentation). We show that the two methods have complementary advantages. We propose a combination of INSEG and OUTSEG and preliminary results indicating that our final system may be able to handle unconstrained handwriting. Finally, we present an application of "siamese" neural networks to signature verification.

7.3.1 Isolated Character Recognition

The simplest task that we have addressed is that of isolated character recognition, which involves the combination of modules shown in Fig. 7.6. The specifications of the neural feature extractor (a TDNN up to its second to last layer) are summarized in Table 7.1. We report on several experiments with the same neural feature extractor but different classification layers. We always use one output neuron per character interpretation or class. The desired output values for a given interpretation $char_i$ are -1 for all of the neurons except the ith, which has a desired output of 1. We train the network with backpropagation.

The experiments are performed with data collected in the cafeteria of an AT&T facility. Approximately 250 writers contributed to at least one complete set of the 10 digits, the 26 uppercase letters, and the 26 lowercase letters. We address the writer-independent task, meaning that different sets of writers are used for training and testing.

In [1], we report results obtained on digits and uppercase letters. A set of approximately 12,000 examples is used for training and 2500 examples for testing. A TDNN with 36 outputs in its classification layer is trained to

Table 7.1. TDNN Feature Extractor for Character Recognition

Input length	Subsampling steps	δ/m 1st layer	δ/m 2nd layer	δ/m 3rd layer	δ/m 4th layer
90	3.3.2	8 / 7	6 / 10	4 / 16	5 / 24

The subsamplings steps are time subsampling steps from one layer to the next. The time scale thus is contracted by a factor of 18. The kernels have dimensions δ by m (Fig. 7.3). The neural feature extractor has 34,660 connections in its unfolded version (Fig. 7.3) but only 3106 independent weights, due to "weight sharing" (Sec. 7.2.2).

recognize either digits or uppercase letters. This means that, during training, no error is backpropagated from the uppercase letter neurons if a digit is presented, and vice versa. At utilization time the a priori information about whether the character to be recognized is a digit or an uppercase letter is needed. For instance, the neuron with maximum activation among the digit output neurons is selected if the character to be recognized is known to be a digit. We obtain an error rate on the test set of 3.4% error (2.3% if tested on digits only and 3.8% if tested on uppercase letters only).

In [2], we propose a training algorithm (the "emphasizing scheme") that enables atypical patterns, such as characters written by left-handed people, to be learned. The method is a simple way of improving the information-theoretic learning efficiency. It consists, during backpropagation training, in presenting more often the least predictable patterns (i.e., the ones with the largest squared losses). The method is related to minimax learning procedures (Sec. 7.2.6) such as optimum margin classification [18] and boosting techniques [30]. Using this method, the error rate is reduced from 3.4 to 2.8%. Even more importantly, the variation in error rate for different writers is substantially decreased.

In [3], we present results on lowercase letters. In this study, we propose a "super-supervised" learning technique, the purpose of which is to detect undesirable outliers (mislabeled or meaningless patterns introduced by accident into the database). During the supervised learning session, which utilizes the class labels stored in our database, a human "super-supervisor" double checks the labels of the patterns that are hardest to learn and are therefore suspicious. We show that the "emphasizing scheme" works best when the data are cleaned by a "super-supervisor." A TDNN with the same feature extraction layers as specified in Table 7.1, but with only 26 neurons in the classification layer (for the lowercase letters), is trained with approximately 9500 lowercase letters and tested with 2000 letters from different writers. Our final error rate of 6.9% is considerably better than the initial error rate of 11% error that is obtained with standard backpropagation and without cleaning the data.

In [2, 5], we address the problem of writer adaptation. No matter how good a writer-independent system becomes, there always will be too many variations in writing styles to ensure very high recognition accuracy for all writers. It is therefore important to allow a given writer to fine-tune the system to his own handwriting. We have developed techniques that allow the same neural feature extractor to be kept and retrain only the classification module. The simplest method consists in using a nearest-neighbor classifier instead of the last layer of the network. A selection of prototypes is stored in the representation of the neural feature extractor. Patterns are classified according to the class of their nearest prototype. When the system fails to recognize a pattern, an additional prototype is generated. New classes can be introduced, allowing for customizations of the recognizer by adding new sets of symbols and gestures. Performing nearest-neighbor classification in the representation of the neural feature extractor, as opposed to the network input representation, is advantageous for two reasons: few patterns are required for adaptation (usually only one or two), thanks to the robustness of the representation; and little storage is required, thanks to the compactness of the representation. Using an optimal margin classifier instead of the nearest-neighbor classifier, the same accuracy is retained while recognition speed is improved and fewer prototypes are needed. This enables us to reach very high recognition accuracies for most writers (less than 1% error) with fast adaptation and at the expense of no recognition speed degradation.

7.3.2 Hand-Printed Word Recognition

In [6, 7], we address the harder task of recognizing hand-printed words that are not a priori segmented into letters. This task involves the combination of modules shown in Fig. 7.7. We report results obtained with two methods of recognition-based segmentation. The methods are designed to work in the "run-on mode," where there is no constraint on the spacing between characters. While both methods use a neural network recognition engine and a graph-algorithmic postprocessor, their approaches to segmentation are quite different.

The experiments are carried out on data from a large variety of writers which were collected in the cafeteria of an AT&T facility. We collected approximately 9000 one- to five-letter words, which we separated into 8000 for training and 1000 for testing. Short words (one to three letters) are random combinations of all letters, and longer words are legal English words. Another set of 600 English words of any length (from an 80,000-word dictionary) also is used for testing. Our best result is 11% word error rate on that last set, using lexical checking.

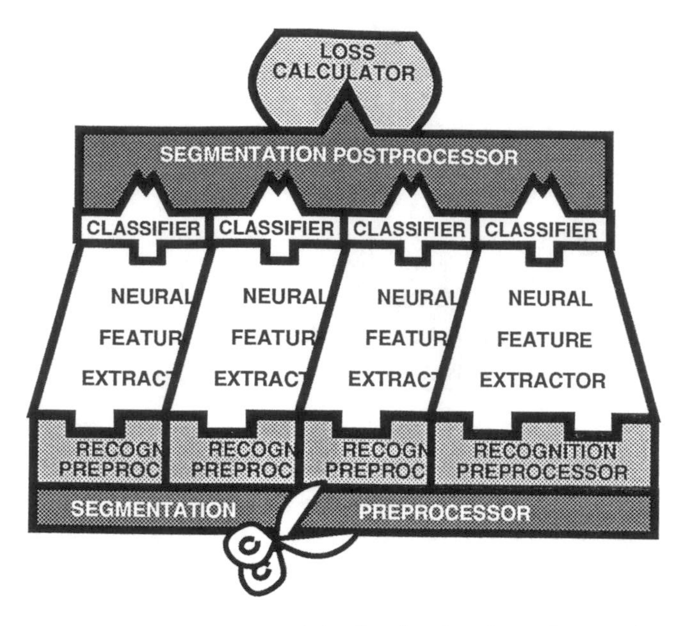

Fig. 7.7. The architecture of the hand-printed word recognizers.

INSEG

We call our first method INput SEGmentation, or INSEG, because of the key role played by the segmentation preprocessor. The INSEG technique is inspired by earlier work in optical character recognition [14]. A set of heuristics is used to determine likely segmentation points, or "tentative cuts," For the recognition of on-line hand-printed characters, pen-lifts are very natural tentative cuts. Our segmentation preprocessor uses both pen-lifts and spaces to cut the data stream and define a set of "tentative characters," usually a small superset of the valid characters (Fig. 7.8).

An interpretation graph is built with the recognition scores of all tentative characters (Fig. 7.8). The transition probabilities are simply set to the inverse of the preceding node fan-out and do not include information about the relative position of the tentative characters [31] nor the frequencies of letter successions in English [32]. The segmentation postprocessor searches the K-best paths in the graph. In the case of the figure, the best path is "UFT." But if the word is written by an English-speaking person, we would rather select the second-best path "LIFT," which is a valid English word. In practice, we use up to 20 best paths and try to match them with the closest English words using the "ispell" program [33]. The words thus obtained are reordered using the interpretation graph. The valid English word with the highest score is selected. Using as a recognition engine the TDNN described in Sec. 7.3.1, we obtain a word error rate on the short-word data of 29% without lexicon and 23% with lexicon.

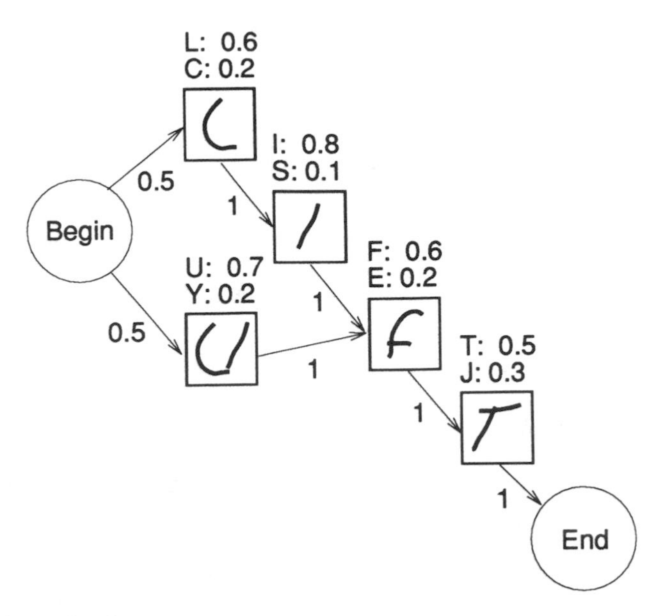

Fig. 7.8. Example of an interpretation graph produced by the INSEG system. The first two best scores are shown on top of each box. The transition weights indicated below the arrows are just the inverse of the preceding node fan-out. The best path "UFT" has a score of 0.10. The second-best path "LIFT" has a score of only 0.07, but it will be preferred when lexical checking is used.

We have found that better recognition rates are obtained by optimizing the overall system: recognizer and segmentor. This is done by retraining the TDNN with both examples of valid characters and counterexamples of characters. These counterexamples correspond to mistakes of the segmentation preprocessor, such as incorrect splitting of characters or incorrect groupings of strokes. If the correct interpretation is not the best path in the interpretation graph, the segmentation postprocessor pursues its search until the nth path gives the correct interpretation. All of the incorrect paths up to the $(n-1)$th are used as counterexamples to retrain the network. With such retraining, much improvement is obtained. The word error rate without lexicon drops to 18% and with lexicon to 15%, on the short-word test set.

Our best performances are obtained using retraining and by combining two different neural networks with a voting scheme: a TDNN and a two-dimensional convolutional network [34]. On the short-word test set, we reach, without lexicon, a word error rate of 13% and, with lexicon, 10%. On the long-word test set, we reach an error rate of 23% without lexicon and 11% with lexicon. It should be emphasized that the test sets are not cleaned from mislabeled and meaningless patterns that are introduced during data collection. The error rates we obtain are more than three times smaller than the error rate obtained by a commercial recognizer tested on the same data.

OUTSEG

The INSEG system is well suited for the recognition of characters separated by pen-lifts. For connected handwriting, we prefer an alternative segmentation technique which does not presume any segmentation points prior to recognition. We call this other technique OUTput SEGmentation, or OUTSEG.

The segmentation preprocessor presents to the recognizer a window of the input signal. The window is shifted in time to show the next tentative character. Therefore, the task of the segmentation preprocessor is trivial.

Scores for the different tentative characters are obtained as an ordered sequence of score vectors (Fig. 7.3). Both the location and the interpretation of the characters are determined by the segmentation postprocessor which therefore handles the segmentation per se. Hence the name OUTSEG.

How does the segmentation postprocessor proceed? If no character is present in the input window, all of the network outputs are below a certain activation threshold. Conversely, if a character is present in the input window, there is a high activation value for the neuron corresponding to the correct interpretation. We avoid, however, the very delicate problem of tuning thresholds by filtering the sequence of output scores. We use a digital filter that implements a model of character duration and of duration of the spacing between characters. In Fig. 7.9, we show the sequence of network outputs and the interpretation path selected by our postprocessor. Several high scores are filtered in this path by the duration modeling.

We first use a TDNN trained on isolated characters only, similar to the one described in the previous section. We obtain with that network a very large error rate: more than 80% word error, with or without lexicon, on the short-word database.

As in the case of the INSEG method, we have found that it is important to retrain the TDNN with entire words. With retraining, the network learns both to recognize characters and to detect transitions between characters by giving a low score to all outputs when the window is not centered on a character. We use for that purpose the position-invariant training technique proposed in [35, 36]. We also have tried to optimize the overall system, recognizer plus segmentor, by backpropagating errors through the duration model, with similar success. After retraining, the word error rate on the short-word data reduces to 21% without lexicon and 17% with lexicon. On the long-word data we obtain 49% without lexicon and 21% with lexicon.

The error rates of OUTSEG are about twice as small as that of the commercial recognizer tested on the same data. They are, however, not quite as good as that of INSEG on this task. This is understandable since uppercase printed letters are relatively easy to segment with a good heuristic segmen-

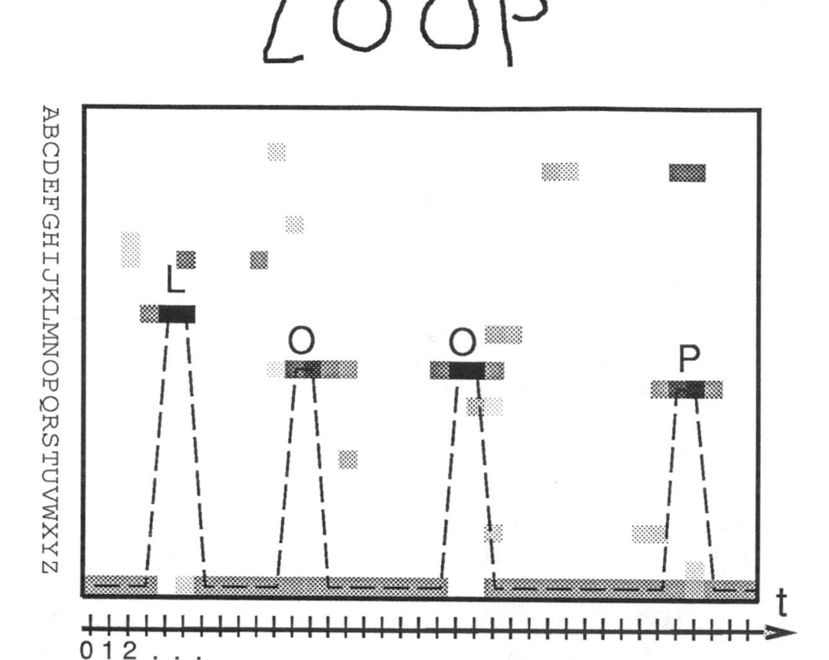

Fig. 7.9. Example of an output from the OUTSEG system. Each row in the matrix represents the score of one output unit for a different input window position in time. Each column represents the scores for all units for one particular position of the time window. The darker the boxes, the more confident the unit is in its interpretation. The last row in the matrix is 1 minus the sum of all the other scores in a given column. It represents the probability that no character is detected. The dashed line indicates the best path found by the postprocessor, using a duration model of the characters and the character transitions.

tation preprocessor. In more than 95% of the words in our database, letters are separated by spaces and/or pen-lifts. Although we do not have direct evidence, the experiments performed in [37] are an indication that, on cursive handwriting, for which good tentative cuts are difficult to determine, OUTSEG would work better than INSEG.

Combination of INSEG and OUTSEG

We would like our end system to be able to recognize printed characters, both cursive script and also mixed styles. Currently existing systems are good at recognizing either printed letters or cursive. We propose a way of combining INSEG and OUTSEG into a single system that has the good features of both systems and permits mixed styles to be handled. We call this new method INOUTSEG.

To extend INSEG to cursive recognition, one would need to introduce more tentative cuts than pen-lifts, perhaps minima of the speed of the pen. But, multiplying the number of tentative cuts is unnecessary for the recognition of hand-printed characters, and can only result in performance degradation. It is also unclear whether tentative cuts other than pen-lifts could be found. On the other hand, OUTSEG can spot character locations without making use of heuristic tentative cuts, but it makes more mistakes than INSEG when characters are nicely separated by pen-lifts.

Our idea is to rely on the INSEG preprocessor to determine tentative cuts in input space that are reported in the OUTSEG output space (see Fig. 7.10). We then detect inconsistencies between the INSEG and OUTSEG segmentations. When OUTSEG finds multiple characters between two consecutive tentative cuts, additional "virtual cuts" are introduced to separate them. The list of tentative cuts and virtual cuts thus obtained is subsequently used to build an interpretation graph, such as the one described in Fig. 7.8. This INOUTSEG graph is filled with the recognition scores of INSEG and OUTSEG combined with a voting scheme.

In the example of Fig. 7.10, the INSEG system uses a neural network trained to recognized uppercase and lowercase letters, and the OUTSEG system uses a network trained to recognize cursive words. The figure shows that OUTSEG cannot recognize the uppercase letter "M" and makes an insertion error at the end of the word. On the other hand, INSEG cannot separate the connected letters "ed." The combined system, however, correctly recognizes the word. Our system still is under development, and experimental results will be reported elsewhere.

7.3.3 Signature Verification

The problem of signature verification is quite different from that of character recognition. For handwriting recognition, the correct interpretation must be discovered, not the identity of the writer. For signature verification, the correct interpretation (writer name) is given, but what needs to be discovered is the identity of the writer (good guy or forger?).

Most approaches to this problem rely on human expertise to devise discriminatory similarity measures. In our approach, we train a "siamese" neural network (see Secs. 7.2.5 and 7.2.7 and Fig. 7.11) to learn a similarity measure from examples.

In [4], we address the signature verification task with the problem of reducing credit card fraud in mind. The credit card holder would give several samples of his own signatures at the bank, and then an encrypted pattern of his signature would be stored on his credit card. At the retail site, his signature would be matched against the pattern stored on his card.

The experiments are carried out on data collected among the staff at AT&T and NCR, and on data collected at Wright State University. A total of approximately 3300 signatures was obtained, including genuine

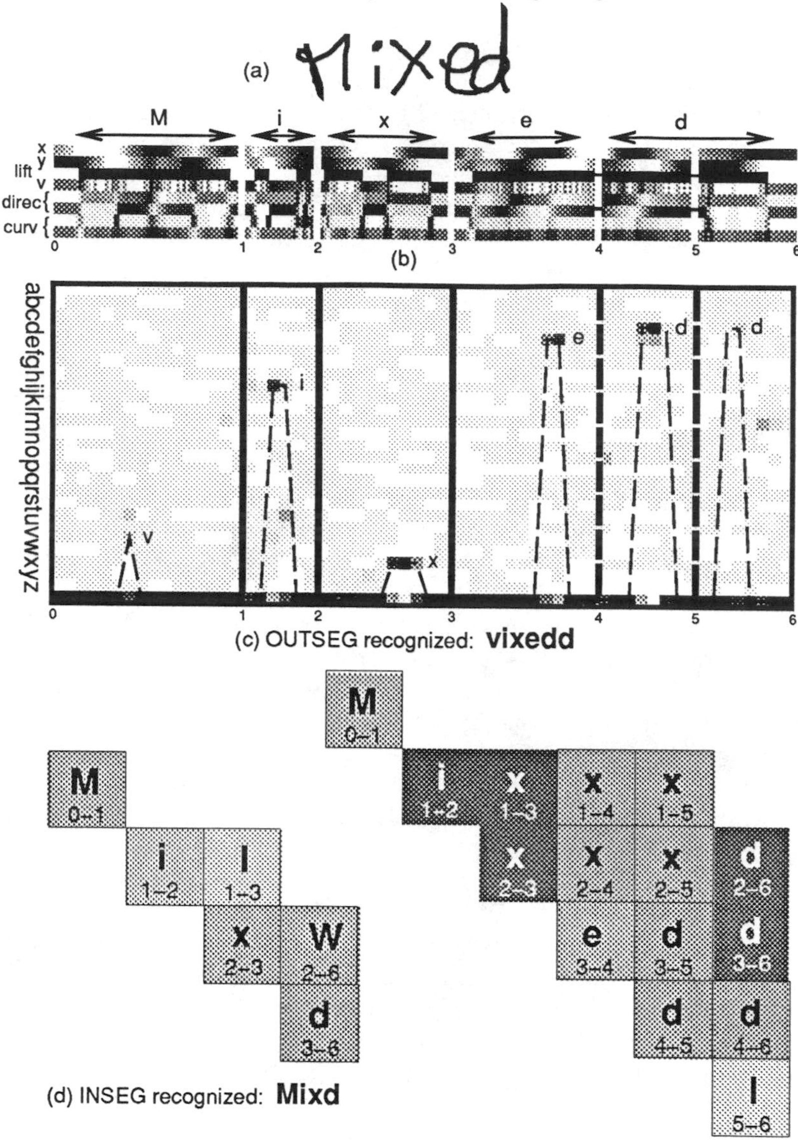

Fig. 7.10. Example of combination of input and output segmentation. (a) The original word. (b) The data as presented to the OUTSEG network. (c) The sequence of output scores obtained by the OUTSEG network. (d) The INSEG interpretation graph. (e) The INOUTSEG interpretation graph, including INSEG tentative cuts, OUTSEG virtual cuts, and the scores of both INSEG and OUTSEG combined with a voting scheme. In (b) and (c), the tentative cuts selected by INSEG (coinciding with some pen-lifts) are shown as bold vertical lines; the virtual cuts provided by OUTSEG are shown as vertical dashed lines. In (d) and (e), arrows are not represented; each box is associated with the tentative character delineated by two cuts whose numbers are at the bottom of the box; in each box, we show only the interpretation associated to the best score whose intensity is indicated with the grey shading (the darker, the larger).

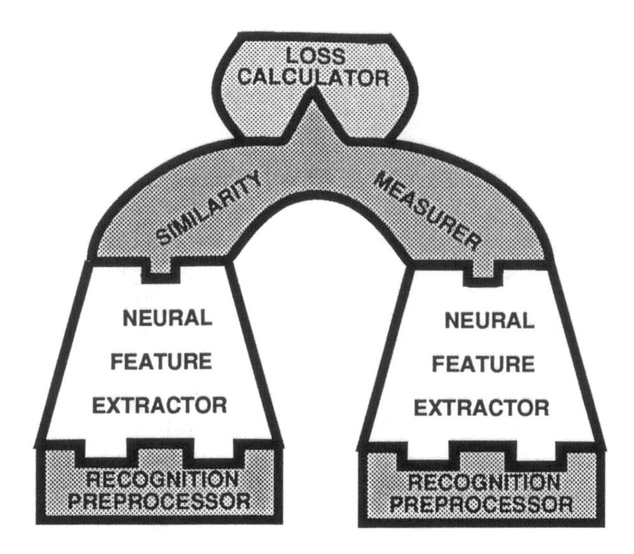

Fig. 7.11. Architecture of the siamese neural network.

signatures and forgeries. The amount of effort put into producing forgeries varied but reflected rather well the spectrum of real-life data since many forgers attempting to use somebody else's credit card produce effortless forgeries and may even use their own signature.

Training of the "siamese" network is performed by presenting pairs of signatures. We train the neural feature extractor to produce a representation of the signatures such that there is a small angle between two genuine signatures from the same person, and a large angle between two distinct signatures and between a signature and its forgery. The angle is measured by the "similarity measurer" (Fig. 7.11) by calculating the cosine between the output vectors of the two neural feature extractors (see Sec. 7.2.5). If both signatures are produced by the same person (genuine signatures), the desired cosine is 1, otherwise it is negative or 0. The exact value of the desired value does not influence the performance significantly.

During utilization, the cosine similarity measure is replaced by a Gaussian similarity measure (Sec. 7.2.5). Credit card holders usually do not contribute to the data that are used to train the neural feature extractor. They provide only a few examples of their signatures, which are encoded as the representations of the outputs of the neural feature extractor. We build a Gaussian model of these patterns, whose mean becomes the reference prototype for the user's signature and whose covariance matrix is used to determine the width and orientation of the Gaussian window.

The siamese network fits the problem quite nicely. After training, a model signature can be encoded in the representation of the neural feature extractor. It is therefore both compressed significantly and encrypted, which is suitable for storage on a credit card magnetic strip. During verification, the signature to be checked is processed at the retail place by an identical neural feature extractor and matched to its model. The model can be updated constantly with examples of successfully verified signatures.

In [4], several network architectures are tried. The networks are trained with 7700 signature pairs, 50% being pairs of genuine signatures from the same person, 40% genuine forgery pairs, and 10% pairs of different signatures. The networks then are tested on a separate set of signatures from different people comprising about 500 genuine signatures and 400 forgeries. For each person, a model signature in the neural feature extractor representation is built from six signatures. The model is tested with six other genuine signatures and forgeries.

By varying a rejection threshold on the output of the system, one can monitor the trade-off between accepting too many forgeries and rejecting too many genuine signatures. Our best network permits detecting 80% of the forgeries while rejecting only 5% of the genuine signatures.

7.4 Conclusion

We presented a neural network approach to solve several problems in on-line handwriting recognition. Our design choices make consistent use of the sequential nature of the data, in both the preprocessing and the neural network architecture (a TDNN). The initial success of our writer- independent isolated character recognizer motivated us to address tasks of greater difficulty. We reported here on applications to signature verification and the recognition of hand-printed words with no spacing constraints between characters. On this last task, our system outperforms a widely distributed commercial recognizer tested on the same data. While we still are improving our first system, we already are developing the next generation, which will try to recognize mixed styles including hand-printed letters and cursive.

Acknowledgments. The work described in this overview was conducted in Larry Jackel's department at Bell Labs and is the result of the joint effort of many researchers, including Paul Albrecht, Yoshua Bengio, Jim Bentz, Bernhard Boser, Léon Bottou, John Denker, Donnie Henderson, Wayne Hubbard, Yann Le Cun, Annick Leroy, Cliff Moore, Craig Nohl, Howard Page, Ed Pednault, Doug Riecken, Eduard Säckinger, Roopack Shah, Vladimir Vapnik, and Anne Weissbuch. Collaborations with our colleagues at ETH-Zürich, and our colleagues at Bell Labs doing research on Optical Character Recognition and Multimedia Communications are gratefully acknowledged. The simulations were performed on the neural network simulator *SN* of Neuristique Inc., written by Léon Bottou and Yann Le Cun, and on the network simulator *v–lisp* written by Bernhard Boser.

References

[1] I. Guyon, P. Albrecht, Y. Le Cun, J. Denker, W. Hubbard (1991) Design of a neural network character recognizer for a touch terminal. *Pattern Recognition* **24**(2)

[2] I. Guyon, D. Henderson, P. Albrecht, Y. Le Cun, J. Denker (1992) Writer independent and writer adaptive neural network for on-line character recognition. In: *From Pixels to Features III*, S. Impedovo (Ed.) (Elsevier, Amsterdam)

[3] N. Matić, I. Guyon, L. Bottou, J. Denker, V. Vapnik (1992) Computer-aided cleaning of large databases for character recognition. *Proc. ICPR*, Amsterdam, (IEEE, New York)

[4] J. Bromley, J. Bentz, L. Bottou, I. Guyon, L. Jackel, Y. Le Cun, C. Moore, E. Sackinger, R. Shah (1993) Signature verification with a siamese time delay neural network. In: *Applications of Neural Networks to Pattern Recognition*, I. Guyon, P.S.P. Wang (Eds.) (World Scientific, Singapore)

[5] N. Matić, I. Guyon, J. Denker, V. Vapnik (1993) Writer adaptation for on-line handwritten character recognition. In: *ICDAR'93*, Tokyo

[6] M. Schenkel, H. Weissman, I. Guyon, C. Nohl, D. Henderson (1992) Recognition-based segmentation of on-line hand-printed words. In: *Advances in Neural Information Processing Systems 5 (NIPS 92)*, (Morgan Kaufmann, San Mateo, CA)

[7] H. Weissman, M. Schenkel, I. Guyon, C. Nohl, D. Henderson (1993) Recognition-based segmentation of on-line run-on hand-printed words: Input vs. output segmentation. *Pattern Recognition*, submitted

[8] C.C. Tappert, C.Y. Suen, T. Wakahara (1990) The state of the art in on-line handwriting recognition. *IEEE Trans. PAMI* **12**(8):787–808

[9] P. Gallinari (1990) A neural net classifier combining unsupervised and supervised learning. In: *Proc. International Neural Network Conference*, Vol. 1, Paris, July 1990 (IEEE, New York), pp. 375–378

[10] K.J. Lang, G.E. Hinton (1988) A time delay neural network architecture for speech recognition. Technical Report CMU-cs-88-152, Carnegie-Mellon University, Pittsburgh PA

[11] A. Waibel, T. Hanazawa, G. Hinton, K. Shikano, K. Lang (1989) Phoneme recognition using time-delay neural networks. *IEEE Trans. Acoust. Speech Signal Proc.* **37**:328–339

[12] Y. Le Cun, B. Boser, J.S. Denker, D. Henderson, R.E. Howard, W. Hubbard, L.D. Jackel (1990) Back-propagation applied to hand-written zipcode recognition. *Neural Comput.* **1**(4)

[13] O. Matan, C.J.C. Burges, Y. Le Cun, J. Denker (1992) Multi-digit recognition using a Space Dispacement Neural Network. In: *Advances in Neural Information Processing Systems 4*, Denver, J.E. Moody et al. (Eds.) (Morgan Kaufmann, San Mateo, CA)

[14] C.J.C. Burges, O. Matan, Y. Le Cun, D. Denker, L.D. Jackel, C.E. Stenard, C.R. Nohl, J.I. Ben (1992) Shortest path segmentation: A method for training neural networks to recognize character strings. In: *IJCNN'92*, Vol. 3, Baltimore (IEEE, New York)

[15] L.-Y. Bottou (1988) Master's thesis, EHEI, Universite de Paris 5, Paris, France

[16] D.E. Rumelhart, G.E. Hinton, R.J. Williams (1986) Learning internal representations by error propagation. In: *Parallel distributed processing: Explorations in the microstructure of cognition*, Vol. I, (Bradford Books, Cambridge, MA), pp. 318–362

[17] Y. Le Cun (1989) Generalization and network design strategies. In: *Connectionism in Perspective*, R. Pfeifer, Z. Schreter, F. Fogelman, L. Steels (Eds.) (Elsevier, Zurich, Switzerland)

[18] B. Boser, I. Guyon, V. Vapnik (1992) A training algorithm for optimal margin classifiers. In: *Fifth Annual Workshop on Computational Learning Theory* (ACM, Pittsburgh, PA), pp. 144–152

[19] L.R. Rabiner (1989) A tutorial on Hidden Markov Models and selected applications in speech recognition. In: *Proc. IEEE*, Vol. 77-2 (IEEE, New York)

[20] R.P. Lippmann (1989) Review of neural networks for speech recognition. *Neural Comput.* **1**(1):1–38

[21] J.S. Bridle (1990) Alpha-nets: A recurrent "neural" network architecture with a hidden markov model interpretation. *Speech Commun.* **9**(1):83–92

[22] H. Bourlard, C. Wellekens (1990) Links between markov models and multilayer perceptron. *IEEE Trans. Pattern Anal. Machine Intell.* **12-12**:1167–1178

[23] E. Levin, R. Pieraccini, E. Bocchieri (1993) Time-warping network: A neural approach to hidden markov model based speech recognition. In: *Applications of Neural Networks to Pattern Recognition*, I. Guyon, P.S.P. Wang (Eds.) (World Scientific, Singapore)

[24] E. Sackinger, J. Bromley (1991) Neural-network and k-nearest neighbour classifiers. Technical Report TM 51323-919819-01, AT&T Bell Labs.

[25] P. Baldi, Y. Chauvin (1992) Neural networks for fingerprint recognition. *Neural Comput.*, to appear

[26] I. Guyon, V. Vapnik, B. Boser, L. Bottou, S.A. Solla (1992) Structural risk minimization for character recognition. In: *Advances in Neural Information Processing Systems 4*, Denver, J.E. Moody et al. (Eds.) (Morgan Kaufmann, San Mateo, CA)

[27] E.B. Baum, D. Haussler (1989) What size net gives valid generalization? *Neural Comput.* **1**(1):151–160

[28] V. Vapnik (1982) *Estimation of Dependences Based on Empirical Data* (Springer-Verlag, New York)

[29] L. Bottou (1991) *Une approche théorique de l'apprentissage connexioniste; Application à la reconnaissance de la parole.* PhD thesis, University of Paris XI, Paris, France

[30] H. Drucker, R. Schapire, P. Simard (1993) Boosting performance in neural networks. In: *Applications of Neural Networks to Pattern Recognition*, I. Guyon, P.S.P. Wang (Eds.) (World Scientific, Singapore)

[31] E. Pednault (1992) A hidden markov model for resolving segmentation and interpretation ambiguities in unconstrained handwriting recognition. Technical Report TM 11352-920929-01, AT&T Bell Labs.

[32] T. Fujisaki, H.S.M. Beigi, C.C. Tappert, M. Ulkeson, C.G. Wolf (1992) On-line recognition of unconstrained handwriting: a stroke-based system and its evaluation. *Preprint*

[33] R.E. Gorin et al. (1991) *UNIX*^{TM} man-page for ispell, version 3.0.06 (beta). 09/17/91.

[34] A. Weissbuch, Y. Le Cun (1992) Private communication

[35] J. Keeler, D.E. Rumelhart, W-K. Leow (1991) Integrated segmentation and recognition of hand-printed numerals. In: *Advances in Neural Information Processing Systems 3*, R. Lippmann et al. (Eds.) (Morgan Kaufmann, San Mateo, CA), pp. 557–563

[36] J. Keeler, D.E. Rumelhart (1992) A self-organizing integrated segmentation and recognition neural net. In: *Advances in Neural Information Processing Systems 4*, J.E. Moody et al. (Eds.) (Morgan Kaufmann, San Mateo, CA), pp. 496–503

[37] D. Rumelhart et al. (1992) Integrated segmentation and recognition of cursive handwriting. In: *Third NEC Symposium Computational Learning and Cognition* (Princeton, New Jersey) (to appear)

8

Topology Representing Network in Robotics

Kakali Sarkar and Klaus Schulten[1]

with 6 figures

Synopsis. We consider the visually guided control of the grasping movements of a highly hysteretic five- joint pneumatic robot arm. For this purpose we apply a modified version of the so-called *topology representing network* algorithm, a vector quantization algorithm that also learns to represent neighborhood relationships. The notion of neighborhood relationships allowed us to average the behavior of neurons which represent similar tasks, both during the training and in generating control signals in the mature state. Based on visual information provided by two cameras, the robot learns to position and orient its end effector properly for the object to be grasped. For simplicity, we consider the grasping of cylindrical objects only. The control is comprised of two stages. In the first stage, the end effector approaches the side of the cylinder facing the robot base; and in the second stage, the end effector grasps the cylinder. Training of the first stage involves a brief episode of supervised learning to prime the network. The control is achieved through a visual feedback loop: for both stages of the motion the system detects the error to target and applies a linear correction. This correction is achieved through a training that yields a vector-quantized representation of a zero-order signal of joint pressures and a first-order correction through Jacobian tensors which relate the error, expressed in terms of camera coordinates, to correct joint pressures. The network is trained satisfactorily after about 300 trial movements, with a residual average error of 1.35 camera pixels. Besides a demonstration of the technical feasibility of control through *topology representing networks*, this chapter provides a tutorial for technical applications of such networks. The algorithm behind a *topology representing network*, its training and employment for task control, is described in complete detail to provide the reader with a comprehensive view of this important class of neural networks in the context of a technical application.

---

[1]Department of Physics/Beckman Institute, University of Illinois, Urbana, IL 61801, USA.

8.1 Introduction

In the early days of research in neurocomputing, networks were seen as devices that were capable of computing logic functions [1]. Such a mechanistic view of neurocomputing became popular mainly because of the fact that computation traditionally was viewed in light of logic gates and switching algebra. However, we have gradually come to know the bottlenecks of the traditional deterministic computer; we observe that the human brain can easily outperform today's supercomputers in tasks where it processes multidimensional analogue data and probabilistic, noisy information. It is now generally believed that an understanding of boolean logic and switching algebra may not enhance our perspective about neuronal information processing in the brain. The quest for a theoretical framework to quantify the underlying computation process has brought computer scientists, physicists, and biologists together. Vigorous research efforts during the last two decades have helped to develop a different perspective about neurocomputing. This interdisciplinary effort has resulted in many promising real-world applications such as speech processing [2], optimization [3], complex control systems [4, 5], and more.

Grasping of objects is one of the most common tasks frequently performed by human beings. Even though this seems to be easy and often spontaneous to most of us, from the control system perspective grasping is complicated. The object to be grasped has to be identified in the environment by its location and by other features. Then the trajectory of the arm movement has to be planned in such a way that it does not collide with any obstacle. Recently, many efforts have been made [6–10] to understand the control mechanism of such complex maneuvers and to make use of these fundamental control techniques to develop viable artificial neural control systems. In this chapter we focus mainly on the control of the execution of grasping motions, assuming an extremely simplified solution for the recognition of the target and the arm's current posture: we provide a set of suitable light-emitting diodes (LEDs) on the arm and the target in an otherwise darkened space.

Nevertheless, the problem of executing motions to grasp a cylinder placed in all possible positions and orientations in a robot's workspace is a difficult one. The motion must involve at least five degrees of freedom and be sufficiently precise. The precision must be achieved for an arm that is subject to random and hysteretic behavior. In fact, in the present case, the controlled arm is driven pneumatically with effectors which are subject to strong hysteresis and oscillations as characterized in [11, 12]. The required control only can be achieved when the network, besides learning the control signals for a sufficiently fine set of arm postures, also learns tensors which allow the arm to linearly correct deviations from the target due to hysteresis and other effects.

The corresponding control problem, in principle, can be formulated in terms of a table look-up algorithm that provides for each target cylinder a table entry which produces the suitable air pressures to move the arm. As was already stated, the entries of the table need to be a set of pressures to move the five degrees of freedom of the arm (see Section 2) as well as a tensor, the Jacobian connecting the deviation from the target, expressed as a vector of five coordinates, to the vector of pressures driving the arm (see Sec. 3). Obviously, such a table look-up program cannot be arbitrarily fine. However, even a coarse grid of, say, 10 points along each coordinate for a five-dimensional space leads to a very large number (100,000) of table entries. Obviously, an optimal choice which, for a given number of entries, produces the smallest error is very desirable. An important ingredient of the criterion stated is the probability distribution of arm postures under normal working conditions. The neural networks used in our study obey such criterion in that they assign their table entries as a result of a training in which arm postures are requested with a frequency distribution which matches that occuring in normal working conditions. In fact, the algorithm allows life-long learning such that the table entries can be continuously adjusted to the work experience.

The problem to optimally assign a finite number of table entries to a continuous space, often of very high dimension, is called the *vector quantization problem*. The neural network algorithm adopted here provides a solution for vector quantization as discussed in [13]. However, there is another important attribute of the control problem that also must be captured by the look-up algorithm in order to be efficient, namely the topology of the control space. This implies that the table entries develop threads between each other which connect entries assigned to arm postures which are very close to each other. These threads serve two purposes, one during training and one after training. The threads can be employed when the table entries are generated, i..e., when the networks are trained. Entries connected through threads contain similar information, and, hence, they can share the improvements to their entries during the training period. The result is a dramatic decrease of the training period since any training episode is shared by many table entries. A particularly important aspect of the sharing of information among table entries is that this feature makes the system much less sensitive to the initial, usually random, entries in the table. In many instances, when table entries are trained separately, convergence to a suitable control program depends on the initial table entries, i.e., the radius of convergence of the training algorithm is not infinite. However, the sharing of table entry updates increases the radius of convergence enormously, as was demonstrated in [14].

The threads between entries are also very beneficial after training, when the system is used to control the arm. The threads allow one to average the control signals (pressures) to the arm over table entries connected through a thread. Such an average improves performance at the early stages of training and can also increase the accuracy of the control: if N units are

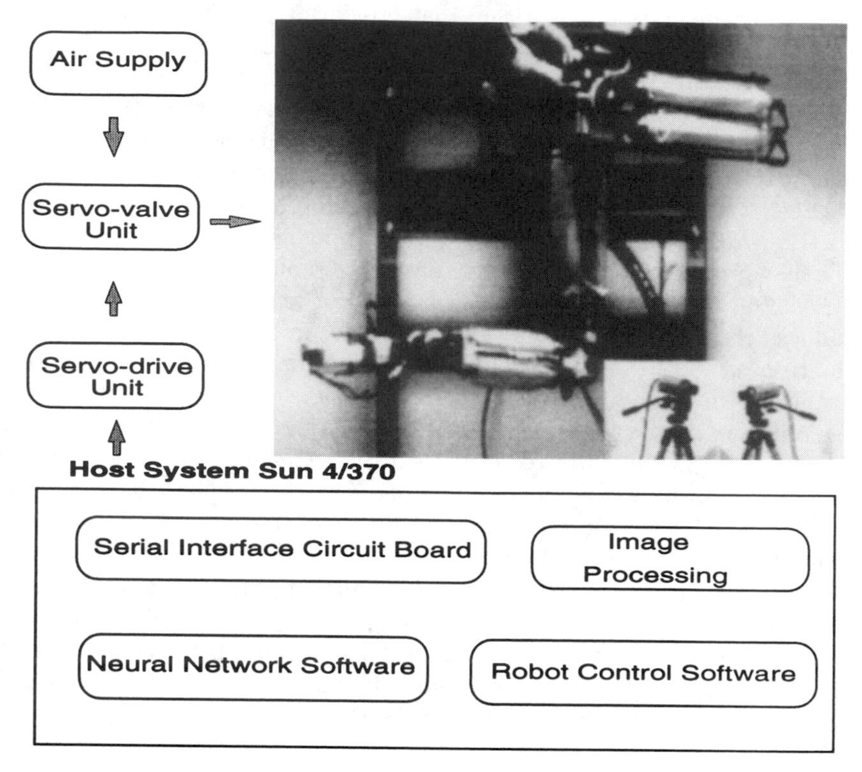

Fig. 8.1. Block diagram of the SoftArm robot system [11].

pooled, each with an error ϵ, the error after averaging (assuming, for the sake of simplicity, that the table entries are coded for exactly the same posture) is $\epsilon/\sqrt{N}$.

The threads between the table entries reflect the topology, i.e., neighborhood relationships, of the control space. In the present case, the topology of the control space is obviously that of $\mathbb{R}^5$ since all arm postures required to grasp a cylinder form a manifold embedded in the five-dimensional Euclidean space. In fact, in the algorithm presented below, the threads between the table entries are never actually established. Rather, we use the Euclidean metric to establish a closeness ranking among table entries and use this ranking instead of threads. However, in many cases, a dimension or metric is not obvious and needs to be established while a system is confronted with training tasks. In an early neural network scheme for control based on Kohonen networks [15, 16], such a dimension needed to be specified beforehand. Theses schemes preserved the given dimension (topology) in that they assigned table entries to the task space while keeping the threads, e.g., those representing a two-dimensional grid, intact. Examples addressing the control of robots in computer simulations are found in

[17, 10, 18, 19, 14]. A comprehensive presentation of these networks in a variety of applications, ranging from brain maps to robot control, can be found in [20]. This textbook also discusses at length the statistical mechanical analysis of the convergence properties of the network and fluctuations of the network's table entries. A particularly interesting application of these networks to visual brain maps can be found in [21].

When we attempted to apply neural network algorithms to control real, i.e., not simulated, robot arms, we established that networks with an a priori topology, like generalized Kohonen networks, are not optimal. Instead, we appended the vector quantization scheme described in [13] with Hebbian rules which provided the required threads between table entries. The resulting topology representing the network had been introduced in [22] and discussed at length in [23]. The network has been applied succesfully to control an industrial robot with precise response to control signals [24, 25] and also to a pneumatically driven robot [11], the same as the one employed in the present study.

In this chapter we present an extension of our previous work [11] on the control of a pneumatic robot arm by incorporating a control mechanism for the grasping of cylinders of arbitrary orientation. In the following section we first characterize the control problem describing the arm geometry and the ideosyncracies of the pneumatic actuators of the robot arm used. In Sec. 3 we present the topology representing network algorithm employed for control. The section provides all of the algorithmic steps involved in complete detail, but it does not explain the algorithm exhaustively as is done in [23]. However, the detailed presentation of the algorithm in the present contribution might be considered by many readers a better explanation of topology representing networks than any general exposition. In Sec. 4 we demonstrate how the algorithm, after training, performs grasping motions.

8.2 Problem Description

The robot–camera system is shown schematically in Fig. 8.1. This system has been described in detail in [11]. The robot contains a pneumatic arm with five joints. At each joint, two or four rubber tubes are connected by chains across sprockets. The rubber tubes are supplied with compressed air from an air compressor. When differential air pressures are supplied to the tubes, differing equilibrium lengths result, which induce a rotation of the joint to a new equilibrium point.

There are five servo drive units for five joints, each of which takes signals from the host computer and sends current output to the servo valve unit. The servo valve unit then converts this electrical signal to pressure information, i.e., it controls the pressures inside the rubber tubes by opening or closing the electrical valves. Two cameras observe the location of the end

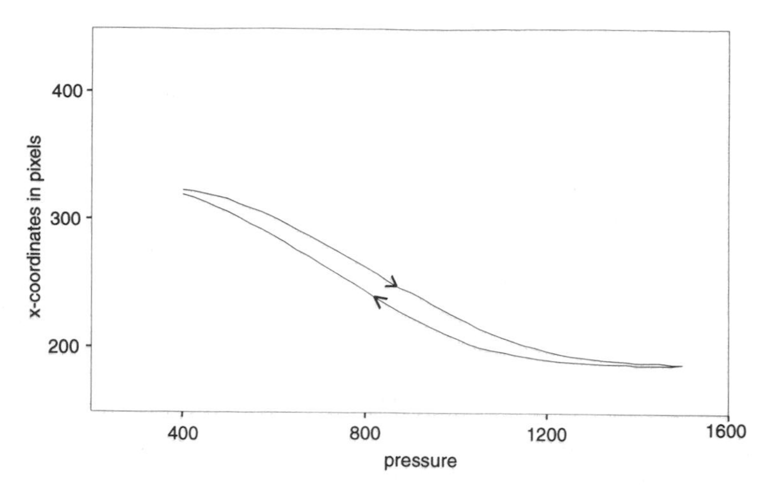

Fig. 8.2. Pressure versus position plot for joint 1. Hysteretic behavior of joint 1, of the softarm. The pressure difference in the agonistic and antagonistic tubes of joint 1 was first increased and then decreased.

effector or the cylinder to be grasped and send back the information to the host computer, which then finds the image coordinates in pixels with the help of two parallel image processors.

The servo drive units can be used to control the robot arm in two modes, a pressure-control mode and a position-control mode [11]. The present work has been carried out in the pressure-control mode. The relation between the joint pressures and position is highly nonlinear and also exhibits hysteresis. When the pressure is increasing, the pressure–position relation follows a particular path, but it follows a different path while the pressure is decreasing again. Figure 8.2 shows such type of behavior for joint 1.

The end effector of the robot arm is a two-fingered one and is presented schematically in Fig. 8.3. The movement of the end effector is controlled by the fourth and fifth joints. Each joint produces a motion which is a combination of rotational motions about the axes XX' and YY'. Pure rotation about XX' and YY' also can be produced, but each of them is a function of both the fourth and fifth joint pressures.

In the present work, we consider the grasping of cylindrical objects only. In order to grasp such an object, several issues need to be addressed. First, the point of grasping should be very close to the center of mass of the cylinder. If the center of mass is far from the chosen grasping position, the generation of undesirable torques makes it difficult to hold the cylinder. The angle between the axis of symmetry(ZZ') of the cylinder and that of the end effector(XX') is another important factor. The end effector should be placed perpendicular to the symmetry axis of the cylinder. In other words, axis ZZ' should be perpendicular to the plane containing axes XX'

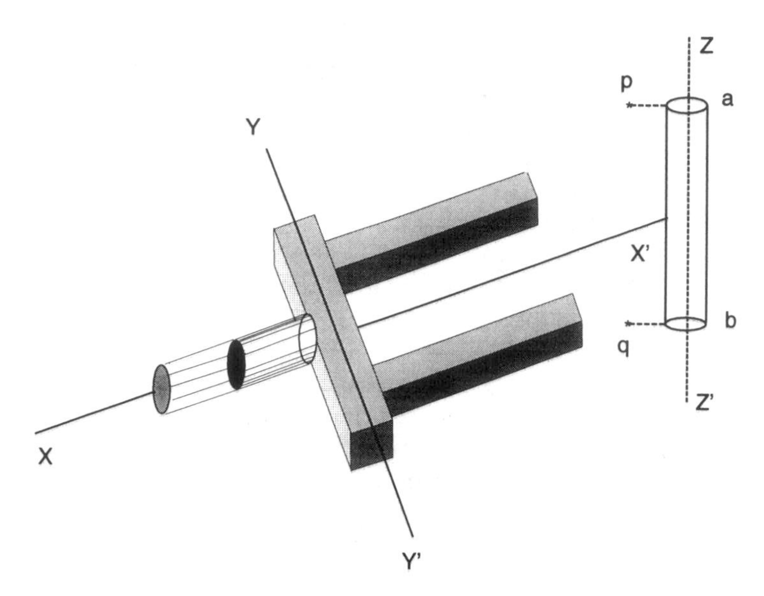

Fig. 8.3. A sketch of the end effector (gripper) and the cylinder to be grasped.

and YY'. These two aspects have played the role of prime significance in all of our grasping algorithms.

8.3 Topology Representing Network Algorithm

The visually controlled motions for grasping cylinders placed in the arm's workspace are carried out in two stages: In the first stage, the arm's gripper is placed in front of the cylinder at a proper orientation as shown in Fig. (8.3); in the second stage, the arm moves toward the center of the cylinder and actually grasps it by closing the gripper's fingers. The training procedures of each stage will be described separately below. Control of the first stage is by far the more difficult problem.

8.3.1 Training of First-Stage Motion

The goal of the first stage of the grasping motion is to generate a set of pressures in the arm's tubes which place and orient the gripper in front of the cylinder in a configuration suitable to carry out the second stage of the grasping motion and actually grasp the cylinder. We refer to the suitable configuration reached at the end of the first grasping stage as the *target configuration*. This configuration is realized through application of a set of vectors to the tubes of the arm which are collected in a pressure vector $\mathbf{P}$.

The target position for the initial placement of the gripper is determined as follows: As is shown in Fig. 8.3, we fix two lights at the positions p and q such that the line joining p and q is coplanar as well as parallel to the cylindrical axis ab. The images of these lights give the representation of the endpoints of another imaginary cylinder of the same size as the original, which, however, is placed at a small distance in front of the original one. The lights appear in the two cameras at points characterized by the coordinates $(u_1, u_2, u_3, u_4)^T$ and $(u_5, u_6, u_7, u_8)^T$. As a result, the position of the target is characterized through an eight-dimensional vector $\mathbf{u}_{target} = (u_1, u_2, u_3, u_4, u_5, u_6, u_7, u_8)^T$. The set of all vectors $\mathbf{u}_{target}$ in the robot's workspace form the so-called feature space $V \subset \Re^8$. We seek a training procedure which, for the first stage of the grasping motion, develops a map $\mathbf{u}_{target} \in V \rightarrow \mathbf{P}(u_{target}) \in \mathcal{F}$ which assigns to $\mathbf{u}_{target}$ the proper pressure vector for $\mathbf{P}$, positioning and orienting the gripper in front of the cylinder.

The robot arm is moved through ten tubes which pairwise act in an agonistic–antagonistic manner to rotate the arm's joints. The sum of the two pressures in each agonist–antagonist tube pair determines the stiffness of the motion. In the present study, the total pressure for each joint was kept constant during the operation of the system. As a result, the arm was moved through five independent pressures, one for each joint. The corresponding pressure vector $\mathbf{P}$ is then five-dimensional and the space $\mathcal{F}$ of joint pressures is then embedded in $\Re^5$.

The goal of the training of the N neurons controlling the first stage of the grasping motion is to develop first a set of Voronoi cells covering the feature space V with centers $\mathbf{w}_k \in V$, $k = 1, 2, \ldots N$, and then to develop a map $V \rightarrow \mathcal{F}$. The latter map is established through local affine maps in each of the Voronoi cells, i.e., in the Voronoi cell assigned to neuron k, through

$$\mathbf{P}(\mathbf{u}_{target}) = \mathbf{P}_k + A_k \cdot (\mathbf{u} - \mathbf{w}_k), \tag{8.1}$$

where $\mathbf{P}_k$ and A_k are constants (a vector and a tensor) which are acquired through the training.

As was stated earlier, the neurons actually achieve their control through averaging their output $\mathbf{P}(\mathbf{u}_{target})$. The average involves the neurons that have Voronoi cells adjacent to each other in the feature space V. To determine the corresponding average, one first needs to determine a ranking among the neurons which describes which neuron's Voronoi cell contains the target vector $\mathbf{u}_{target}$, which Voronoi cell is second closest, third closest, etc. Such ranking is achieved as follows: One determines for each neuron k, $k = 1, 2, \ldots N$, the distance

$$\mathbf{D}_k(\mathbf{u}_{target}) = \| \mathbf{u}_{target} - \mathbf{w}_k \| \tag{8.2}$$

and then determines a ranking $k_0, k_1, \ldots k_{N-1}$ such that

$$D_{k_m}(\mathbf{u}_{target}) \leq D_{k_n}(\mathbf{u}_{target}) \qquad \text{for } m < n.$$

One then defines

$$k(r, \mathbf{u}_{target}) = k_r \tag{8.3}$$

$$r(\ell, \mathbf{u}_{target}) = m\,, \quad \text{where} \quad k_m = \ell. \tag{8.4}$$

This ranking can be employed to achieve the desired averaging. We choose for this purpose the functional form

$$\overline{\mathbf{P}}(\mathbf{u}_{target}) = \sum_{k=1}^{N} \alpha(r(k, \mathbf{u}_{target})) \times \left[\mathbf{P}_{k(r,\mathbf{u}_{target})} + A_{k(r,\mathbf{u}_{target})} \cdot (\mathbf{u}_{target} - \mathbf{w}_{k(r,\mathbf{u}_{target})})\right] \tag{8.5}$$

with

$$\alpha(r) = e^{-r/10} \tag{8.6}$$

The softarm poses a challenging control problem due to drift in the relationship between pressures applied to the arm's joints and the resulting arm configuration. This drift manifests itself on various time scales; on a very short timescale it is characterized by the hysteretic behavior of the arm shown in Fig. 8.2. On longer time scales a drift arises due to temperature sensitivity and dependence on time of usage of the mechanical characteristics of the arm's tubes. Finally, over the lifetime of the softarm the characteristics of the tubes are subject to wear. The long time changes can be overcome by retraining the arm. In fact, the algorithms for training and control of the arm are essentially identical, such that retraining can be realized during actual usage of the softarm.

The hysteretic properties of the softarm require that one linearly corrects the arm posture to reduce the error $d = ||\mathbf{x} - \mathbf{x}_{target}||$, where $\mathbf{x}$ characterizes the current arm posture and $\mathbf{x}_{target}$ is the desired posture. As was specified above, and for the second-stage gripper movement further below, the posture is characterized by certain vectors of camera coordinates such that d is measured in units of camera pixels. The corrections of arm postures seek to reduce the error d below a tolerance

$$\text{tol}(t) = 0.1 + 100 \cdot e^{-t/120}\,\text{pixels}\,. \tag{8.7}$$

Here t counts the number of training steps. The tolerance is chosen large at the beginning of the training and reduces towards a small final value.

Obviously, one cannot enforce an overall accuracy of less than a camera pixel. In fact, the remaining final average error measures a little less than a pixel for each network, and a little over one pixel for the two networks controlling stage-one and stage-two movements combined (see Sec. 4). To reduce the error d below the tolerance [Eq. (8.7)] usually requires several linear correction steps. Accordingly, the control system linearly corrects the arm posture repeatedly until the tolerance is met. In the course of the training, when the tolerance is already at a small value, e.g., after 200 training steps, the system typically requires eight correction moves, whereas it requires only about two to three such moves after training is completed.

The final result of a training procedure is optimal quantities $\mathbf{w}_k$ and $\mathbf{P}_k$, A_k for all N neurons k. At the beginning of the training, these quantities need to be assigned initial values. In many cases [10, 14], the initial values of quantities to be acquired are chosen randomly. However, such choices lead to long learning periods that are particularly unfavorable in cases where "real-world" systems are trained. In the present case, the robot arm requires about 30 s for a single training step, a period that can lead to long overall training times. Furthermore, the radius of convergence of a training procedure [14] might not be infinite, such that some initial assignments, will not lead to convergence. Averaging as in Eq. (8.5) increases the radius of convergence [14], but the radius need not necessarily become infinite. A finite radius of convergence would require that the initial values of $\mathbf{w}_k$ and $\mathbf{P}_k$, A_k be chosen closer to the correct values. For this reason and, in particular, to speed up the overall training period, we acquired initial values in a supervised learning scheme. The learning was continued, after a brief phase, in an unsupervised form. For the sake of a more systematic exposition of the training schemes chosen, it is more suitable to present first the unsupervised learning scheme adopted here and then the supervised scheme, even though the schemes were applied in the opposite order.

Unsupervised Learning Scheme

The unsupervised learning scheme consists of several hundred training steps, each of which results in an update of the quantities $\mathbf{w}_k$ and $\mathbf{P}_k$, A_k. The values of these quantities before the learning step are defined as $\mathbf{w}_k^{old}$ and $\mathbf{P}_k^{old}$, A_k^{old} and after the learning step as $\mathbf{w}_k^{new}$ and $\mathbf{P}_k^{new}$, A_k^{new}.

We now outline how any particular step proceeds. The learning steps are numbered $t = 1, 2, \ldots$, and each learning step consists of ten substeps.

1. A cylinder is placed in a new, usually randomly chosen position in the workspace of the arm. To ascertain that the cylinder is actually placed in the workspace, one often adopts a "split brain" procedure [24], having the robot itself position the cylinder, but then "forgetting" the control signals (joint pressures in the present case). The cameras detect the cylinder and provide the vector $(v_1, \ldots, v_8)^T$ characterizing the cylinder position. For the following we define

$$\mathbf{v}_{target} = (v_1, v_2, v_3, v_4, v_5, v_6, v_7, v_8)^T . \quad (8.8)$$

Actually, the position $\mathbf{v}_{target}$ used for the stage-one motion does not coincide with the cylinder position, but rather is a position between the robot base and the cylinder, close to the cylinder as defined above.

2. The closeness ranking $k(r, \mathbf{v}_{target})$ of the neurons and its inverse $r(k, \mathbf{v}_{target})$ is determined, as described in Eqs. (8.3) and (8.4) above: $k(0)$ is the index of the neuron with its $\mathbf{w}_k^{old}$ closest to $\mathbf{v}_{target}$, $k(1)$ is the index of the neuron with its $\mathbf{w}_k^{old}$ second closest to $\mathbf{v}_{target}$, etc. Conversely, $r(119)$ is the rank of the neuron with index 119, i.e., $r(119) = 5$ implies that the particular neuron 119 has its $\mathbf{w}_{119}^{old}$ sixth closest to $\mathbf{v}_{target}$.

3. The vectors (weights) $\mathbf{w}_k^{old}$ are updated according to

$$\mathbf{w}_k^{new} = \mathbf{w}_k^{old} + \gamma_w \left(r(k, \mathbf{v}_{target}), t\right) \cdot \left(\mathbf{v}_{target} - \mathbf{w}_k^{old}\right) . \tag{8.9}$$

γ_w is a function that decays exponentially with the number t of the learning step as well as with the closeness rank $r(k, \mathbf{v}_{target})$

$$\gamma_w(r, t) = \epsilon \cdot e^{-r/\sigma} e^{-t/\lambda} \tag{8.10}$$

with $\epsilon = 0.7$, $\sigma = 5$, and $\lambda = 100$.

4. The pressure that is supposed to move the robot arm toward the target $\mathbf{v}_{target}$ then is determined according to the averaging procedure [Eq. (8.5)]

$$\begin{aligned} \overline{\mathbf{P}}(\mathbf{v}_{target}) &= \sum_{k=1}^{N} \alpha\left[(k, \mathbf{v}_{target}]\right] \\ &\times \left[\mathbf{P}_{k(r,\mathbf{v}_{target})} + A_{k(r,\mathbf{v}_{target})} \cdot \left(\mathbf{v} - \mathbf{w}_{k(r,\mathbf{v}_{target})}\right)\right] . \end{aligned} \tag{8.11}$$

5. The pressure [Eq. (8.11)] is applied to the robot arm's tubes and the robot moves its gripper. The resulting gripper configuration is detected by the cameras and the vector of camera coordinates $\mathbf{v}_i \in V$ is supplied. This motion was termed in our previous studies [11] the *coarse movement* of the arm.

6. The values $\mathbf{P}_k^{old}$ then are updated according to

$$\mathbf{P}_k^{new} = \mathbf{P}_k^{old} + \gamma_p \left(r(k), t\right) \cdot \left[\overline{\mathbf{P}}(\mathbf{v}_{target}) - \mathbf{P}_k^{old} - A_k(\mathbf{v}_i - \mathbf{w}_k)\right] , \tag{8.12}$$

where $\overline{\mathbf{P}}(\mathbf{v}_{target})$ is the pressure determined in substep 4 and

$$\gamma_p(r, t) = \epsilon' \cdot e^{-r/\sigma} e^{-t/\lambda} \tag{8.13}$$

with $\epsilon' = 0.8$.

7. The system now determines an improved vector of pressures which attempt to correct the remaining differences between $\mathbf{v}_{target}$ and $\mathbf{v}_i$:

$$\overline{\mathbf{P}}_{fine} = \overline{\mathbf{P}}(\mathbf{v}_{target}) + \sum_{r=0}^{S} \alpha(r) \left[A_{k(r)} \cdot (\mathbf{u} - \mathbf{v}_i)\right] , \tag{8.14}$$

where $\overline{\mathbf{P}}(\mathbf{v}_{target})$ is again the pressure determined in substep 4 and $\alpha(r)$ is given in Eq. (8.6).

8. The pressure $\overline{\mathbf{P}}_{fine}$ is applied to the arm's tubes and the robot arm assumes a new gripper position. This position is detected by the cameras and corresponding camera coordinates $\mathbf{v}_f$ are supplied. This motion had been termed *fine movement* in our previous studies [11].

9. The system employs the remaining error between $\mathbf{v}_f$ and $\mathbf{v}_{target}$ to update the tensors A_k according to

$$A_k^{new} = A_k^{old} + \gamma_j(r,t) \cdot (\Delta\mathbf{P} - A_{k(r)}^{old}\Delta\mathbf{v}).\Delta\mathbf{v}^T \|\Delta\mathbf{v}\|^{-2}, \qquad (8.15)$$

where

$$\gamma_j(r,t) = \epsilon'' \cdot e^{-r/\sigma} e^{-t/\lambda} \qquad (8.16)$$

with $\epsilon'' = 0.01$ and where we defined $\Delta\mathbf{P} = \overline{\mathbf{P}}_{fine} - \overline{\mathbf{P}}(\mathbf{v}_{target})$, $\overline{\mathbf{P}}(\mathbf{v}_{target})$ as again being the pressure vector of substep 4, and $\Delta\mathbf{v} = \mathbf{v}_f - \mathbf{v}_i$.

10. The system determines the error $d = \|\mathbf{v}_f - \mathbf{v}_{target}\|$ between the present gripper position and the target position. In the case where d exceeds the tolerance [Eq. (8.7)], another correction move is executed and, accordingly, the system carries out steps 7–9 again; otherwise, the system goes to the next step. In the case where steps 7–9 are executed once more, one first redefines $\mathbf{P}_k^{new} \rightarrow \mathbf{P}_k^{old}$ and $A_k^{new} \rightarrow A_k^{old}$.

11. The unsupervised learning scheme either terminates when a set number of steps has been executed or starts another round of substeps, beginning with substep 1 above.

Supervised Learning Scheme

The supervised learning scheme described now was employed to obtain better starting values for the quantities $\mathbf{w}_k$ and $\mathbf{P}_k$, A_k, which specify how the neurons k, $k = 1, 2, \ldots N$ control the initial stage of the grasping motion. The supervised learning scheme defines a sequence of target camera coordinates $\mathbf{v}_{target}$ by actually moving the gripper to the respective configuration and communicating the respective pressures to the learning scheme. The procedure, applied in the first $n_{sup} = 50$ steps of the learning scheme, is as follows:

1. A random pressure vector $\mathbf{P}_{target}$ is chosen.

2. $\mathbf{P}_{target}$ is applied to the tubes of the arm and the arm moves to a new position. The gripper configuration is detected by the cameras and the corresponding camera coordinates $\mathbf{v}_{target}$ are supplied.

3. The closeness ranking $k(r, \mathbf{v}_{target})$, $r(k, \mathbf{v}_{target})$ of the neurons is determined as in the unsupervised scheme.

4. The vectors (weights) $\mathbf{w}_k^{old}$ are updated, as in the unsupervised scheme, according to

$$\mathbf{w}_k^{new} = \mathbf{w}_k^{old} + \gamma_w\left(r(k, \mathbf{v}_{target}), t\right) \cdot (\mathbf{u} - \mathbf{w}_k^{old}), \tag{8.17}$$

where γ_w is as defined in Eqs. (8.3) and (8.4).

5. The pressure vectors $\mathbf{P}_k^{old}$ are updated according to

$$\begin{aligned} \mathbf{P}_k^{new} &= \mathbf{P}_k^{old} + \gamma_p\left(r(k, \mathbf{v}_{target}), t\right) \\ &\quad \times \left[\mathbf{P}_{target} - \mathbf{P}_k^{old} - A_k(\mathbf{v}_{target} - \mathbf{w}_k)\right], \end{aligned} \tag{8.18}$$

where $\gamma_p(r, t)$ is as defined in Eq. (8.9).

6. The system then determines a pressure vector

$$\begin{aligned} \overline{\mathbf{P}}(\mathbf{v}_{target}) &= \sum_{k=1}^{N} \alpha\left[r(k, \mathbf{v}_{target})\right] \cdot \left[\mathbf{P}_{k(r,\mathbf{v}_{target})}\right. \\ &\quad \left. + A_{k(r,\mathbf{v}_{target})} \cdot (\mathbf{v}_{target} - \mathbf{w}_{k(r)})\right] . \end{aligned} \tag{8.19}$$

7. This pressure is applied to the arm's tubes, and, as a result, the arm moves its gripper to a new position.

8. The cameras detect the new gripper position and supply the corresponding camera coordinates $\mathbf{v}_i$.

9. The system now determines an improved vector of pressures which attempt to correct the remaining differences between $\mathbf{v}_{target}$ and $\mathbf{v}_i$:

$$\overline{\mathbf{P}}_{fine} = \overline{\mathbf{P}}(\mathbf{v}_{target}) + \sum_{r=0}^{S} \alpha(r) \cdot \left(A_{k(r)} \cdot (\mathbf{v}_{target} - \mathbf{v}_i)\right) . \tag{8.20}$$

10. The pressure $\overline{\mathbf{P}}_{fine}$ is applied to the arm's tubes and the gripper moves to a new position.

11. The cameras detect the new gripper position and supply the corresponding camera coordinates $\mathbf{v}_f$.

12. The system then updates the tensors A_k^{old} according to

$$\begin{aligned} A_k^{new} &= A_k^{old} + \gamma_j(r(k, \mathbf{v}_{target}), t) \\ &\quad \times \left[(\mathbf{P}_{target} - \overline{\mathbf{P}}_{fine}(\mathbf{v}_{target}) - A_{k(r)}^{old}(\mathbf{v}_{target} - \mathbf{v}_f))\right] \\ &\quad \times (\mathbf{v}_{target} - \mathbf{v}_f)^T \|\mathbf{v}_{target} - \mathbf{v}_f\|^{-2} . \end{aligned} \tag{8.21}$$

Note that both expressions updating $\mathbf{P}_k^{old}$ and A_k^{old}, i.e., Eqs. (8.18) and (8.21), include $\mathbf{P}_{target}$, i.e., knowledge of the pressure which would have guided the arm, except for hysteretic effects, exactly to the target gripper position characterized by $\mathbf{v}_{target}$.

13. The system determines the error $d = ||\mathbf{v}_f - \mathbf{v}_{target}||$ between the present gripper position and the target position. In the case where d exceeds the tolerance [Eq. (8.7)], another correction move is executed, and, accordingly, the system carries out steps 9–12 again; otherwise, the system goes to the next step. In the case where steps 9–12 are executed once more, one redefines first $\mathbf{P}_k^{new} \rightarrow \mathbf{P}_k^{old}$ and $A_k^{new} \rightarrow A_k^{old}$.

14. In the case where n_{sup} training steps have been completed, the system terminates; otherwise, it begins another round of substeps beginning with substep 1 above.

8.3.2 Training of Final Grasping of the Cylinder — Second Stage of Movement

After the gripper has been placed and oriented properly in front of the cylinder (see Fig. 8.3) in the first stage of the movement, the gripper needs to be translated toward the cylinder until the fingers of the gripper enclose the cylinder sufficiently, i.e., until the center of the gripper coincides with the center of the cylinder. This translation is referred to as the *second stage* of the gripper movement. Since this movement does not require rotation of the gripper, only three degrees of freedom are active in the second stage of the movement. This considerably simplifies the control problem which requires, hence, a lower resolution of the neural network representation such that 200 neurons suffice.

The algorithm employed here for control and training of stage-two movement has been described in [11]; for the sake of completeness and consistency of notation, we review the algorithm below.

The aim of the algorithm is to guide the center of the gripper $\mathbf{g}$ to the center of the cylinder. The latter is characterized through two sets of camera coordinates, (c_1, c_2) and (c_3, c_4), corresponding to the image of the gripper center in the left and in the right camera, respectively. For the control of stage-two movement, the map

$$\mathbf{c} \rightarrow \mathbf{p} \tag{8.22}$$

is required, where $\mathbf{c} = (c_1, c_2, c_3, c_4)$ is a four-dimensional vector and $\mathbf{p}$ defines the set of pressures to translationally move the gripper. Since the last two joints of the five-jointed softarm are involved in gripper rotation, they are not required for the second-stage movement and only three pressures need to be specified. Accordingly, the map to be determined is $\Re^4 \rightarrow \Re^3$.

The embedding spaces $\Re^4$ and $\Re^3$ define a (Euclidean) metric $||\cdots||$ that will be employed.

The strategy of the present neural network approach, as outlined in [11], is to represent the relevant three-dimensional manifold Ω of gripper centers $\mathbf{c} \in \Re^4$ through vector quantization involving n neurons, where $n = 200$. The neurons labeled $\ell, \ell = 1, 2, \ldots n$ are to be assigned positions $\omega_\ell \in \Re^4$, which represent the manifold Ω of possible gripper centers. To each of the neurons we also assign a pressure vector $p_\ell \in \Re^3$ and 3×4-tensor a_ℓ. The latter are to be chosen to establish affine maps

$$\mathbf{p}(\mathbf{c}) = \mathbf{p}_\ell + a_\ell \cdot (\mathbf{c} - \omega_\ell), \tag{8.23}$$

which optimally approximate the exact map [Eq. (8.22)] in the Voronoi cell of neuron ℓ in the manifold Ω, i.e., in the space of all gripper centers $\mathbf{c}$ with $||\mathbf{c} - \omega_\ell|| \leq ||\mathbf{c} - \omega_m||$, $m = 1, 2, \ldots n$.

In order to determine the pressure that guides the gripper to the cylinder center $\mathbf{c}_{target}$ in stage two of the movement, one determines, in analogy to the case of stage-one movements, the closeness ranking $\ell(r, \mathbf{c}_{target})$ and, inversely, $r'(\ell, \mathbf{c}_{target})$. As in the case of a stage-one movement, the pressures supplied to the robot arm are actually averages of the pressures [Eq. (8.23)] contributed by neurons of neighboring Voronoi cells. The corresponding averages for the control of stage-two movements are given by

$$\begin{aligned} \overline{\mathbf{p}}(\mathbf{c}_{target}) &= \sum_{\ell=1}^{n} \alpha(r'(\ell, \mathbf{c}_{target})) \\ &\quad \times \left[\mathbf{p}_{\ell(r,\mathbf{c}_{target})} + a_{\ell(r,\mathbf{c}_{target})} \cdot (\mathbf{c}_{target} - \omega_{\ell(r,\mathbf{c}_{target})})\right], \end{aligned} \tag{8.24}$$

where $\alpha(r)$ is as defined in Eq. (8.6).

The final result of the training procedure is optimal quantities ω_ℓ and $\mathbf{p}_\ell$, a_ℓ for all n neurons ℓ. At the beginning of the training procedure these quantities are assigned random values. Stage-two movement control does not require supervised learning to improve the initial values and cuts down the training period; the reason for this is that the three-dimensional posture control of a robot arm with averaging of control signals converges rapidly with an infinite convergence radius, as is demonstrated in [14].

Learning Scheme

The unsupervised learning scheme consists of several hundred training steps, each of which results in an update of the quantities ω_ℓ and $\mathbf{p}_\ell$, a_ℓ. The quantities before the learning step are defined as ω_ℓ^{old} and $\mathbf{p}_\ell^{old}$, a_ℓ^{old}, and after the learning step ω_ℓ^{new} and $\mathbf{p}_\ell^{new}$, a_ℓ^{new}. We now outline how any particular step proceeds. The learning steps are numbered $t = 1, 2, \ldots$, and each learning step consists of nine substeps.

1. A target position $\mathbf{c}_{target}$ is chosen randomly to operate the robot in a "split brain" fashion: a random set of pressures (p_1, p_2, p_3) is applied

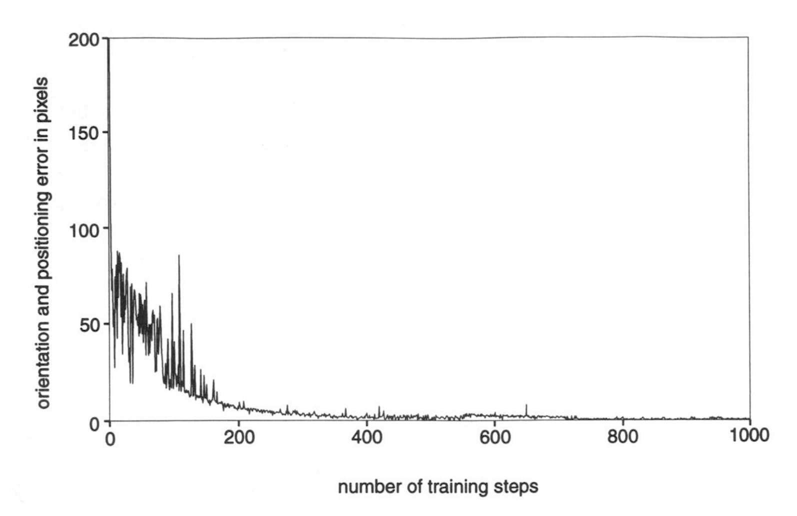

Fig. 8.4. Positioning and orientation error versus number of steps. This figure shows the learning curve for the network controlling the first stage of the gripper movement.

to the tubes of the first three joints of the softarm. The arm moves to a corresponding position. This position is detected through the cameras and communicated to the system in the form of the four-dimensional vector $\mathbf{c}_{target}$. This procedure ascertains that the chosen positions $\mathbf{c}_{target}$ actually belong to the workspace of the arm.

2. The closeness ranking $\ell(r, \mathbf{c}_{target})$ and its inverse $r(\ell, \mathbf{c}_{target})$ are established.

3. The values ω_ℓ^{old} are updated using the expression

$$\omega_\ell^{new} = \omega_\ell^{old} + \gamma_w(r(\ell, \mathbf{c}_{target}), t) \cdot (\mathbf{c}_{target} - \omega_\ell^{old}) . \tag{8.25}$$

Here $\gamma_w(r, t)$ is chosen as

$$\gamma_w(r, t) = e^{-r/\sigma_2} e^{-\sqrt{t}/9}, \tag{8.26}$$

where $\sigma_2 = 5$.

4. The pressure vector $\overline{\mathbf{p}}(\mathbf{c}_{target})$, which is supposed to move the gripper center toward $\mathbf{c}_{target}$, then is determined according to the averaging procedure in Eq. (8.24).

5. This pressure is applied to the tubes of the robot arm and the arm moves the gripper. The resulting position of the gripper center is detected by the cameras and the vector $\mathbf{c}_i$ of camera coordinates is supplied.

6. The values $\mathbf{p}_\ell^{old}$ are then updated according to

$$\begin{aligned}\mathbf{p}_\ell^{new} &= \mathbf{p}_\ell^{old} + \gamma_p'(r(\ell, \mathbf{c}_{target}), t) \\ &\times \left[\overline{\mathbf{p}}(\mathbf{c}_{target}) - \mathbf{p}_\ell^{old} - a_\ell^{old}(\mathbf{c}_i - \omega_\ell^{old})\right],\end{aligned} \tag{8.27}$$

where $\overline{\mathbf{p}}(\mathbf{c}_{target})$ is the pressure vector determined in substep 4 and where

$$\gamma_p'(r(\ell, \mathbf{c}_{target}), t) = \epsilon'' \cdot e^{-r/\sigma_2}\, e^{-\sqrt{t}/9} \tag{8.28}$$

with $\epsilon'' = 0.8$ and $\sigma_2 = 5$.

7. The system now determines an improved vector of pressures which attempt to correct the remaining differences between $\mathbf{c}_{target}$ and $\mathbf{c}_i$:

$$\overline{\mathbf{p}}_{fine} = \overline{\mathbf{p}}(\mathbf{c}_{target}) + \sum_{r=1}^{n} \alpha(r'(\ell, \mathbf{c}_{target})) \cdot a_{\ell(r)} \cdot (\mathbf{c}_{target} - \mathbf{c}_i), \tag{8.29}$$

where $\overline{\mathbf{p}}(\mathbf{c}_{target})$ is again the pressure vector determined in substep 4 and where $\alpha(r)$ is as defined in Eq. (8.6).

8. The pressure $\overline{\mathbf{p}}_{fine}$ is applied to the arm's tubes and the arm assumes a new gripper position. This position is detected by the cameras and the corresponding camera coordinates $\mathbf{c}_f$ are supplied.

9. The system employs the remaining error between $\mathbf{c}_{target}$ and $\mathbf{c}_f$ to update the tensors a_ℓ^{old}:

$$a_k^{new} = a_k^{old} + \epsilon''' e^{-r/\sigma} \cdot a_k^{old}(\mathbf{c}_{target} - \mathbf{c}_f)\Delta\mathbf{c}^T \|\Delta\mathbf{c}\|^{-2} \tag{8.30}$$

with $\epsilon''' = 0.01$, $\sigma = 5$, and $\Delta\mathbf{c} = \mathbf{c}_f - \mathbf{c}_i$.

10. The system determines the error $d = ||\mathbf{c}_f - \mathbf{c}_{target}||$ between the present gripper position and the target position. In the case where d exceeds the tolerance [Eq. (8.7)], another correction move is executed and, accordingly, the system carries out steps 7–9 again; otherwise, the system goes to the next step. In the case where steps 7–9 are repeated, one first redefines $\mathbf{p}_\ell^{new} \rightarrow \mathbf{p}_\ell^{old}$ and $a_\ell^{new} \rightarrow a_\ell^{old}$.

11. The learning scheme either terminates when a set number of steps have been executed or starts another round of substeps, beginning with substep 1 above.

8.4 Experimental Results and Discussion

8.4.1 Robot Performance

Target locations for the training were selected by moving the end effector to a position that was chosen by supplying random pressures to the joints.

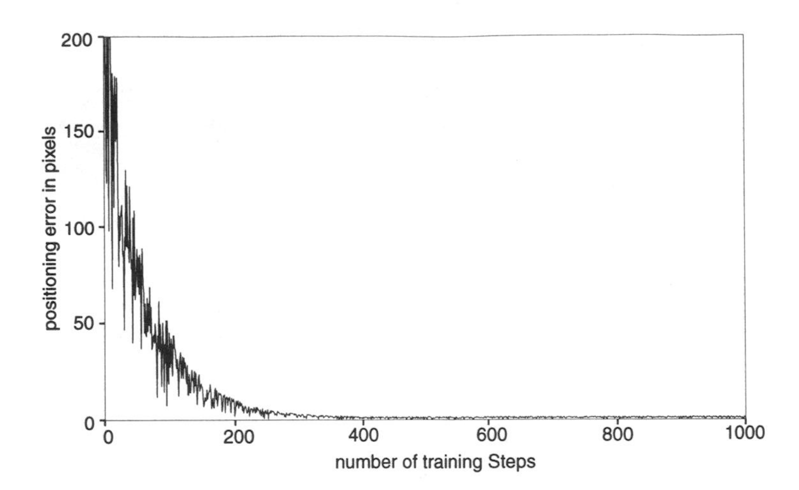

Fig. 8.5. Positioning error of the end effector for the neural network controlling the second stage of gripper movement.

Maximum and minimum pressures for each joint were stated such that the robot arm picked target positions within a workspace of size 375 mm $\times$ 750 mm $\times$ 750 mm.

The camera viewed the resulting position and orientation of two lights that were fixed at positions p and q (Fig. 8.3) and sent the corresponding $\mathbf{v}_{target}$ to the system.

In each learning step, after the target location $\mathbf{v}_{target}$ was chosen, the robot arm went to a particular arbitrarily chosen position from where it tried to reach the target location $\mathbf{v}_{target}$ using one coarse movement and several fine movements.

All of the weights $\mathbf{w}_k$, pressures $\mathbf{P}_k$, and Jacobians A_k initially were assigned randomly. The initial $n_{sup} = 50$ learning steps followed the supervised procedure, introduced in Sec. 3, in which the knowledge of the pressures $\mathbf{P}_{target}$ corresponding to the target positions $\mathbf{v}_{target}$ were provided. After the first 50 steps, the robot started to learn in an unsupervised mode, i.e., the pressures $\mathbf{P}_{target}$ no longer were provided. Each trial, on average, took 30 s to complete. Two networks were trained separately in this way. One network, consisting of 1000 neurons, was employed for stage-one movements which positioned and oriented the gripper in front of the cylinder. For S, introduced in Eqs. (8.14) and (8.20), a value of 400 was chosen. The robot learned a set of five pressures $\mathbf{P}_k$ and a set of 5×8 Jacobian matrices. A smaller network of 200 neurons was employed for second-stage movements leading to grasping. In the later case, only three joints were used, and here the robot learned a set of 3×4 Jacobian matrices in an unsupervised way, as was already described in [11]. The tolerance level for

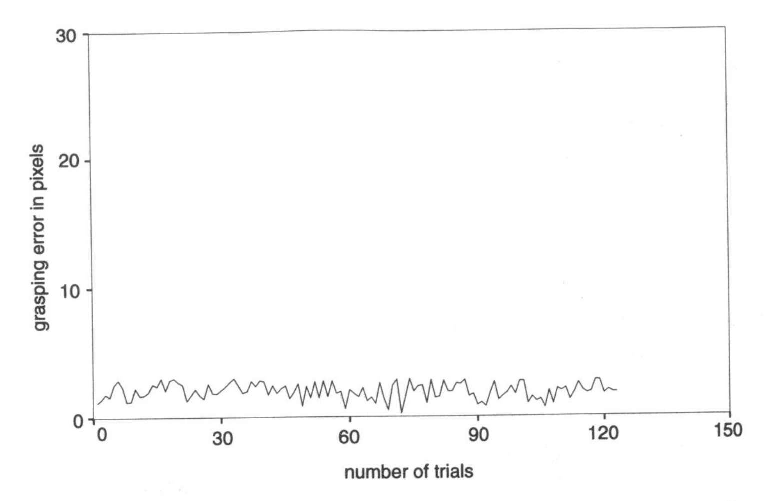

Fig. 8.6. Grasping error versus number of trials; the figure here shows the combined error for both of the networks.

error $(\mathbf{u}_{target} - \mathbf{v}_f)$ for each learning step was an exponential function of time [Eq. (8.7)].

As in Eq. (8.7), in the initial stages the tolerance was set to a high level, and as the network became mature it became lower and lower. Both of the networks took 400 steps to reduce the error for both the positioning and orientation below 3 pixels. Figures 8.4 and 8.5 show error levels for both of the networks after 1000 learning steps. For a mature network, three fine movements were sufficient to reduce the error below the tolerance level.

8.4.2 Combination of Two Networks for Grasping

After the training was completed, the mature networks were tested for grasping a cylinder. The combined network, trained first by the supervised and then by the unsupervised algorithm, was used to place the robot gripper in front of the actual cylinder by sending visual inputs from two lights at positions p and q (Fig. 8.3). After this initial positioning, the visual inputs were changed to the images of the center of line ab. The network consisting of 200 neurons then became activated and the gripper approached that center slowly by small movements. The results for the two networks then were combined and are shown in Fig. 8.6. Figures 8.5 and 8.6 demonstrate that the network is satisfactorily trained after only about 300 trial movements, with a residual average error of 1.35 camera pixels.

8.4.3 Discussion

Control of positioning and grasping movements of robot arms often has been addressed in the literature, in particular by researchers in control theory and artificial intelligence [26]. The major problem with the control theory and the artificial intelligence approaches is that they both depend on the domain knowledge and, therefore, require cumbersome efforts to design the control system. Moreover, these approaches are not robust when one deals with real life, e.g., hysteretic, robots. In this work we have taken a different approach which is based on our understanding of the map-generating mechanism in human brains [21]. Our previous effort to control the positioning of the end effector of a pneumatic robot [11] was successful but limited to a restricted set of target configurations. In the present study we allow arbitrary orientations of a target cylinder to be grasped and thereby have made the problem of grasping control more difficult to accomplish. Nevertheless, the *topology representing network* algorithm along with supervised tuning accomplished control of grasping after only a modest number (300) of training episodes. Presently, we extend this study to network architectures that closely resemble biological motor pathways, in particular those that involve cortical as well as cerebellar components. We also employ a more sophisticated method for visual recognition of target and arm posture.

Acknowledgments. We would like to thank Ted Hesselroth for the algorithms of Sec. 3.2, Hillol Kargupta for helpful discussions about the supervised algorithm, Joerg Walter for the vision system code, and Volker Ehrlich and Benno Puetz for help with Fig. 8.1. The authors express their gratitude to the Carver Charitable Trust for support. Funds for the robot system were provided by the Beckman Institute through the Capital Development Board of the University of Illinois. The computations were carried out in the National Institutes for Health Resource for Concurrent Biological Computings (grant 1 P41 RR05969*01).

References

[1] W. Mc Culloch, W. Pitts (1943) A logical calculus of the ideas immanent in the nervous activity. *Bull. Math. Biophys.*, **5**:115–133

[2] O. Ghitza (1987) Robastness against noise: The role of timing-synchrony measurement. *Proc. Int. Conf. on Acoustics Speech and Signal Processing, ICASSP-87, Dallas*, April 1987

[3] D.W. Tank, J.J. Hopfield (1986) Simple neural optimization networks: An A/D converter, signal decision circuit and a linear programming circuit. *IEEE Trans. Circuits Syst.* **CAS-33**:533–541

[4] K. Furuta, M. Sampei (1988) Path control of a three-dimensional linear motional mechanical system using laser. *IEEE Trans. Indust. Electron.* **35**(1):52–59

[5] L.E. Weiss, A.C. Sanderson, C.P. Neuman (1987) Dynamic sensor-based control of robots with visual feedback. *IEEE J. Robotics Automat.* **RA-3**(5):404–417

[6] M. Kuperstein (1987) Neural model of adaptive hand-eye coordination for single postures. *Science* **239**:1301–1311

[7] M. Kuperstein (1987) Adaptive visual-motor coordination in multijoint robots using parallel architecture. *IEEE Int. Automat. Robotics (Raleigh, NC)*, 1596–1602

[8] J. A. Walter, T. M. Martinetz, K. Schulten (1991) Industrial robot learns visuomotor coordination by means of "neural-gas" network. In: *Proc. Int. Conf. Artificial Neural Networks, Helsinki, 1991* (Elsevier, Amsterdam)

[9] H. Miyamoto, M. Kawato, T. Setoyama, R. Suzuki (1988) Feedback-error-learning neural network for trajectory control of a robotic manipulator. *Neural Networks* **1**:251–265

[10] T. Martinetz, H. Ritter, K. Schulten (1990) Three-dimensional neural net for learning visuo-motor coordination of a robot arm. *IEEE Trans. Neural Networks* **1**:131–136

[11] T. Hesselroth, K. Sarkar, K. Schulten, P.P. van der Smagt (in press) Neural network control of a pneumatic robot arm. *IEEE Trans. Syst., Man Cybernet.*

[12] P. van der Smagt, K. Schulten (1993) Control of pneumatic robot arm dynamics by a neural network. In: *Proc. World Congress on Neural Networks, Portland, OR, July 11–15, Vol. 3*, pp. 180–183

[13] T. Martinetz, S. Berkovich, K. Schulten (1993) "Neural gas" for vector quantization and its application to time series prediction. *IEEE Trans. Neural Networks* **4**:558–569

[14] T. Martinetz, K. Schulten (1993) A neural network for robot control: Cooperation between neurons as a requirement for learning. *Comput. Electr. Engrg.* **19**:315–332

[15] T. Kohonen (1982) Analysis of a simple self-organizing process. *Biol. Cybern.* **44**:135–140

[16] T. Kohonen (1982) Self-organized formation of topologically correct feature maps. *Biol. Cybern.* **43**:59–69

[17] H. Ritter, T. Martinetz, K. Schulten (1989) Topology-conserving maps for learning visuo-motor coordination. *Neural Networks* **2**:159–168

[18] T. Martinetz, H. Ritter, K. Schulten (1990) Learning of visuo-motor coordinaation of a robot arm with redundant degrees of freedom. In: *Parallel Processing in Neural Systems and Computers*, R. Eckmiller, G. Hartmann, G. Hauske (Eds.) (Elsevier, Amsterdam), pp. 431–434

[19] T. Martinetz, K. Schulten (1990) Hierarchical neural net for learning control of a robot's arm and gripper. In: *International Joint Conference on Neural Networks, San Diego, CA, Vol. 2* (Institute of Electrical and Electronics Engineers, New York), pp. 747–752

[20] H. Ritter, T. Martinetz, K. Schulten (1992) *Neural Computation and Self-Organizing Maps* (revised English Edition) (Addison-Wesley, Reading, MA)

[21] K. Obermayer, G.G. Blasdel, K. Schulten (1992) Statistical mechanical analysis of self-organization and pattern formation during the development of visual maps. *Phys. Rev. A* **45**:7568–7589

[22] T. Martinetz, K. Schulten (1991) A neural gas network learns topologies. In: *Artificial Neural Networks*, T. Kohonen, O. Simula, J. Kangas (Eds.) (Elsevier, Amsterdam), pp. 397–402

[23] T. Martinetz, K. Schulten (1996) Topology representing networks. *Neural Networks* **7**:507–522

[24] J. A. Walter, K. Schulten (1993) Implementation of self-organizing neural networks for visuo-motor control of an industrial robot. *IEEE Trans. Neural Networks* **4**:86–95

[25] T. Martinetz, K. Schulten (1993) A neural network with hebbian-like adaptation rules learning visuo-motor coordination of a PUMA robot. In: *Proc. IEEE Int. Conf. Neural Networks (ICNN-93), San Francisco*, pp. 820–825

[26] P. H. Winston (1984) *Artificial Intelligence* (Addison-Wesley, Reading, MA)

Index

S